Variational Principles of Continuum Mechanics with Engineering Applications

Volume 1: Critical Points Theory

Mathematics and Its Applications

Managing Editor:

M. HAZEWINKEL

Centre for Mathematics and Computer Science, Amsterdam, The Netherlands

Editorial Board:

F. CALOGERO, *Universita degli Studi di Roma, Italy*
Yu. I. MANIN, *Steklov Institute of Mathematics, Moscow, U.S.S.R.*
A. H. G. RINNOOY KAN, *Erasmus University, Rotterdam, The Netherlands*
G.-C. ROTA, *M.I.T., Cambridge, Mass., U.S.A.*

Vadim Komkov

Department of Mathematics, Winthrop College, U.S.A.

Variational Principles of Continuum Mechanics with Engineering Applications

Volume 1: Critical Points Theory

D. Reidel Publishing Company

A MEMBER OF THE KLUWER ACADEMIC PUBLISHERS GROUP

Dordrecht / Boston / Lancaster / Tokyo

Library of Congress Cataloging in Publication Data

CIP

Komkov, Vadim.
Variational principles of continuum mechanics with engineering applications

(Mathematics and its applications)
Includes bibliographies and indexes.
Contents: v. 1. Critical Points theory.
1. Continuum mechanics. 2. Calculus of variations. I. Title. II. Series: Mathematics and its applications (D. Reidel Publishing Company)
QA808.2.K66 1985 531 85-28105
ISBN 90-277-2157-2 (v.1)

Published by D. Reidel Publishing Company,
P.O. Box 17, 3300 AA Dordrecht, Holland

Sold and distributed in the U.S.A. and Canada
by Kluwer Academic Publishers,
190 Old Derby Street, Hingham, MA 02043, U.S.A.

In all other countries, sold and distributed
by Kluwer Academic Publishers Group,
P.O. Box 322, 3300 AH Dordrecht, Holland

Printed in The Netherlands

CONTENTS

EDITOR'S PREFACE

Approach your problems from the right end and begin with the answers. Then one day, perhaps you will find the final question.

'The Hermit Clad in Crane Feathers' in R. van Gulik's *The Chinese Maze Murders*.

It isn't that they can't see the solution. It is that they can't see the problem.

G.K. Chesterton. *The Scandal of Father Brown* 'The point of a Pin'.

Growing specialization and diversification have brought a host of monographs and textbooks on increasingly specialized topics. However, the "tree" of knowledge of mathematics and related fields does not grow only by putting forth new branches. It also happens, quite often in fact, that branches which were thought to be completely disparate are suddenly seen to be related.

Further, the kind and level of sophistication of mathematics applied in various sciences has changed drastically in recent years: measure theory is used (non-trivially) in regional and theoretical economics; algebraic geometry interacts with physics; the Minkowsky lemma, coding theory and the structure of water meet one another in packing and covering theory; quantum fields, crystal defects and mathematical programming profit from homotopy theory; Lie algebras are relevant to filtering; and prediction and electrical engineering can use Stein spaces. And in addition to this there are such new emerging subdisciplines as "experimental mathematics", "CFD", "completely integrable systems", "chaos, synergetics and large-scale order", which are almost impossible to fit into the existing classification schemes. They draw upon widely different sections of mathematics. This programme, Mathematics and Its Applications, is devoted to new emerging (sub)disciplines and to such (new) interrelations as exempla gratia:

- a central concept which plays an important role in several different mathematical and/or scientific specialized areas;
- new applications of the results and ideas from one area of scientific endeavour into another;
- influences which the results, problems and concepts of one field of enquiry have and have had on the development of another.

The Mathematics and Its Applications programme tries to make available a careful selection of books which fit the philosophy outlined above. With such books, which are stimulating rather than definitive, intriguing rather than encyclopaedic, we hope to contribute something towards better communication among the practitioners in diversified fields.

Because of the wealth of scholarly research being undertaken in the Soviet Union, Eastern Europe, and Japan, it was decided to devote special attention to the work emanating from these particular regions. Thus it was decided to start three regional series under the umbrella of the main MIA programme.

A long time ago it was certainly time that mechanics (both particle mechanics and continuum mechanics) was a main source of mathematical problems and ideas. By the first half of this century, however, the limits between the two areas of research certainly had become rather tenuous. Also, a standard gap of some 50 years between the developments of theory and applications had developed.

Now things have changed again. A period of gathering in the harvest of some 70 years of intense and often intensely specialised developments has started. Also, the fifty-year gap is fast disappearing and has disappeared completely in some areas. Continuum mechanics is one of the areas profiting from all this. There is also a new awareness that the matter of "what can one really do with all these beautiful theoretical results and ideas" is an important part of research. This book testifies to the fact that abstract results and techniques are useful and can be put to effective use and are, given a good expositor, accessible to, in this case, engineers.

The unreasonable effectiveness of mathematics in science ...

Eugene Wigner

Well, if you know of a better 'ole, go to it.

Bruce Bairnsfather

What is now proved was once only imagined.

William Blake

As long as algebra and geometry proceeded along separate paths, their advance was slow and their applications limited.

But when these sciences joined company they drew from each other fresh vitality and thenceforward marched on at a rapid pace towards perfection.

Joseph Louis Lagrange.

Bussum, July 1985 Michiel Hazewinkel

INTRODUCTION

VARIATIONAL PRINCIPLES OF CONTINUUM MECHANICS

In writing this monograph, I had to consider the basic interplay between mathematics and mechanics. In particular one has to answer some obvious questions in considering the development of a mathematical theory which is primarily oriented towards an applied science. A majority of engineers or physicists would have given an obvious answer concerning the role of mathematics. It is used for solving problems. Modern physicists are not quite so certain that this is a primary role of mathematics, even of mathematical physics. First of all, mathematics provides an abstract language in which one can attempt to state precisely some physical laws. Secondly, mathematics is used as a source of physical concepts. I have always believed in the continued interplay of mathematical and physical ideas. Important physical concepts usually led in the past to an enrichment of the mathematical ideas. Vice versa, a concept which occurs naturally in diverse areas of mathematics must have an important physical interpretation.

We seem to be witnessing a rebirth of the classical attitudes. The fundamental works of Truesdell, Noll, Coleman, and Gurtin in continuum mechanics; the works of Lichnerowicz and Hermann in modern physics, and the unexpected application of the Attiya-Singer index theory to quantum mechanics, the "rigorous" theories of Feynman integration-all point towards a new era of physical interpretation of mathematical concepts. This monograph attempts to contribute in a modest way toward this general trend.

The author realizes how hard it is for an engineer to absorb new mathematical ideas. At the same time, more and more do the modern mathematical ideas filter into the graduate courses of our engineering colleges. Just how much is to be taken for granted is hard to decide. What sufficed in the 1950-s is insufficient in the 1980-s.

As a compromise, some elementary concepts of functional analysis have been included in the Appendices A and B, Volume I, of this work. This monograph covers only a narrow range of mathematical material which is generally labeled "variational methods" trying to bypass most of the typical 19-th century arguments which are the backbone of most "mathematical methods for ..." expository materials and textbooks.

Volume 2 deals with related topics. Algebraic approach (Lie group, Lie algebras), invariance theory; modern theories of sensitivity; connection with problems of control theory and optimization theory.

It is impossible to treat all important aspects of the interaction between physics, engineering and mathematics even when we relate it exclusively to the variational approaches.

The author selected some topics, omitted others (perhaps even more important) and relied on some mathematical developments, while at best paying only a lip service to others. Some developments in the theory of design optimization, sensitivity, group theoretic and model theoretic methods grow so fast that any research monograph is bound to be slightly out of date by the time it is published.

The author has no intention of producing an encyclopedic text, or trying to complete with such texts, but concentrates specifically in Volume 1 of this work on the critical point theory and its applications to continuum mechanics.

CHAPTER 1

ENERGY METHODS, CLASSICAL CALCULUS OF VARIATIONS APPROACH - SELECTED TOPICS AND APPLICATIONS

1.1 The Energy Methods. -- Some Engineering Examples

The points of view of Hamilton, Lagrange, Gauss, Hertz, and Lord Rayleigh emphasized the concept of energy rather than force. The equations of motion of the system are not derived by consideration of equilibrium of forces acting on the system, possibly including the Newtonian and d'Alambert inertia forces. Instead, the primary role is played by energy considerations. As an example of such an approach, a condition of stable equilibrium under static loads is replaced by the condition of a local minimum for the potential energy of a mechanical system. Instead of solving the equations of motion of a vibrating system to find its natural frequencies, it is possible to consider the mean values of potential and kinetic energies, or to minimize an appropriate energy functional.

It was Rayleigh who first discovered that in the natural mode of vibration of an elastic system the equal distribution of average kinetic and potential energies caused the frequency of the system to be minimized. Moreover, any other assumed motion, satisfying the boundary conditions (but not necessarily obeying any physical laws!) will result in a higher value of the fundamental frequency.

This type of a problem is traditionally "solved" by techniques of classical analysis best illustrated by an example of a vibrating string fastened at the ends. Let the string have length ℓ, mass density ρ (per unit length), and be subjected to approximately constant tention T. One could carry out the usual analysis based on Newton's laws of motion to determine the equations of motion and the corresponding natural frequency. As an exercise we shall carry out the details of a heuristic analysis in the general form still offered in many engineering texts.

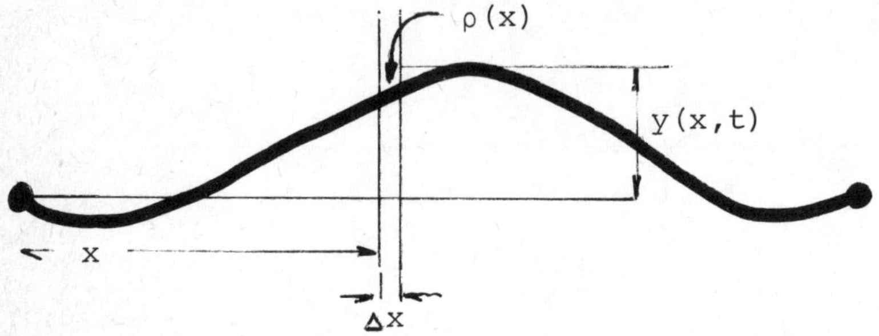

Figure 1

We wish to find the deflection function y(x,t) obeying boundary conditions:

$$y(o) \equiv 0, \quad y(\ell) \equiv 0$$

These boundary conditions are independent of time.

At this point I will repeat a heuristic argument found in many engineering texts, and one which I have heard in my young days in the classroom. We look through a magnifying glass at a small segment of the string which now "looks almost like a segment of a line." (Remark: We have just assumed "spatial differentiability" of the displacement function!)

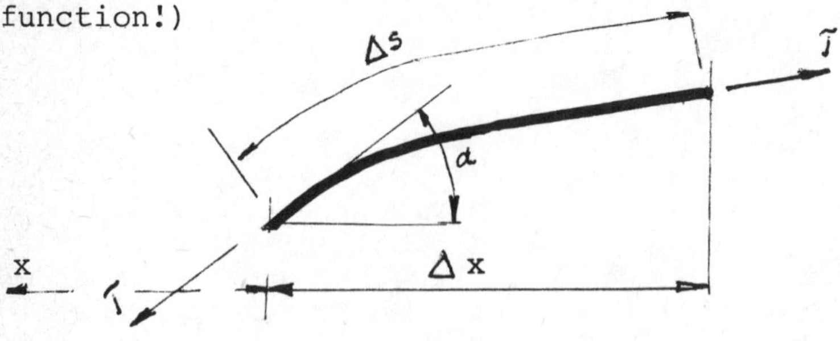

Figure 2

The force of gravity is ignored and we assume that only the tension forces acting on the ends of this segment "are of any importance." Balancing the y components of forces (per unit length of the string), we have

$$[T \sin\alpha/\Delta s]_{x+\Delta x} - [T \sin\alpha/\Delta s]_{x} \cong \rho \frac{\partial^2 y}{\partial t^2} \cdot \Delta x ,$$

where $\cong$ means "is approximately equal."

If α "is small" we substitute $\tan\alpha = \frac{\partial y}{\partial x}$ for $\sin\alpha$, and use the mean value theorem for

$$\frac{[T \frac{\partial y}{\partial x} (1 + (\frac{\partial y}{\partial x})^2]^{\wedge}_{x+\Delta x} - [T \frac{\partial y}{\partial x} (1 + (\frac{\partial y}{\partial x})^2]^{\wedge}_{x}}{\Delta x}$$

to write:

$$\rho \frac{\partial^2 y}{\partial t^2}\Big|_{x=\xi} = \frac{\partial}{\partial x}(T \frac{\partial y}{\partial x} (1 + (\frac{\partial y}{\partial x})^2))\Big|_{x=\xi}$$

$x \leqq \xi < x + \Delta x$, and observe that ξ is quite arbitrary. Then we assume that $1 + (\frac{\partial y}{\partial x})^2$ is "very close to unity," and that "T is a constant function of x." Hence, putting $T/\rho = c^2$, we finish by writing the classical equation of the one-dimensional wave propagation

$$\frac{\partial^2 y}{\partial x^2} = \frac{1}{c^2} \frac{\partial^2 y}{\partial t^2}, \tag{1.1}$$

describing the dynamic behavior of the string.

Now, let us look at our argument closely to answer a simple question, "what kind of a (genuine physical) string motion is really described by this equation?" To realize what can possibly cause a string not to behave in a fashion which fits into our equation, we should list all the suspicious sounding statements which were quoted in the inverted commas.

Now, assumming that everything is in order, we proceed to separate variables by writing $y(x,t) = X(x) \cdot \theta(t)$. We suggest that the reader should stop here and consider the physical implications of writing $y(x,t) = X(x) \cdot \theta(t)$. What are the physical implications of this mathematical assumption? These are certainly nontrivial. Let us overlook this point, and simply assume that the solution of the problem $y(x,t)$ can be written in the separated form

$$y(x,t) = X(x) \cdot \theta(t) \tag{1.2}$$

It follows easily that

$$\frac{X''(x)}{X(x)} = \frac{1}{c^2}\frac{\theta''(t)}{\theta\ (t)} = \text{constant};$$

We denote this constant by $-\Lambda^2$. Then $\theta(t)$ satisfies the equation

$$\theta'' + \omega^2\ \theta = 0\ ,\ \text{where}\ \omega = \Lambda C,$$

and $\theta = A_o \sin\ (\omega t)$ solves it. ω is interpreted as the natural angular velocity. The constant A_o is the amplitude. The natural frequency is given by $f = \omega/2\pi$.

A more sophisticated approach would consider the average values of potential and kinetic energies.

$$T = \frac{1}{2} \int_o^{\ell} \rho\ (\frac{dy}{dt})^2\ dx$$

Suppose that the string is vibrating with frequency f, where $\omega = 2\pi f$. Integrating over a complete cycle of vibration gives a formula for average kinetic energy $\bar{T}$ and potential energy $\bar{V}$.

$$\bar{T} = \frac{1}{4} \rho \, \omega^2 \int_0^{\ell} y^2 dx,$$

$$\bar{V} = \frac{1}{4} E \int_0^{\ell} \left(\frac{dy}{dx}\right)^2 dx,$$

where E denotes Young's modulus.

We replace the constant tension assumption by the equivalent assumption stating that the cross-sectional area of the string is constant and Young's modulus is constant.

Denoting by $\langle f, g \rangle$ the $L_2 \, [0,\ell]$ product, i.e., $\langle f, g \rangle = \int_0^{\ell} f(x) \cdot g(x) \, dx$, we rewrite the formula for ω_1^2 in this notation.

$$\omega_1^2 = \frac{E}{\rho} \, \frac{\langle Ay_1, y_1 \rangle}{\langle y_1, y_1 \rangle},$$ where A stands for the operator:

$$A \equiv - \frac{d^2}{dx^2} .$$

According to Rayleigh's principle, the natural mode $y_1(x)$ minimizes ω_1^2. That is, among all possible shapes $\eta(x)$ which are physically admissible, and satisfy $\eta(0) = \eta(\ell) = 0$, the natural mode $y_1(x)$ minimizes ω_1^2. Hence, ω_1^2 may be regarded as a functional depending on the shape and distribution of weight of the vibrating string $\eta(x)$, i.e., $\omega_1^2 = \omega_1^2(\eta)$. One may investigate this dependence and the <u>sensitivity</u> of ω_1.

If we know how to differentiate ω_1^2 with respect to η, the necessary condition for the minimum of ω_1^2 becomes $\frac{d(\omega_1^2)}{d\eta} = 0$. According to the rules of Fréchet differentiation given later in the Appendix A,

$$\frac{d}{d\eta}(\omega_1^2(\eta)) = \frac{d}{d\eta}\left(\frac{\langle A\eta,\eta\rangle}{\langle\eta,\eta\rangle}\right) = \frac{2}{\langle\eta,\eta\rangle}(A\eta - \omega_1^2\eta) = 0$$

This is really the equation for the vibrating system in its fundamental mode determining the best design. The equation of motion is given by $d(\omega_1^2)/dy(x) = 0$, where ω is regarded as a function of the unknown displacement. That is, we claim that the derivative of the Rayleigh quotient with respect to y(x) is equal to zero.

It is interesting to observe that the entire discussion considering equilibrium of forces acting on the string has been bypassed, and replaced by the simple-minded statement that the first derivative must be equal to zero when a differentiable function (functional) assumes a local minimum.

This is a deep observation contrasting the points of view of Newton and Huygens, but also providing an insight into modern viewpoint of classical and continuum mechanics.

The important aspect of our simple example is the replacement of analysis concerning vectors (forces, displacements, etc.) by an analysis concerning some extremal properties of a functional (a function whose range is a subset of real or complex numbers). Specifically, we differentiate the energy functional with respect to some vector. In examples offered in this chapter, this is a straightforward operation. However in some example discussed in later chapters concerning continuum mechanics, it is not clear what "differentiation" means or how to compute such abstract derivatives. One needs some preparation in functional analysis. Elementary concepts of functional analysis as needed in critical point theory are given in Appendices A and B of this volume of the monograph.

First, let us illustrate the energy approach by an elementary example of a direct application of the energy method.

Consider a mass m attached to two points by linear springs and subjected to the force of gravity, as shown on Figure 3.

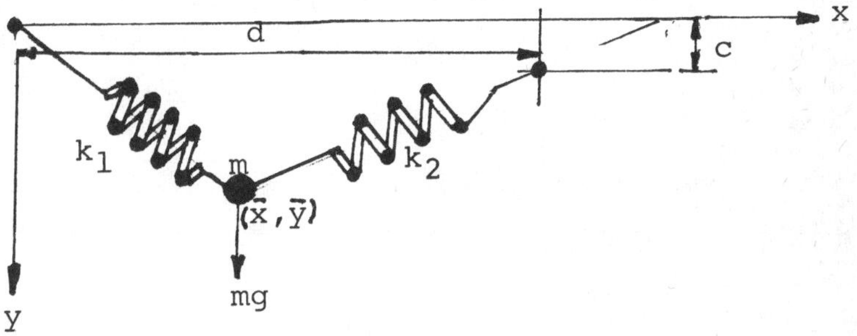

Figure 3

The spring constants are k_1, k_2 as indicated. Find the position of equilibrium.

Instead of attempting to balance the forces and moments acting on the mass m, as is done in an elementary statics course, we shall use the principle of minimum potential energy.

We choose coordinates x, y as indicated on Figure 3 so that the force of gravity acts along a line parallel to axis y. The potential energy is given by

$$V = \frac{1}{2} K_1(\bar{x}^2 + \bar{y}^2) + \frac{1}{2} \cdot K_2[(d-\bar{x})^2 + (\bar{y}+c)^2] + mg\bar{y}.$$

Note that $\bar{y}$ appears to be negative on the Figure A-1; however the positive y-direction is downward. There is no reason why the positive direction should be up!

V attains a minimum only if

$$\frac{\partial V}{\partial \bar{x}} = \frac{\partial V}{\partial \bar{y}} = 0, \text{ or } \frac{\partial V}{\partial \bar{x}} = K_1\bar{x} - K_2(d-\bar{x}) = 0$$

$$\frac{\partial V}{\partial \bar{y}} = K_1\bar{x} + K_2(\bar{y} + c) + mg = 0 \quad .$$

Solving for $\bar{x}$ and $\bar{y}$, we obtain the coordinates of the equilibrium point.

$$\bar{x} = \frac{K_2 d}{K_1 + K_2}, \quad \bar{y} = \frac{mg + K_2 c}{K_1 + K_2} \quad .$$

We check that forces acting on the mass m are in balance (i.e., their sum is equal to zero). Indeed they are.

Somewhat similar problems can be regarded as exercises.

1. Compute the equilibrium position of the mass suspended as shown by three linear springs by minimizing the potential energy of this system.

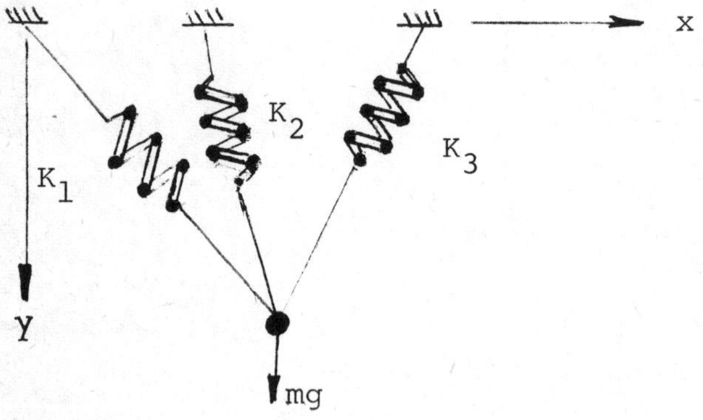

Figure 4

(Observe that this problem is not statically determinate!)

2.

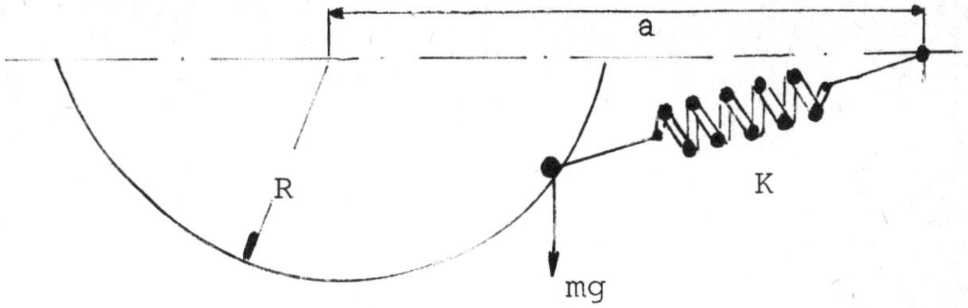

Figure 5

A mass is constrained to a circular path (as shown). A linear spring attracts the mass towards a point A with a > R. Find the equilibrium position.

1.2 <u>Examples From Structural Mechanics</u>
<u>The Theorems of Castigliano and Betti</u>

Suppose that external forces are applied to a structure which deflects elastically. The deflections are "small" and are linearly dependent on the applied forces, i.e., if the forces applied are given by an n-vector, the m-dimensional deflection vector is of the form

$$\underset{\sim}{q} = A\underset{\sim}{f}$$

where A is an m x n matrix. Let us assume Hooke's law. Castigliano's theorem asserts that the deflection δ in the direction of a force f_i at the point of application of that force is equal to the derivative with respect to that force of the total strain energy of the structure:

$$\delta_i = \frac{\partial U}{\partial f_i}$$

This formula leads to direct computational results based on approximate formulas for energy forms.

The approximate formulas for strain energy are given below. The strain energy of a uniform elastic bar in pure tension, or compression, is

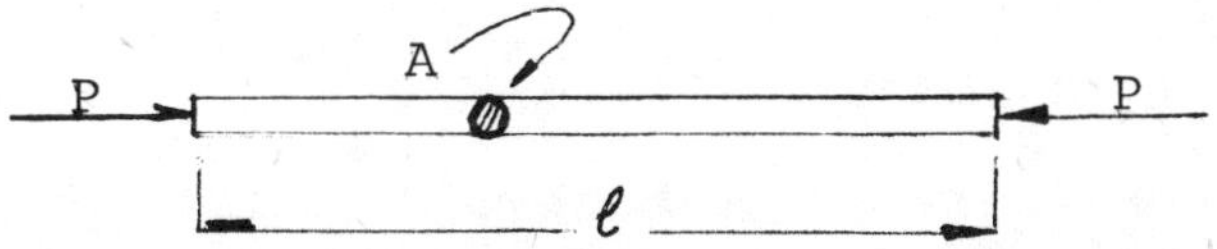

Figure 6

$$U = \frac{P^2 \ell}{2AE},$$

where

P is the force.
A is the cross-sectional area.
E is Young's modulus.
ℓ is the length.

For a pin-jointed structure containing n-members, the analogous formula is

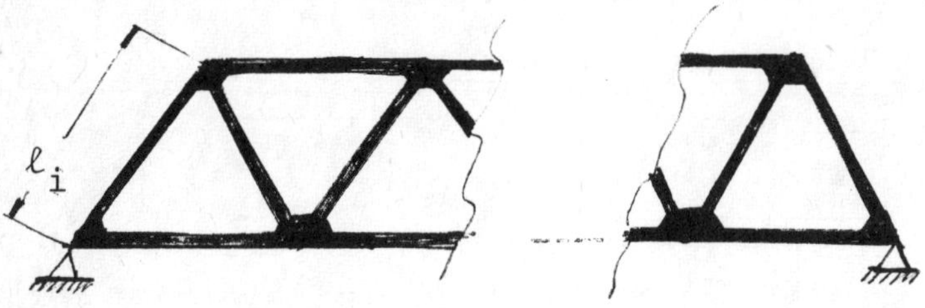

Figure 7

$$U = \sum_{i=1}^{n} \frac{S_i^2 \ell_i}{2EA_i}$$

A_i is the area of cross section of the i-th member.
ℓ_i is the length of the i-th member.

S_i is the force transmitted by the i-th member.

E is the Young's modulus.
For a single beam in bending, the strain energy is

$$U = 1/(2E) \int_0^{\ell} \left(\frac{M^2(x)}{I(x)}\right) dx$$

M(x) is the bending moment.
I(x) is the moment of inertia of the cross-sectional area about the neutral axis.

The simple shear of a beam:

$$U = 1/(2\,G) \int_0^{\ell} \left(\frac{V^2(x)}{A(x)}\right) dx,$$

where
V is the shear force (per unit length).
A is the cross sectional area.
G is the elastic shear modulus.

Pure torsion of a bar:

$$U = 1/2 \int_0^{\ell} \left(\frac{T^2}{JG}\right) dx$$

T is the torque.
J is the polar moment of inertia of the cross sectional area about the center of torsion.

1.3 An Example of Application

It is necessary to determine the deflection at the free end of a cylindrical rod bent into a quarter-circle shape. The load P applied to the free end is perpendicular to the plane of the axis of the rod as shown on the sketch below.

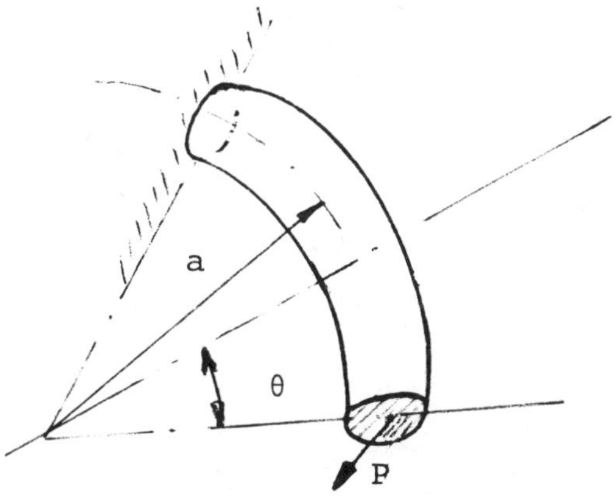

Figure 8

Solution

The strain energy is the sum of the bending and torsional energy. The shear energy due to the direct shear is ignored.

Hence

$$U = \int_0^{\ell} \frac{M^2 dx}{2\,E\,I} + \int_0^{\ell} \frac{T^2 dx}{2\,G\,J}$$

where M is the bending moment and T the Torque. E,I,G,J are constants.

The potential energy is given by the difference between work performed and the strain energy.

In this case

$$V = -P \cdot \delta_p + U ,$$

where δ_p denotes the deflection in the direction of the force P computed at the point where P is applied. V(P) is minimized only if $\frac{\delta V}{\delta P} = 0$, or if $\delta_p = \frac{\delta U}{\delta_P}$ which is the form of Castigliano's theorem used to solve this problem.

Hence,

$$\delta_P = \frac{\partial}{\partial P} \{\int_0^{\ell} \frac{M^2(P)}{2\,E\,I}\, dx + \int_0^{\ell} \frac{T^2(P)}{2\,G\,J}\, dx\}$$

The bending moment is:

$$M(P) = Pa \sin\theta \,.$$

The torque (twisting moment) is

$$T(P) = Pa(1 - \cos\theta).$$

We compute

$$\delta_p = \int_0^{\ell} \{\frac{M \frac{\partial M}{\partial P}}{EI} + \frac{T \frac{\partial T}{\partial P}}{GJ}\, dx\} =$$

$$a \int^{\Pi/2} [\frac{Pa^2}{EI} \sin^2\theta\,]\, d\theta + a \int_0^{\Pi/2} \frac{Pa^2}{GJ} (1-\cos\theta)^2\, d\theta =$$

$$\frac{\pi}{4} Pa^3 \left[\frac{1}{EI} + \left(3 - \frac{8}{\pi}\right) \frac{1}{GJ}\right] .$$

Exercise

Compute the deflection of the quarter-circle shaped bar in Figure 8 if the force P acts in the plane of the axis of the bar.

1.4 Betti's Theory

Castigliano's theorem states that if we measure the deflection at a point of application of a concentrated generalized force Q_i, the corresponding deflection q_i in the direction of Q_i is given by

$$q_i = \frac{\partial V (Q)}{\partial Q_i}$$

It was Betti who first observed a seemingly paradoxical result. The formula "works" independently of the magnitude of the generalized force Q_i. Application of an infinitesimal generalized force Q_i at any point of the structure can be used to compute the defection at that point and in the direction of Q_i. This is the so-called "dummy load" approach, where one computes the effects of Q_i, then sets $Q_i = 0$ in the resulting formulas. The method is best illustrated in the case of simple bending of a beam.

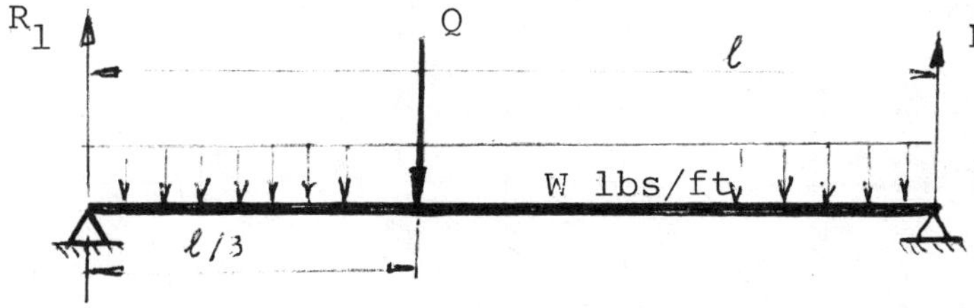

Figure 9

Let us suppose that we wish to find a deflection at the point $x = \frac{1}{3} \ell$ for the simply supported beam subjected to a uniformly distributed load, as shown on Figure 9.

We apply load Q at $x = \ell/3$, and compute the bending moment

$$M(Q)\begin{cases} = \frac{W\ell x}{2} - \frac{Wx^2}{2} + \frac{2}{3}Qx, \text{ if } 0 \le x \le \ell/3; \\ = [\frac{2x}{3} - (x - \ell/3)] \cdot Q \quad \text{for } \frac{\ell}{3} \le x \le \ell \end{cases}$$

Then

$$\delta_Q = \frac{1}{EI}\int_0^\ell M(Q)\frac{\partial M(Q)}{\partial Q}\,dx.$$

Computing the required displacement we proceed to the limit as $Q \to 0$.

$$\delta_{\ell/3} = \lim_{Q\to 0}\delta_Q = \frac{1}{EI}\left\{\int_0^{\ell/3}\left(\frac{w\ell x}{2} - \frac{wx^2}{2}\right)\cdot\frac{2x}{3}\,dx\right\}$$

$$+ \int_{\ell/3}^{\ell}\left(\frac{w\ell x}{2} - \frac{wx^2}{2}\right)\left[\frac{2x}{3} - (x-\ell/3)\right]dx = \frac{11}{972}\,\frac{w\ell^4}{EI},$$

which is the correct answer.

Of course, the method is quite general and the purpose of the specific computation performed above was to illustrate the general idea.

At this point another observation could be made

$$\lim\left[M(Q)\frac{\partial M(Q)}{\partial Q}\right] = \lim_{Q\to 0} M(Q)\lim_{Q\to 0}\frac{\partial M(Q)}{\partial Q} = M(x)\cdot m(x,\xi).$$

Here $M(x)$ denotes the bending moment due to the actual generalized forces applied to the system, while $m(x,\xi)$ denotes the bending moment at $\hat{x}$ due to a unit load positioned at the point $x=\xi$, where we wish to compute the deflection.

Referring to the example illustrated on Figure 9, we compute

$$M(x) = \left(\frac{w\ell\hat{x}}{2} - \frac{w\hat{x}^2}{2}\right)$$

$$m(\hat{x}, \xi = \ell/3) = \frac{2\hat{x}}{3} \text{ if } 0 \le \hat{x} \le \ell/3$$

$$\frac{2\hat{x}}{3} - (\hat{x}-\ell/3) \text{ if } \ell/3 \le \hat{x} \le \ell, \text{and}$$

$$\int_0^{\ell} \frac{M(\hat{x}) \cdot m(\hat{x}, \ell/3)}{EI} d\hat{x} = \frac{11}{972} \frac{w\ell^4}{EI},$$

as before.

1.5 The Kinetic Energy Terms

Kinetic energy for a body with mass density $\rho(x,y,z)$ can be written as

$$T = \frac{1}{2} \int_{\Omega} (\rho\ (x,y,z) \cdot \nu^2)\, dxdydz,$$

where $\underset{\sim}{\nu}\ (x,y,z)$ is the velocity vector and $\nu^2 = ||\underset{\sim}{\nu}\ ||^2 = \underset{\sim}{\nu} \cdot \underset{\sim}{\nu}$.

For a rigid body whose motion is observed by an observer located at the origin of a coordinate system, we denote by ν the velocity of a point in a relative to the origin; Γ, the position vector; and ω, the rotation vector.

The kinetic energy expression can be rewritten in the form

$$T = \frac{1}{2} \int_{\Omega} \rho\ \underset{\sim}{\nu} \cdot (\omega \times \Gamma)\ dxdydz$$

For a rotation about a fixed point (say the origin) we have

$$T = \frac{1}{2}\ (I_{xx}\omega_x{}^2 + I_{yy}\omega_y{}^2 + I_{zz}\omega_z{}^2)$$

where $I_{xx} = \int_{\Omega} [\rho(x,y,z)\cdot(y^2+z^2))]\ dxdydz$, and I_{yy}, I_{zz} are defined similarly.

For a discrete system of masses, integrals are replaced by finite sums.

$$T = \frac{1}{2} \Sigma\, m_i \upsilon_i^{\,2}$$

or

$$T = \frac{1}{2} \Sigma\, m_i \upsilon_i \cdot (\underset{\sim}{\omega} \times \underset{\sim}{r}_i)$$

For a rigid body rotating about a fixed point or for a system of discrete masses we can use an identical formula:

$$T = \frac{1}{2} \underset{\sim}{\omega} \cdot \underset{\sim}{L}, \text{ where } \underset{\sim}{\omega} = \{\omega_x, \omega_y, \omega_z\}$$

$\underset{\sim}{L}$ is the angular momentum.

If [I] denotes the matrix of the moments of inertia with respect to x,y,z axes,

$$[I] = \begin{bmatrix} I_{xx} & I_{xy} & I_{xz} \\ I_{yx} & I_{yy} & I_{yz} \\ I_{zx} & I_{zy} & I_{zz} \end{bmatrix}$$

then $L = [I] \cdot \underset{\sim}{\omega}^T$ (i.e., $L_x = I_{xx}\omega_x + I_{xy}\omega_y + I_{xz}\omega_z$, etc.),

with $I_{xy} = \int_\Omega (\rho \cdot x \cdot y)\, dxdydz, \ldots$

1.6 Some Finite Dimensional Examples of Energy Principles

Two points of view, representing different approaches to discretizing structural and mechanical systems are (1) matrix structural analysis and (2) finite element representation.

In the matrix structural analysis, "lumped parameter" approach is adopted from the start. A structure is considered as an assembly of pieces whose behavior is represented by a finite set of constants plus a set of modal conditions and force-displacement relations.

In contrast, the finite element technique adopts a continuum point of view. The displacement (or strain) field obeys certain differential equations, which relate the displacements to applied loads. From the "practical engineer's" point of view, the structure is subdivided into fairly simple elements which amounts to a piecewise polynomial approximation for the displacement field.

These developments are well known to all practicing engineers with highly sophisticated numerical techniques and computer software available for implementation to specific problems. (The best known software program is the NASTRAN code, now routinely available in the Western world.)

Generally, some variational principle is used in simplifying the numerical computation.

The basic steps in matrix structural analysis consist in determining the so-called stiffness and mass matrices.

As an example of both approaches, consider a beam which is an element of a much larger structure. All forces acting on the beam are transmitted to it through the joints situated at both ends. We adopt a local x-y coordinate system, the x-axis, coinciding with the initial axis of the beam.

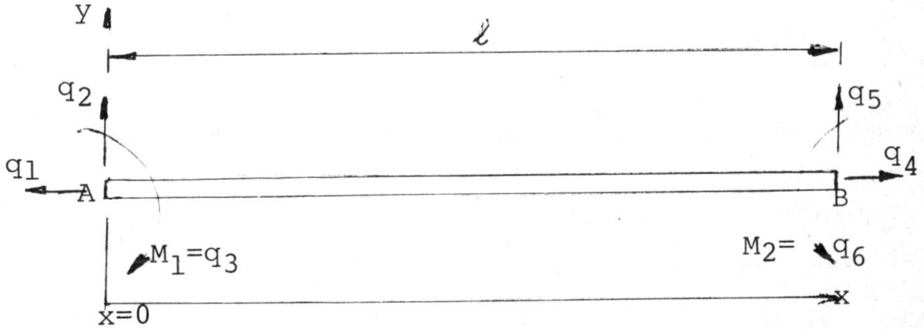

Figure 10

The y-displacement of a point with coordinate $\hat{x}$ due to the system of loads illustrated on Figure 10 is

$$y(\hat{x}) = \frac{q_2}{\ell^3}(2\hat{x}^3 - 3\ell\hat{x}^2 + \ell^3) - \frac{q_5}{\ell^3}(2\hat{x}^3 - 3\ell x^2)$$

$$+ \frac{q_3}{\ell^2}(\hat{x}^3 - 2\ell\hat{x}^2 + \ell^2\hat{x}) + \frac{q_6}{\ell^2}(\hat{x}^3 - \ell x^2)$$

The strain energy is given by

$$V = \frac{E}{2}\int_0^\ell \left[A\left(\frac{q_1}{\ell} - \frac{q_4}{\ell}\right)^2\right] dx + \frac{E}{2}\int_0^\ell \{[I(\frac{q_3}{\ell^3}(12\hat{x} - 6\ell)$$

$$- \frac{q_5}{\ell^3}(12\hat{x} - 6\ell) + \frac{q_3}{\ell^2}(6\hat{x} - 4\ell) + \frac{q_6}{\ell^2}(6\hat{x} - 2\ell))]^2\} \, dx,$$

where A(x) is the cross-sectional area and I(x) is the moment of inertia of the cross-sectional area about the neutral axis of bending.

The strain energy V can be rewritten in the matrix form

$$V = \frac{1}{2}\{\underset{\sim}{q}^T \cdot \mathbb{K} \cdot \underset{\sim}{q}\},$$

where $\underset{\sim}{q} = \{q_1 \; q_2 \; \dots \; q_6\}^T$, is the generalized displacement column vector, q^T the corresponding row vector and the symmetric matrix $\mathbb{K}$ is given by:

$$\mathbb{K} = \begin{bmatrix} A\ell^2 & 0 & 0 & -A\ell^2 & 0 & 0 \\ 0 & 12I & 6\ell I & 0 & -12I & 6\ell I \\ \text{symmetric} & & 4\ell^2 I & 0 & -6\ell I & 12\ell^2 I \\ \text{about the} & & & A\ell^2 & 0 & 0 \\ \text{diagonal} & & & & 12I & -6\ell I \\ 0 & 6\ell I & & & & 4\ell^3 I \end{bmatrix}$$

if A and I are constant.

In a truss all bending effects are neglected and only the forces q_1 and q_4 indicated on Figure 10 are supposed to influence the energy of the structure.

The truss' stiffness matrix is given by

$$\mathbb{K} = \begin{bmatrix} 1 & 0 & 0 & -1 & 0 & 0 \\ & 0 & 0 & 0 & 0 & 0 \\ & & 0 & 0 & 0 & 0 \\ & & & 1 & 0 & 0 \\ & \text{symmetric about the diagonal} & & & 0 & 0 \\ & & & & & 0 \end{bmatrix}$$

In a similar manner we compute the kinetic energy term

$$T = \frac{1}{2} \int_0^{\ell} \rho \, A[(\frac{ds}{dt})^2 + (\frac{dw}{dt})^2] \, dx,$$

where ρ is the material density, s is the displacement in the y-direction.

The kinetic energy admits a matrix representation

$$T = \frac{1}{2} (\dot{\mathbf{q}}^T \cdot \mathbb{m} \cdot \dot{\mathbf{q}}),$$

where $\mathbb{m}$ is the so-called mass matrix.

$$\mathbb{m} = \frac{\rho A \ell}{420} \begin{bmatrix} 140 & 0 & 0 & 70 & 0 & 0 \\ & 156 & 22\ell & 0 & 54 & -132\ell \\ & & 4\ell & 0 & 13\ell & -3\ell \\ \text{symmetric} & & & 140 & 0 & 0 \\ & & & & 156 & -22\ell \\ & & & & & 4\ell \end{bmatrix}$$

If bending effects are neglected (i.e., for a truss), the matrix m simplifies to:

$$m = \frac{\rho A \ell}{6} \begin{bmatrix} 2 & 0 & 0 & 1 & 0 & 0 \\ & 2 & 0 & 0 & 1 & 0 \\ & & 0 & 0 & 0 & 0 \\ & \text{symmetric} & & 2 & 0 & 0 \\ & & & & 2 & 0 \\ & & & & & 0 \end{bmatrix}$$

For a free, conservative motion (i.e., a free vibration), the equations of motion are determined by envoling Hamilton's principle. The first variation $\delta \int_o^T (T - V)\, dt$ must vanish. This implies that the Lagrangian integral, otherwise known as the action integral for the structural system, is stationary in the space of kinematically admissible variables Q ($\underset{\approx}{q} \in Q$). Hence, the behavior of the structure is determined if one applies this principle.

This is, of course, easier said than done. We recall that we assigned a local coordinate system to each element of the structure. The displacements of the nodes, at which such elements are joined, must be expressed in a global coordinate system to which all local coordinate systems are related by means of a transformation matrix S.

For each element of our finite decomposition, we introduce a rotation matrix S rotating the local coordinates to become parallel to a global coordinate system $\hat{q}^i$.

$$Sq^i = \hat{q}^i.$$

If we introduce a generalized global coordinate vector $\underset{\approx}{x}$, the coordinates $\hat{q}^i$ are related to $\underset{\approx}{x}$ by a matrix $\underset{\approx}{B}$ consisting only of zeros and ones.

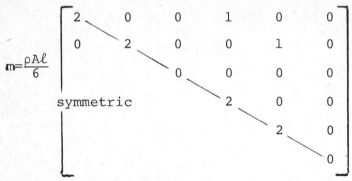

$$\hat{q}^i = B^i x.$$

The strain energy of an i-th element written in the local coordinates is given by:

$$V^i = \frac{1}{2} = q^i \cdot K^i \cdot q^{i^T}$$

where K^i is the stiffness matrix for the i-th element.
Therefore:

$$V^i = \frac{1}{2}\{q^i S^i K^i S^{i^T} q^{i^T}\} = \frac{1}{2}\{\underset{\sim}{x} \cdot B^i S^i K^i S^{i^T} B^{i^T} \underset{\sim}{x}^T\},$$

and

$$V = \sum_i V^i = \frac{1}{2}\{\underset{\sim}{x} \cdot \Sigma(B^i S^i K^i S^{i^T} B^{i^T}) \cdot \underset{\sim}{x}\} = \frac{1}{2}(\underset{\sim}{x} \cdot K_g \cdot \underset{\sim}{x}^T)$$

is the total strain energy of the structure.

An example of a two-bar truss, given below, is copied from the monograph by the author, E.J. Haug and K. Choi.

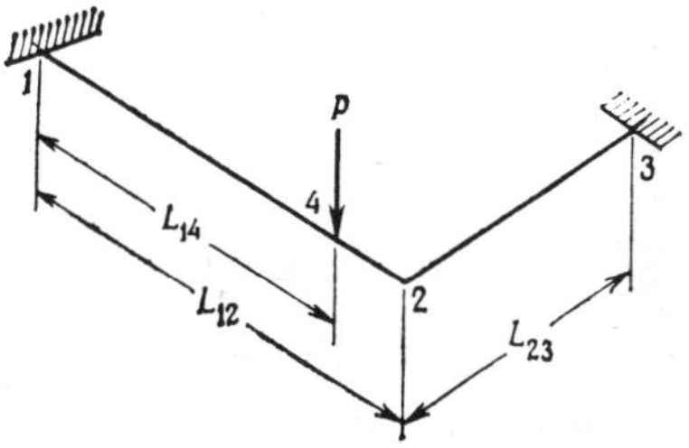

Figure 11

$$K = \begin{bmatrix} \left(\frac{12E_{12}I_{12}}{L_{12}^{3}} + \frac{12E_{23}I_{23}}{L_{23}^{3}}\right) & -\frac{6E_{12}I_{12}}{L_{12}^{2}} & -\frac{6E_{23}I_{23}}{L_{23}^{2}} \\ \frac{-6E_{12}I_{12}}{L_{12}^{2}} & \left(\frac{4E_{12}I_{12}}{L_{12}} + \frac{G_{23}J_{23}}{L_{23}}\right) & 0 \\ \frac{-6E_{23}I_{23}}{L_{23}^{2}} & 0 & \left(\frac{4E_{23}I_{23}}{L_{23}} + \frac{G_{12}J_{12}}{L_{12}}\right) \end{bmatrix},$$

$$S = \begin{bmatrix} P\left(\frac{3L_{14}^{2}}{L_{12}^{2}} - \frac{2L_{14}^{3}}{L_{12}^{3}}\right) \\ P\left(\frac{L_{14}^{3}}{L_{12}^{2}} - \frac{L_{14}^{2}}{L_{12}}\right) \\ 0 \end{bmatrix}.$$

It appears that linear algebra suffices in the analysis of discretized structural systems and specifically in the analysis of trusses.It may not suffice when such discrete ,that is finite dimensional representation,is not accurate.

In a discussion of structural systems with distributed parameters one needs to introduce some concepts of modern analysis.For the sake of completeness some relevant concepts of functional analysis are provided in the appendices A and B.

Before we proceed with what is called "the modern approach"we may indulge in a brief review of some classical concepts and terminology.Most engineers encounter these concepts in their undergraduate mechanics courses.Most mathematicians and theoretical physicists bypass all of this material in their training due to the eagerness of their instructors to reach quickly "the really important" modern ideas.

1.7 A Review of Some Concepts and Examples from Classical Mechanics

This brief outline of some basic concepts of classical mechanics is certainly no substitute for a serious study of this subject. Landau and Lifshitz [4] or Goldstein [2] provide an excellent introduction to this subject for an original treatise by a famous contribution to this subject read Hertz [16], Routh [6] or Lord Rayleigh.

For an unusual critique read Ernst Mach [7].

An example of the minimum potential energy law:

The von Mises catastrophe.

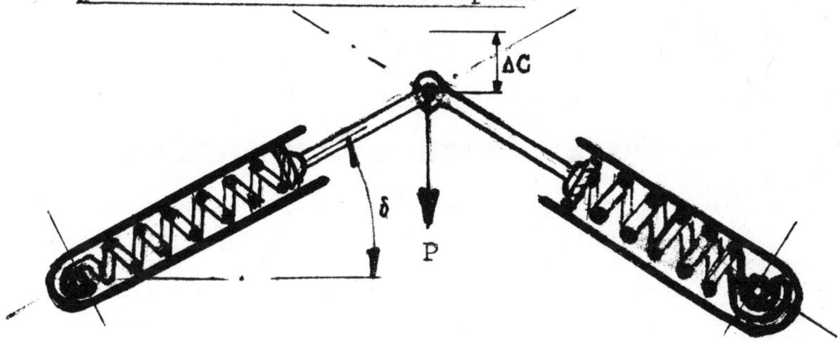

Figure 12

An elastic bar truss is subjected to a load P as shown on Figure 12

$$\Delta c = \frac{\ell}{\sin\delta} - \frac{\ell}{\sin\delta_o}$$

where the angle δ_o corresponds to P=0. u denotes the vertical displacement due to the force P.

We have $P\Delta c = \left(\frac{\ell}{\sin\delta}\right)P - \left(\frac{\ell}{\sin\delta_o}\right)P.$

The force in each bar is

$$N = \frac{P}{2\cos\delta} = EA\left(1 - \frac{\sin\delta_o}{\sin\delta}\right)$$

Note that $N = E\,A\,\varepsilon$ is Hooke's law.
EA is replaced by the spring's constant k if linear springs are used instead of elastic bars.

Therefore $P = P(\delta)$

$$P = 2\,E\,A\,(1 - \frac{u}{\ell}\tan\delta_o)\,[1/\sqrt{\tan^2\delta_o + (1-\frac{u}{\ell}\tan\delta_o)^2}$$

$- \cos\delta]$. We plot P versus u.

The laws of minimum potential energy imply that the quantity

$$U = V - Pu = [E\,A\,c_o\{1-\cos\delta_o\sqrt{\tan^2\delta_o+(1-\frac{u}{\ell}\tan\delta_o)^2}\}^2$$

$- Pu]$ is minimized.

The "usual",or "expected" P-u graph

Figure 13

Extrema of U lie on the P-u curve indicated below.

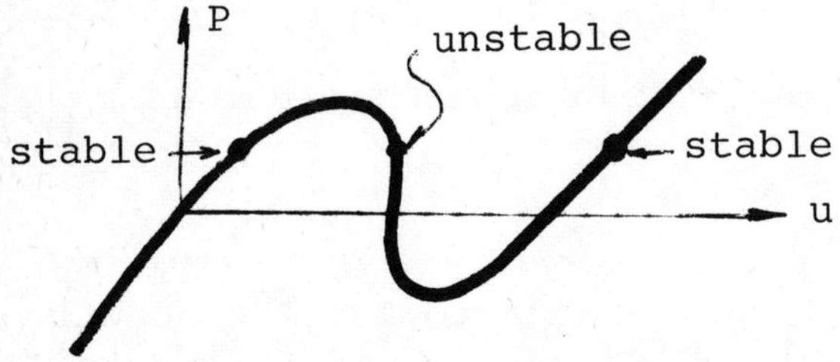

Figure 14

Cyclic loading patterns indicate discontinuous jumps from one branch of the P-u curve to another. The dotted lines represent unstable equilibrium positions satisfying $\frac{\partial U}{\partial u} = 0$ but on this part of the curve U(u) is not a minimum of the potential energy.

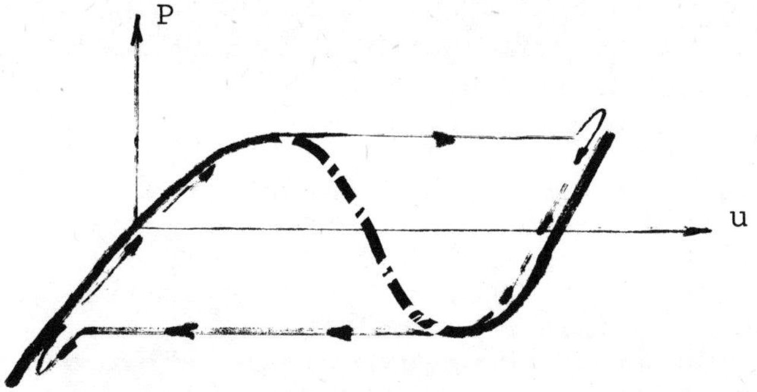

Figure 15

Some concepts and terminology of classical mechanics.

In most elementary texts on engineering mechanics the student is introduced to configurations or states of a mechanical system through the analysis of forces and moments that "act on the system".To carry out this analysis we have to be certain that all forces and moments acting on the system have been accounted for and that all constraints and interactions with other,external, systems have been replaced by appropriate forces and moments.

Let us deal,for the time-being with final-dimensional cases only.We perceive an idealized model of a mechanical system whose configuration is completely determined by specifying n-quantities such as linear displacements at certain points within the system,or,perhaps the temperatures, pressures,or velocities measured at specific times times and at specific spatial locations.Our mathematical model is such that the knowledge of these quantities $\{q_1,q_2,\dots q_n\}$ completely describes all the information we need concerning the current configuration,that is the state of the system.Other quantities such as forces,accelara tions,entropy,various energy forms may be directly computed if the state of the system is known.

It sounds reasonable that given the displacements the corresponding forces may be derived from the laws of physics and from "constitutive equations" but on occasions it is not easy to attain this objective.An appropriate technique is known in the jargon of engineering statics as "the free body analysis".A philosophical difficulty has been pointed out by H.Hertz in [16] .We do not have a satisfactory definition of the concept of force,particularly of a static force.Hertz suggested that we should introduce a generalized displacement vector,but that we should delay the introduction of the concept of a generalized force. Let us assume that n-parameters $q_1,q_2,\ldots q_n$, determine uniquely the configuration of the system,or that all physically possible(admissible) configurations can be represented by such ordered n-tuples,with the usual definition of addition of ordered n-tuples and of multiplication by a scalar.Generally such definitions reflect our physical interpretation of the mathematical model.

The linear vector space V of the ordered n-tuples $\{q_1,q_2,\ldots q_n\} = \underset{\sim}{q}$ over the real or comlex number field will be called the space of general displacements of the system and any vector $q \in V$ will be called a general displacement vector. We observe that q_i-s are not assumed to have necessary the physical dimension of length.

The space of generalized forces is defined as V^*,that is as the topological dual of V.Hence, F is a generalized force acting on the system if for any $\underset{\sim}{q} \in V$, $< F,q >$ represents a functional, which is a continuous linear map from V into the real (or complex) numbers.It is easy to show that the topological dual of an n-dimensional linear vector space is again an n-dimensional linear vector space.Without introduction of basis it is harder to show that every continuous,linear map from V into the reals can be written in the form

$$W(\underset{\sim}{q}) = \sum_{i=1}^{n} F_i q_i = \langle \mathbf{F}, \mathbf{q} \rangle$$

for every $\mathbf{q} \in V$. (This is a consequence of the Lax-Milgram theorem quoted in this volume in the Appendix B. Write $F_i = A\phi_i$, $\phi \epsilon V^*$, where A is the operator stated in that theorem.) The component of F_i in the bilinear product $\langle F, q \rangle$ is called the component of generalized force corresponding to the generalized displacement q_i. If q_i has the physical dimension of length (inches, cm., ...) then F_i will have the physical dimension of force (lbs., Kgs,...). If q_i is dimensionless (say radians), then F_i will have the dimension of energy (lbs ·inches, Kg.· cms.,...).

So far we did not insist that the generalized displacements q_i are linearly independent. In fact, there may be a number of constraints, say m<n constraints of the form

$$\phi_1(q_1,q_2,\ldots,q_n) = 0 \ ,$$
$$\phi_2(q_1,q_2,\ldots,q_n) = 0 \ ,$$
$$\vdots$$
$$\phi_m(q_1,q_2,\ldots,q_n) = 0 \ ,$$

which may allow us to eliminate m generalized displacements. Without any loss of generality let us assume that $q_{n-m+1}, q_{n-m+2},\ldots, q_n$ have been eliminated, and that $q_1,q_2,\ldots,q_{n-m}$ are the remaining generalized displacements. Let us also assume that the remaining displacements $q_1,\ldots,q_{n-m}$ are linearly independent. Then n-m is called the number of degrees of freedom of the system.

Example: The motion of a double pendulum. (i.e. a pendulum suspended from another pendulum). The masses move in a plane.

This system is illustrated on figure 16.

The position of the system is uniquely determined by the two angles θ_1, θ_2. (Hence we have a system with two degrees of freedom.) Taking θ_1 and θ_2 as the generalized displacements, we see that the corresponding generalized forces are the restoring moments M_1 and M_2. The work performed by the moments acting on the displacements θ_1, θ_2 is given by

$W = \sum_{i=1}^{2} M_i\theta_i$. The total (negative) work performed by raising the masses from the position $\theta_1 = 0, \theta_2 = 0$ is computed as $V = -mg\ell\,(2\cos\theta_1 + \cos\theta_2)$. This is the potential energy. The position of the system corresponding to $\theta_1 = 0$ and $\theta_2 = 0$ coincides with the extremal value of the functional $V(\theta_1,\theta_2)$.

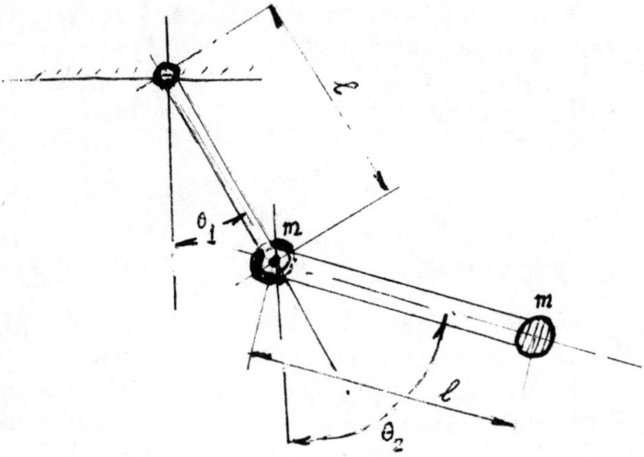

Figure 16

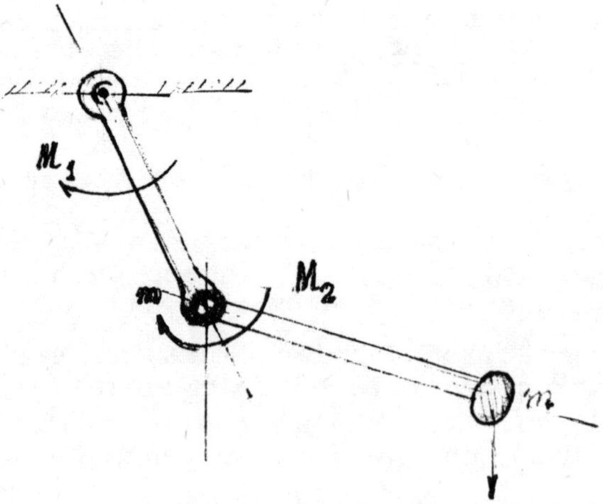

Figure 17

(1.8) A Formal Introduction To Generalized Coordinates

Consider a system which may be completely described by n independent physical quantities: $q_1, q_2, \ldots, q_n$. For example, for a single particle mechanical system, we need 3 coordinates to describe its position, such as x, y, z-Cartesian coordinates, or R, θ, z-cylindrical polar coordinates. These quantities q_i as we have indicated, will be our "generalized position coordinates," or simply, the generalized coordinates.

In order to describe a system of m particles in a 3-dimensional space, we need 3m coordinates. For example, to describe the positions of m-mass points, we need 3m coordinates. However, if we also have r-geometric constraints, then we only need 3m-r coordinates. The number 3m-r will be called the number of degrees of freedom of the system.

The physical meaning of these additionally given relationships will be an imposition of some constraints on the general motion of the system. Without attempting to define it we shall consider an "infinitesimal change of the system," compatible with the constraints. The virtual displacements will be such infinitessimal changes (denoted by $\delta q_1, \delta q_2, \ldots \delta q_n$) in the values of the generalized coordinates $q_1, q_2, \ldots q_n$.

Note: A rigorous definition requires some knowledge of the mathematical model theory.

(1.9) The "δ" Notation (a heuristic approach).

The δ-symbol as presented here was first introduced around 1790 by Lagrange who had no scrupules in considering infinitesimal displacements.

The meaning of the "δ" notation may be best explained as follows: Let us consider a continuous and differentiable function of the position coordinates:

$$F(q_1, q_2, \ldots, q_n)$$

with some holonomic constraints:

$$r_1(q_1,q_2,q_3,\ldots q_n) = 0 \, ,$$

$$r_2(q_1,q_2,q_3,\ldots q_n) = 0 \, ,$$

$$\cdot$$

$$\cdot$$

$$\cdot$$

$$r_k(q_1,q_2,q_3,\ldots q_n) = 0 \, .$$

The constraints are defined to be holonomic if the equations expressing the conditions of constraint could be put in a form involving only the coordinates. For example, if the derivatives of the coordinates are involved in the constraint conditions, and they can not be eliminated by quadratures, such constraints are nonholonomic.

As an example we could consider a function $V(x,y)$ expressing the potential energy of a single particle moving in the x,y plane while a constrain $r(x,y) = x^2 + y^2 - 1 = 0$, expresses the fact that its position is restricted to the circumference of the unit circle:

$$x^2 + y^2 = 1.$$

Let's assume that in fact we have computed the potential energy of this particle at the point P with coordinates $x = \frac{\sqrt{2}}{2}$, $y = \frac{\sqrt{2}}{2}$. Now we wish to investigate how the potential energy would change if an infinitesimal displacement took place along the path:

$$x^2 + y^2 = 1$$

This displacement will be called a "virtual displacement," while the infinitesimal change in the value of the function V will be called the "variation of V" and denoted by δV.
We denote the infinitesimal changes in the coordinate values by the symbols:

$$\delta q_1, \ \delta q_2, \ \ldots \ \delta q_n \ .$$

We accept the chain rule of differentiation

$$\delta V = \frac{\delta V}{\delta q_1} \cdot \delta q_1 + \frac{\delta V}{\delta q_2} \cdot \delta q_2 + \ldots \frac{\delta V}{\delta q_n} \cdot \delta q_n \approx \sum_{i=1}^{n} \frac{\delta V}{\delta q_i} \cdot \delta q_i \; .$$

The quantity

$$Q_i = \frac{\delta V}{\delta q_i} \approx \frac{\partial V}{\partial q_i}$$ (if δq_i is "small") will be

called the generalized force corresponding to the virtual displacement along generalized coordinate q_i.

If q_i's are Cartesian coordinates whose physical interpretation is length, Q_i's have the dimension of force. In this case the word "generalized" can be safely omitted. We observe that if constraints have been imposed δq_1, δq_2, $\ldots \delta q_n$ are not arbitrary. In our previous

example the particle could move only along the circumference of the unit circle, forcing the condition

$$\frac{\delta x}{\delta y} = \frac{\cos\,(x,s)}{\cos\,(y,s)} = \frac{\cos\,\pi/4}{\cos\,\pi/4} = 1 \; .$$

(Note: cos (x,s) means the cosine of the angle between the x-axis and the direction of displacement ds). This is true in general, and we can express the compliance with constraints by relationships:

$$\delta q_1 = t\alpha_1 \, , \; \delta q_2 = t\alpha_2 \, , \ldots , q_n = t\alpha_n \, ,$$

where α_i is the cosine of the angle formed between the q_i coordinate and the direction in which the particle is free to move, and t is an (undefined) infinitesimal parameter.

1.10 The principle of virtual work

A commonly used, but undefined concept (until the advent of non standard analysis) which is most useful and appeals to the intuition is the concept of virtual displacement.

Let us consider a mechanical system with N degrees of freedom, whose displacements can be located in a Euclidean n-space R^n with Cartesian coordinates $x_1, x_2 \ldots x_n$.

The system is subjected to m-holonomic constraints, that is to constraints that can be reduced to the form

$$(1.6)\quad f_\alpha(x_i, t) = 0 \qquad \alpha = 1,2,\ldots m \quad i = 1,2,\ldots r.$$

Then $\delta x = (\delta x_1, \delta x_2, \ldots, \delta x_n)$ will denote the virtual displacement of the system, which is heuristically described as an infinitessimal change in the values of coordinates $x_1, x_2, \ldots, x_n$ that is consistent with constraints. That is $\delta \underset{\sim}{x} = \underset{\sim}{x}' - \underset{\sim}{x}$ where $\underset{\sim}{x}' = (x_1', x_2', \ldots, x_n')$ are

Cartesian coordinates of the system that satisfies the constraints, and which are infinitesimally close to the coordinates $\underset{\sim}{x}$. Of course the "definition" makes no sense when $x \in \mathbb{R}^N$, and we are clearly talking about a non-standard model $^*\mathbb{R}^N$ of $\mathbb{R}^N$. We only need to say at this point that we have the mathematical tools to make this entire discussion rigorous.

However, various statements which will follow in these notes will appear in a slightly unfamiliar form. The reason for that is that when "infinitesimals" are introduced, extreme care has to be taken making rigorous statements about the functions,sets and relations in an arbitrary model. For example $f(\delta x) = 0$ may have entirely different meaning then the one we ascribe to it "intuitively". It could mean ≈ 0, or it could mean $f(\delta x) = o(\delta x)$, or it could really mean "$= 0$", since in the language used to describe the properties of the real line it is impossible to define any of the concepts, or symbols introduced above except "equal to zero".

The usual statement is that the virtual work due to reaction forces generated by the constraint is zero (See an engineering text such as for example[26]-pages 102-103)

$$(1.7) \quad \sum_{i=1}^{n} R_j \delta x_j = 0$$

where

$$R_j = \sum_{\alpha=1}^{N} \frac{\partial f_\alpha}{\partial x_j} \quad . \qquad (1.8)$$

There is a temptation to introduce a parameter t (say time) and rewrite (1.8) as

$$\sum_{j=1}^{N} R_j \frac{\partial x_i(t)}{\partial t} = 0 \qquad (1.9)$$

subject to constraints $f_\alpha(x_j, t) = 0$.

Unfortunately, the system (1.9) is more complex to manipulate and difficult to interpret for static equilibrium of forces, where the principle of virtual work is most useful.

The difficulty appears in our interpretation of the meaning of R_j. Is it computed at the point $\tilde{x}_0$, or at the point $\tilde{x}_0 + \delta\tilde{x}_0$? As a simple example we could consider the reaction forces arising when we restrict the motion of a rigid body by pivoting it at some point p.

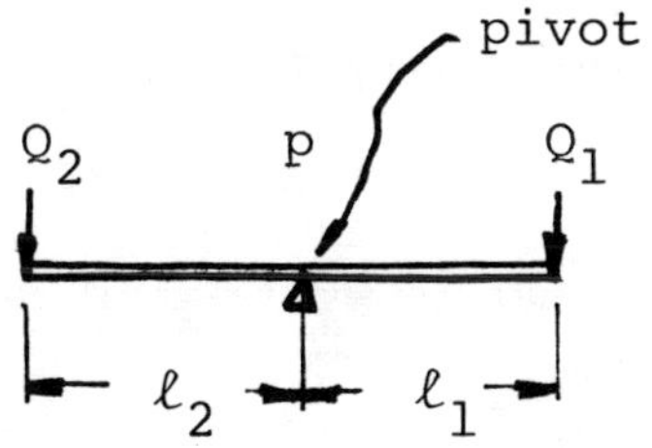

Figure 18

Examining this particular case it should become clear that the principle of virtual work states in the form (1.8) that

$$\lim_{\delta x_j \to 0} \left(\frac{\sum_{j=1}^{N} R_j \delta x_j}{\| \delta x \|} \right) = 0$$

or in the o,O (little o, big O) notation that

$$\sum_{j=1}^{N} R_j \delta x_j = o\|\delta x\|^{-1} .$$

Realizing this fact we can keep the "δ" notation, and derive-without the fear of blunder-almost all well known variational principles of classical mechanics as a consequence of the virtual work principle.

Newton's second law applied to a system of N-masses, or to an equivalent mechanical system, is described by a system of second order differential equations

$$m_j \ddot{x}_j = X_j + R_j ,$$

where X_j are the external forces and R_j are the forces due to the constraints. It can be considered a consequence of the virtual work principle

$$(1.10) \qquad \sum_{j=1}^{N} (m_j \ddot{x}_j - X_j - R_j)\cdot \delta x_j = 0.$$

If no work is performed by the constraints, then the equation (1.10) becomes

$$(1.11) \qquad \sum_{j=1}^{N} (m_j \ddot{x}_j - X_j)\cdot \delta x_j = 0$$

(Or more correctly $= o(\|\delta \underset{\sim}{x}\|^{-1})$.

Introducing the kinetic energy functional

$T = \frac{1}{2} \sum_j m_j \dot{x}_j^2$ we can rewrite the relation (1.11) as

$$(1.12) \quad \sum_j \left(\frac{d}{dt}\left(\frac{\partial T}{\partial \dot{x}_i}\right) - X_j\right) \delta x_j = 0,$$

and upon a coordinate change

$$x_j = x_j(\underset{\sim}{q},t) , \quad \underset{\sim}{q} = (q_1,q_2,\ldots,q_n) ,$$

$$(1.13) \quad \sum_{i=1}^{N} \left(\frac{d}{dt}\left(\frac{\partial T}{\partial \dot{q}_i}\right) - \frac{\partial T}{\partial q_i} - Q_i\right)\cdot \delta q_i = 0,$$

where

$$Q_i = \sum_{j=1}^{N} X_j \frac{\partial x_i}{\partial q_j} ,$$

i.e.,

$$(1.4) \quad < \frac{d}{dt}\left(\frac{\partial T}{\partial \dot{q}_i}\right) - \frac{\partial T}{\partial q_i} - Q_i , \ \delta q_i > = 0 ,$$

which is a more general form than one previously obtained by setting to zero the Gateaux derivative of the Lagrangian functional in the direction of an arbitrary displacement vector, which is consistent with constraints.

A special case when the forces Q_i are derived from a potential V, i.e.

$$(1.5) \quad Q_i = -\frac{\partial V}{\partial q_i}, \quad i=1,2,\ldots,n,$$

results in a more familiar formulation in terms of the Lagrangian L, which in this case is $L = T - V$.

(1.11) <u>The Hamiltonian, and an introduction of a momentum variable.</u>

The problem of extremizing a functional

$$L(q_i, \dot{q}_i, t) = \int_{t_0}^{t_1} L dt$$

may be replaced by an equivalent problem of finding solutions to the so called canonical equations by first introducing a transformation

$$(\tilde{q}, \tilde{\dot{q}}) \to (\tilde{q}, \tilde{p})$$

defined by

$$(1.16) \qquad p_i = \frac{\partial L}{\partial \dot{q}_i} .$$

We introduce a functional called the Hamiltonian

$$(1.17) \quad H = \sum_i p_i \dot{q}_i - L .$$

(See [2], or [1]).

The virtual work principle (1.11) becomes after substituting (1.16) and (1.17)

$$\delta \int_{t_0}^{t_1} L\, dt \approx \int_{t_0}^{t_1} \{ \sum_{i=1}^{N} [\dot{q}_i - \frac{\partial H}{\partial p_i}]\, \delta p_i + $$

$$(\dot{p}_i + \frac{\partial H}{\partial q_i})\, \delta q_i \}\, dt + \sum_{i=1}^{N} p_i \delta q_i \Big|_{t_1}^{t_2} . \qquad (1.8)$$

If we assume that the virtual displacements are time independent. i.e. the constraints imposed on the system were time independent, then the additional assumptions of existence of all derivatives appearing in (1.18) and the previously discussed denseness hypothesis concerning δq_i, δp_i, and the additional assumption that

$$(1.19) \qquad (\Sigma p_i \delta q_i)_{t=t_1} = (\Sigma p_i \delta q_i)_{t=t_2} \qquad \text{does}$$

lead to the canonical equations of Hamilton

$$(1.20) \qquad \dot{q}_i = \frac{\partial H}{\partial p_i}, \qquad \dot{p}_i = -\frac{\partial H}{\partial q_i}, \quad i=1,2,..,N$$

An interpretation of all symbols in formulas (1.16) - (1.20) could be attained by discussing the motion of a spherical pendulum in the usual r,θ,φ spherical coordinate system.

If the Lagrangian L and consequently the Hamiltonian H does not explicitely depend on time, then one can deduce from the canonical equations (1.20) that $H(p,q)$ is constant along any trajectory of motion for the system. A simple computation shows that $L \doteq T - V$ implies that $H = T + V = h = \text{const.}$ H is easily identified as the total energy of the system. $H = h = \text{constant}$ implies that the total energy of the system is conserved.

If we also make the assumption that the kinetic energy is a homogeneous quadratic form in $\dot{q}_i$ that is

$$(1.21) \qquad T = \tfrac{1}{2} \sum_{i,j=1}^{N} a_{ij}\dot{q}_i\dot{q}_j \quad ,$$

where

$$a_{ij}(q) = \frac{\partial^2 T}{\partial \dot{q}_i \partial \dot{q}_j} ,$$

then we can rewrite the Euler-Lagrange equation (1.18) in the form

$$\Delta \sum_{A}^{B} ds = 0 \quad ,$$

$$ds^2 = \sum_{i,j=1}^{N} b_{ij} d\dot{q}_i d\dot{q}_j \ , \ b_{ij} = 2(E-V)a_{ij} \qquad (1.22)$$

and Δ is the total variation which is given by

$$\Delta q_i = q_i(t + \Delta t) + \delta q_i(t + \Delta t) - q_i(t)$$

reducing the "strong" minimization problem (that is with assumptions concerning the existence of all strong derivatives) to a geodesic problem in Riemannian geometry.

Some examples and questions

1) We need to check that the Hamiltonian $H(p,q)$ is invariant along any trajectory of motion if H does not explicitely depend on the time variable.

2) Consider a torsional system consisting of two flywheels mounted on a flexible elastic shaft, as illustrated on figure 19.

The total kinetic energy is

$$(1.23) \qquad T = \tfrac{1}{2}I_1(\dot{\theta}_1)^2 + \tfrac{1}{2}I_2(\dot{\theta}_2^{\,2}) \quad , \ I_1, I_2 > 0.$$

We change variables by denoting

$$q_1 = \sqrt{I_1}\ \theta_1 \ , \quad q_2 = \sqrt{I_2}\ \theta_2 \ ,$$

and we introduce the following simple metric

$$ds^2 = b_1 \cdot (dq_1)^2 + b_2 \cdot (dq_2)^2 \ , \qquad b_1, b_2 \geq 0.$$

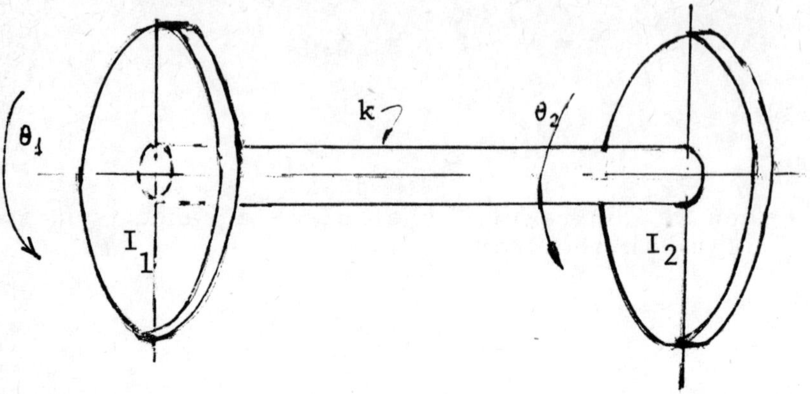

The kinematic configuration manifold is a torus.

Figure 19

If each shaft is regarded as a linear torsional spring then the potential energy can be assumed to be equal to the strain energy, and it is given by the formula

$$(1.24)\quad V=\tfrac{1}{2}k(\theta_1-\theta_2)^2 = \tfrac{1}{2}k(q_1/I_1-q_2/I_2)^2 \;.$$

Related questions
1. Name in this example the coefficients b_{ij} of the formula (1.22).
2. What do the geodesics look like on the kinematic manifold illustrated on figure 19 ?
3. When is the motion of this flywheel system represented by a periodic function of time?

(1.12) A note on a possible variant of the duality of the Lagrangian and the Hamiltonian formalisms arising from the Legendre transformation(2. 16).

Symmetrising the Lagrangian, and offering a symmetric analogue of Noble's formalism has been suggested in classical mechanics(the monograph of Corben and Stehle [1] and in the paper by Komkov and Valanis [27] on the hidden variable approach to the thermodynamics of solids.

Let us again rewrite the Euler-Lagrange equations in the form:

$$\frac{\partial L}{\partial \underset{\sim}{p}} = 0 = a\underset{\sim}{u} - f(\underset{\sim}{u})$$

$$\underset{\sim}{u} \in H_1$$

$$a : H_1 \to H_2 \text{ (linear)}$$

$$f : H_1 \to H_2,$$

where $\underset{\sim}{p} = \frac{\partial L}{\partial (a\underset{\sim}{u})}$.

$$L = < a\,\underset{\sim}{u},\ \underset{\sim}{p} >_\Omega - < f(\underset{\sim}{u}), \underset{\sim}{p} >_\Omega + B(\underset{\sim}{u}, \underset{\sim}{p})_{\partial\Omega}$$

where B is a bilinear form defined on $\partial\Omega$. Of course this is not the way it is done in physics, where one starts with the Lagrangian $L(\underset{\sim}{u}, a\underset{\sim}{u}, t)$.

a is usually the operator $\frac{d}{dt}$. Then one transforms the system to the Hamiltonian formalism by the use of Legendre transformation

$$\underset{\sim}{p} = \frac{\partial L}{\partial \dot{\underset{\sim}{u}}}, \quad H = < p,\ \dot{u} > - L.$$

We shall denote by $[L]$ the commutator $[L] = [L]_\Omega + [L]_{\partial\Omega}$

For the time being let us assume homogeneous boundary conditions that is $B(\mathbf{u},\mathbf{p})_{\partial\Omega} \equiv 0$, and set

$[L]_{\partial\Omega} \equiv 0$. The commutator $[L]_\Omega$ is defined by the formula:

$$[L]_\Omega = < a\underset{\sim}{u},\ \underset{\sim}{p} >_\Omega - < \underset{\sim}{u}, a^* \underset{\sim}{p} >_\Omega + C$$ where C is an arbitrary constant. If a and a^* are true adjoints, then by definition

$[L]_\Omega \equiv C$ along any trajectory of the system.

We notice that for conservative systems replacing L by $([L]-L)$ interchanges the roles played by the generalized coordinates $\underset{\sim}{u}$ and by the generalized momenta $\underset{\sim}{p}$, in the manner reminiscent of the duality theory for electrical networks.

We shall demonstrate the validity of this statement on a completely trivial example.

We consider the motion of a harmonic oscillator $m\ddot{x} + Kx = 0$. The Lagrangian is

$$L = \frac{1}{2}(m^{\frac{1}{2}}\dot{x})^2 - \frac{1}{2}(k^{\frac{1}{2}}x)^2, \quad m>0,\ k>0.$$

$$t \in [0,\infty],\ x \in C^1(-\infty, +\infty).$$

Ignoring the initial or boundary conditions assigned to this harmonic oscillator, we formulate

$-L + [L] = p\dot{x} + x\dot{p} - L$ (recalling that $(\frac{d}{dt})^* = -\frac{d}{dt}$); $\bar{L} = -L + [L] = -\frac{1}{2}\left(\frac{\dot{p}}{k^{\frac{1}{2}}}\right)^2 + \frac{1}{2}\left(\frac{p}{m^{\frac{1}{2}}}\right)^2$

Since $[L]$ = Const. along any trajectory, the minimum of the action integral $\int_0^T L(x,\dot{x})dt$ corresponds to the maximum of the dual integral $\int_0^T \bar{L}(p,\dot{p})\,dt$.

Hence, $\bar{L}$ satisfies the Euler-Lagrange equations of the form:

$$m\ddot{p} + kp = 0,$$

which in this case is a triviality. As an exercise we could consider a similar problem, where the result is not quite so trivial.

Consider the oscillator with a variable mass $m(t)$.

Identifying $\mathcal{a} \cdot = m(t) \frac{d}{dt} (\cdot)$,

$$\mathcal{a}^{*}\cdot = - \frac{d}{dt}[m(t)\ (\cdot)\ ]$$

We have $[L] = \langle m\dot{x}, \hat{p} \rangle + \langle x, (m\hat{p})^{\cdot} \rangle + C$.

$\bar{L} = [L] - L$ is simply equal to $C - L$ along a trajectory.

However the generalized momentum $\hat{p}$ is not

is not the usual linear momentum $m\dot{x}$, since $\hat{p} = \frac{\partial L}{\partial(\mathcal{a}x)} = \dot{x}$. The trajectory can be established by requiring $\frac{\partial \bar{L}}{\partial x} = 0$, i.e.

$$\frac{d}{dt}(m(t)\dot{p}(t)) + k\ p(t) = 0. \qquad (*)$$

Clearly, this equation could have been obtained by multiplying $(m\ddot{x} + kx)$ by p and then integrating by parts, but the conclusion that the equation (*) is valid needs to be justified since x(t) was not any arbitrary function, but one satisfying the original equation of motion. One needs to do some hand waving at this point to justify the validity of the adjoint equation (*) without an appropriate functional analytic reasoning.

<u>An example</u>: The Lagrangian for a two-body problem.

We assume that the interaction forces between the two bodies (or particles) depend only on the distance between them.

Hence, we assume that the potential is a function of the distance between the particles $\mathbf{r}_{12} = \vec{\mathbb{R}}_1 - \vec{\mathbb{R}}_2$.

We denote this distance by r_{12}, $r_{12} = |\mathbf{r}_{12}|$.

$$r_{12} = +\sqrt{(x_1-x_2)^2 + (y_1-y_2)^2 + (z_1-z_2)^2} = |\vec{\mathbb{R}}_1 - \vec{\mathbb{R}}_2|,$$

in some previously chosen system of Cartesian coordinates. $U = U(r_{12})$ is the potential.

The Lagrangian is given by

$$L = \frac{1}{2} m_1 || \dot{R}_1 ||^2 + \frac{1}{2} m_2 || \dot{R}_2 || - U(\mathbf{r}_{12})$$

$\mathbb{R}_1$ denotes the position vector for the first particle, $\mathbb{R}_2$ for the second particle with respect to some (arbitrarily chosen) coordinate system. The origin of the Cartesian coordinate system is supposedly fixed.

Let the origin move with the motion of the particles in such way that it coincides with the center of mass of the system. In the new coordinates (ξ,η,ζ), $\xi = 0$, $\eta = 0, \zeta = 0$, corresponds to $\vec{\mathbb{R}} = m_1\mathbb{R}_1 + m_2\mathbb{R}_2$. Let $m = \frac{m_1 m_2}{m_1+m_2}$ be the "reduced mass".

The "new" Lagrangian in the ξ,η,ζ coordinate system is

$$\bar{L} = \frac{1}{2} m|\dot{\mathbf{r}}_{12}(\xi,\eta,\zeta)|^2 - U(\mathbf{r}_{12}) \text{ with } r_{12} =$$

$|\mathbf{r}_{12}|$, as before.

Thus, the Lagrangian L for a two-body problem has been reduced to the Lagrangian $\bar{L}$ for a single particle in the so called "reduced" coordinate system.

A research suggestion:

Could a similar "trick" work for a 3-body, or perhaps for an n-body system?

1.12 A <u>more complex</u> <u>example</u>: Coupled bending and torsional vibrations of a long beam.

<u>Notation and basic definitions</u>

We shall consider the vibration of a beam governed by the pair of differential equations:

$$(1.25) \quad L_1(y,\theta) = \frac{\partial^2}{\partial x^2} [E(x)I(x)\frac{\partial^2 y(x,t)}{\partial x^2}] + \rho(x)A(x)$$

$$\frac{\partial^2 y(x,t)}{\partial t^2} - \rho(x)A(x)e(x)\frac{\partial^2\theta(x,t)}{\partial t^2} = \phi_1(x)u_1(t)$$

$$L_2(y,\theta) = \frac{\partial^2}{\partial x^2}[E(x)C_w(x)\frac{\partial^2\theta(x,t)}{\partial x^2}] -$$

$$\frac{\partial^2}{\partial x^2}[G(x)C(x)\ \theta(x,t)] - \rho(x)A(x)e(x)\frac{\partial^2 y(x,t)}{\partial t^2}$$

$$- \rho(x)I_0(x)\ \frac{\partial^2\theta(x,t)}{\partial t^2} = \phi_2(x)u_2(t).$$

The symbols used above have the following physical meaning:

$y(x,t)$ - the transverse displacement;

$$\theta, y \in H_o^2([0,\ell] \times [0,T]) = H_o^2(\Omega)$$

$$\phi_i(x) \in L_2^o\ [0,\ell],\ u_i(t) \in H^{-2}(\Omega).$$

$\theta(x,t)$ - the angle of rotation of the cross-section;

$e(x)$ - the distance from the centroid to the center of torsion;

$y_g(x,t)$ - the transverse displacement of the centroid of the cross-section; $(y_g = y - e\theta)$

$A(x)$ - the cross-sectional area

$I_p(x)$ - the polar moment of inertia of the cross-section about the centroid;

$I_0(x)$ - the polar moment of inertia with respect to the shear center $(I_0 = I_p + Ae^2)$

$E(x)$ - Young's modulus;
$\rho(x)$ - the material density;
C_w - the modulus of warping rigidity;

(EC_w is the warping rigidity of the beam)
G - modulus of torsional rigidity;
GC - denotes the torsional rigidity, and
C is the torsional constant, satisfying $T_t = GC\frac{\partial\theta}{\partial x}$
where T_t is the twisting moment due to shear stresses;

$\phi_1(x)u_1(t)$ - are the forces applied to the beam;

$\phi_2(x)u_2(t)$ the first one has the physical dimension of force per unit length, the second- of moment per unit length.

We assume that the applied force (i.e. the inhomogeneous term) is of the form $\phi_i(x)u_i(t)$, i= 1,2, where $\phi_i(x)$ is either a bounded and square integrable, (hence absolutely integrable function of x) on the interval $[0,\ell]$, or is either the Dirac delta function or its derivative concentrated on a finite number of points of $[0,\ell]$. That is, ignoring derivatives of Dirac delta, we have

$$\phi_j(x) = \sum_{i=1}^{K} a_{ij}\delta(x - \zeta_i) + \sigma_j(x), \; j = 1, 2,$$

where a_{ij} are constants, $\sigma_j(x)$ is a uniformly bounded square integrable function on $[0,\ell]$.

There may be a restriction of the type:

$$\sum_{i=1}^{K} |a_{ij}| + \int_0^{\ell} |\sigma_j(x)|dx \leq C$$ for some specified constant C, in the a priori choice of constants a_{ij}, ζ_i. $u_m(t)$ is a piecewise continuous function on $[0,T]$, satisfying $||u_i||_m < 1$. u_i, i = 1,2, are to be determined to optimize the energy level as postulated in the control problem. ($||\cdot||_m$ is the usual Euclidean m-norm.)

We shall consider here only the problem of constant E, ρ, I, A, e, hence of constant I_0, C_w on $[0,\ell]$. This is done for reasons of simplicity.

The basic discussion can be easily generalized to the case where EI, ρA, EC_w, GC, ρI_0, e are arbitrary piecewise continuous functions of x. The modified equations $(1\tilde{\underline{a}})$, $(1\tilde{\underline{b}})$ are:

$$(1.25^a)\quad L_1(y,\theta) = EI\frac{\partial^4 y(x,t)}{\partial x^4} + \rho A\,\frac{\partial^2 y(x,t)}{\partial t^2} - \rho Ae\,\frac{\partial^2\theta(x,t)}{\partial t^2} = \phi_1(x)u_1(t).$$

$$(1.25^b)\quad L_2(y,\theta) = EC_w\frac{\partial^4\theta(x,t)}{\partial x^4} - GC\frac{\partial^2\theta(x,t)}{\partial x^2} - \rho Ae\,\frac{\partial^2 y(x,t)}{\partial t^2} - \rho I_0\frac{\partial^2\theta(x,t)}{\partial t^2} = \phi_2(x)u_2(t).$$

The beam has an axis of symmetry, here the x-axis, with the y axis passing through the shear center, and the x-axis coinciding with the centroidal axis, as illustrated on figure 20.

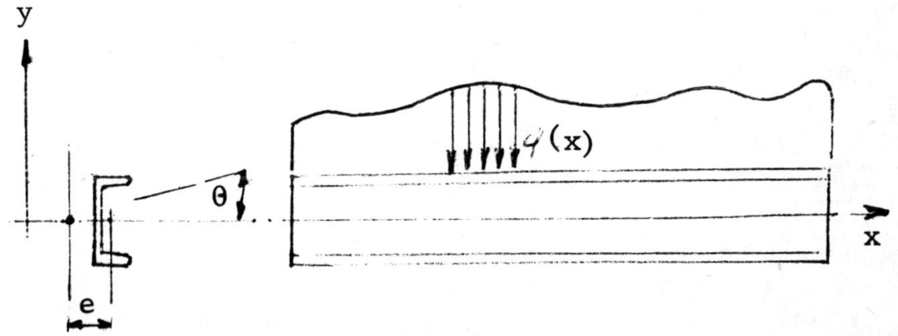

Figure 20

If a load w(x) is distributed along the x-axis of the beam it is causing a torque of magnitude w(x) · e to be distributed along the beam. Let us now assume that the only load applied is the inertia of the beam, and the beam is vibrating freely. Linearlized equations of a free motion are:

$$L_1(y,\theta) = EI\frac{\partial^4 y}{\partial x^4} + \rho A\frac{\partial^2 y}{\partial t^2} - \rho Ae\frac{\partial^2 \theta}{\partial t^2} = 0 \quad (1\underline{^{aH}})$$

$$L_2(y,\theta) = EC_w\frac{\partial^4 \theta}{\partial x^4} - GC\frac{\partial^2 \theta}{\partial x^2} - \rho Ae\frac{\partial^2 y}{\partial t^2} - \rho I_0\frac{\partial^2 \theta}{\partial t^2} = 0 \quad (1\underline{^{bH}})$$

We note that if e = 0 the equations (1^{aH}), (1^{bH}), become uncoupled and the free motion of the beam consists of two independent vibrations, one a purely torsional vibration, the other a bending vibration in the y-direction. In general the motion will be a coupled vibration with the energy transfer taking place between the two modes of vibrations.

The total potential energy is:

$$(1.26)\quad V = \frac{EI}{2}\int_0^{\ell}\left(\frac{\partial^2 y}{\partial x^2}\right)^2 dx + \frac{GC}{2}\int_0^{\ell}\left(\frac{\partial\theta}{\partial x}\right)^2 dx$$

$$+\frac{EC_w}{2}\int_0^{\ell}\left(\frac{\partial^2\theta}{\partial x^2}\right)^2 dx,$$

and the total kinetic energy is:

$$(1.27)\quad T = \rho\frac{A}{2}\int_0^{\ell}\left(\frac{\partial y_g}{\partial t}\right)^2 dx + \rho\frac{I_p}{2}\int_0^{\ell}\left(\frac{\partial\theta}{\partial t}\right)^2 dx .$$

where y_g is the y coordinate of the centroid of cross-section of the beam. If we assume the approximate relationship $y_g = y - e\theta$, we obtain

$$\frac{\partial y_g}{\partial t} = \frac{\partial y}{\partial t} - e\frac{\partial\theta}{\partial t},$$

which can be substituted into the equations (1.25) The equations (1.25) are Euler-Lagrange equation for the action integral L=T-V.

The total energy is defined by the equation:

$$\mathcal{E}(t) = T + V$$

2. <u>The boundary conditions.</u>

We note that warping and longitudinal stresses in the beam are proportional to

$\frac{\partial\theta}{\partial x}$, $\frac{\partial^2\theta}{\partial x^2}$ respectively. As an example we could consider either a bounded and square integrable, (hence absolutely integrable function of x) on the interval $[0,\ell]$, or is either the Dirac delta function or it's derivative concentrated on a finite number of points of $[0,\ell]$. That is

$$\phi_j(x) = \sum_{i=1}^{K} a_{ij}\delta(x-\xi_i)+\sigma_j(x),\ j = 1,2,\text{ where}$$

a_{ij} are constants, $\sigma_j(x)$ is a uniformly bounded square integrable function on $[0,\ell]$.

There may be a restriction of the type:

$$\sum_{i=1}^{K} |a_{ij}| + \int_0^{\ell} |\sigma_j(x)|dx \leq C \text{ for some specified}$$

constant C, in the a priori choice of $a_{ij}, \xi_i, \sigma_j(x)$.

$u_m(t)$ is a piecewise continuous function on $[0,T]$, satisfying $||u_i||m< 1$. u_i, i = 1,2, are to be determined to optimize the energy level as postulated in the control problem.

We shall consider here only the problem of constant E, ρ, I, A, e, hence of constant I_0, C_w on $[0,\ell]$. This is done for reasons of simplicity. The basic discussion can be easily generalized to the case where EI, ρA, EC_w, GC, ρI_0, e are aribitrary piecewise continuous functions of x.

The modified equations (1.25^a),(1.25^b) are:

$$L_1(y,\theta) = EI\frac{\partial^4 y(x,t)}{\partial x^4} + \rho A\frac{\partial^2 y(x,t)}{\partial t^2}$$

$$- \rho Ae\frac{\partial^2\theta(x,t)}{\partial t^2} = \phi_1(x)u_1(t), \text{ etc.}$$

Let us assume the boundary conditions of a simple support (at $x = 0$, or $x = \ell$), which implies restraint against rotation; however the beam is free to warp at the support point. Hence we have at:

$$\left\{\begin{array}{l} x = 0 \\ \text{or} \\ x = \ell \end{array}\right. \quad (2^a) \qquad \left\{\begin{array}{l} \theta = 0 \\ \dfrac{\partial^2\theta}{\partial x^2} = 0 \end{array}\right. \quad (2^b)$$

If the support is of the built-in type, no warping deformation is possible and we have:

$$\left\{\begin{array}{l} \theta = 0 \\ \dfrac{\partial\theta}{\partial x} = 0 \end{array}\right.$$

Combining these with the usual support conditions for w(x,t), we now postulate the boundary conditions of either $(2\underline{a})$ or $(2\underline{b})$ type at the boundary points $x = 0$ and $x = \ell$.
Let b denote either 0, or ℓ, then we have at the contact end points(that is at the points of support for the beam) either:

Simple support

$$\left.\begin{array}{ll} 2^b \left\{\begin{array}{l} y(b,t) \equiv 0 \\ \dfrac{\partial^2 y(b,t)}{\partial x^2} \equiv \end{array}\right. \\ 2^c \left\{\begin{array}{l} \theta(b,t) \equiv 0 \\ \dfrac{\partial^2\theta(b,t)}{\partial x^2} \equiv 0 \end{array}\right. \end{array}\right\} \quad t \in [0,T]$$

or: Built-in end (fixed support)

$$2^d \begin{cases} y(b,t) \equiv 0 \\ \dfrac{\partial y(b,t)}{\partial x} \equiv 0 \end{cases} \qquad 2^e \begin{cases} \theta(b,t) \equiv 0 \\ \dfrac{\partial \theta(b,t)}{\partial x} \equiv 0 \end{cases} \quad t \in [0,T]$$

and the initial conditions

$$\left.\begin{aligned} \theta(0,x) &= \theta_0(x) \\ y(0,x) &= y_0(x) \\ \frac{\partial\theta(0,x)}{\partial t} &= \omega_0(x) \\ \frac{\partial y(0,x)}{\partial t} &= v_0(x), \end{aligned}\right\}$$

where θ_0, y_0, ω, v_0 are continuously differentiable functions of x, whose second derivatives are square integrable and bounded functions on the interval $[0,\ell]$.
The conditions a) b) or c) are natural conditions for the action integral L.

We proceed as in [7] by introducing an inner product $<,>$.

The energy product.

Let $\tilde{Y}_{(1)}$, $\tilde{Y}_{(2)}$ be two solutions of (6) corresponding to vectors $\underset{\sim}{u}_{(1)}$, $\underset{\sim}{u}_{(2)}$ respectively.
We define:

$$<\underset{\sim}{Y}_{(1)},\underset{\sim}{Y}_{(2)}> = \frac{\rho}{2}\left\{\int_0^\ell \left[A(x)\left(\frac{\partial y_{g_1}}{\partial t}\right)\left(\frac{\partial y_{g_2}}{\partial t}\right)\right.\right.$$

$$\left. + I_p\left(\frac{\partial\theta_1}{\partial t}\right)\left(\frac{\partial\theta_2}{\partial t}\right)\right] dx +$$

$$\frac{E}{2}\int_0^\ell \left[I(x)\left(\frac{\partial^2 y_1}{\partial x^2}\right)\left(\frac{\partial^2 y_2}{\partial x^2}\right) + C_w(x)\left(\frac{\partial^2\theta_1}{\partial x^2}\right)\left(\frac{\partial^2\theta_2}{\partial x^2}\right)\right]$$

$$dx + \frac{G}{2}\int_0^{\ell} C(x)\left(\frac{\partial\theta_1}{\partial x}\right)\left(\frac{\partial\theta_2}{\partial x}\right) dx .$$

(Note that in the simple case A, I, I_p, C, C_w are constant, and can be brought outside the integral sign.) It is easily checked that this product satisfies all axiomatic requirements of an inner product.

As a consequence we have the Cauchy-Schwartz inequality:

$$< \underset{\sim}{Y}_{(1)}, \underset{\sim}{Y}_{(2)} >^2 \leq < \underset{\sim}{Y}_{(1)}, \underset{\sim}{Y}_{(1)} > \cdot < \underset{\sim}{Y}_{(2)}, \underset{\sim}{Y}_{(2)} >$$

$$= \varepsilon_1(\underset{\sim}{Y}_{(1)})\, \varepsilon_2(\underset{\sim}{Y}_{(2)}) \qquad (7)$$

where $\varepsilon_i(\underset{\sim}{Y}_{(i)}) = \varepsilon_i(Y_{(i)}(x,t))$, $i = 1,2$; is the total energy corresponding to the vector $\underset{\sim}{Y}_{(i)}$, $i = 1,2$. The equality holds only if for some constant C, it is true that $Y_{(1)} = C\, Y_{(2)}$. We compute the time derivative of the product $< Y_{(1)}, Y_{(2)} >$:

$$\frac{d}{dt} < \underset{\sim}{Y}_{(1)}, \underset{\sim}{Y}_{(2)} > = \sum_{i\neq j}^{2} \{\frac{\rho}{2}\int_0^{\ell}[A(x)\, \frac{\partial y_{g_i}}{\partial t}\, \frac{\partial^2 y_{g_j}}{\partial t^2}$$

$$+ I_p\, \frac{\partial \theta_i}{\partial t}\, \frac{\partial^2\theta_j}{\partial t^2}]\, dx + \frac{E}{2}\int_0^{\ell}[I(x)\, \frac{\partial^2 y_i}{\partial x^2}\cdot\frac{\partial}{\partial t}\left(\frac{\partial^2 y_j}{\partial x^2}\right) +$$

$$C_w(x)\, \frac{\partial^2\theta_i}{\partial x^2}\cdot\frac{\partial}{\partial t}\left(\frac{\partial^2\theta_j}{\partial x^2}\right)]\, dx$$

$$+ \frac{G}{2}\int_0^{\ell}[C(x)\, \frac{\partial\theta_i}{\partial x}\, \frac{\partial}{\partial t}\left(\frac{\partial\theta}{\partial x}\right)]\, dx$$

substituting $y_g = y_i - e\theta_i$, we have after integration by parts (and using boundary conditions of either $2\underline{a}$, $2\underline{b}$ or $2\underline{c}$ type and $2\underline{d}$, $2\underline{e}$):

$$\frac{d}{dt} < \underset{\sim}{Y}_{(1)}, \underset{\sim}{Y}_{(2)} > = \sum_{\substack{i,j=1,2 \\ i \neq j}} \{ \frac{1}{2} \int_0^{\ell} \dot{Y}_i [\rho A \ddot{y}_j -$$

$$\rho A e \ddot{\theta}_j + EI \frac{\partial^4 y_j}{\partial x^4}] \, dx + \frac{1}{2} \int_0^{\ell} \dot{\theta}_i [-\rho I_0 \ddot{\theta}_j - \rho A e \ddot{y}_j$$

$$- GC \frac{\partial^2 \theta_j}{\partial x^2} + EC_w \frac{\partial^4 \theta_j}{\partial x^4}] dx \}$$

$$= \frac{1}{2} \int_0^{\ell} [\dot{y}_1 L_1(y_2, \theta_2) + \dot{y}_2 L_1(y_1, \theta_1)$$

$$+ \dot{\theta}_1 L_2(y_2, \theta_2) + \dot{\theta}_2 L_2(y_1, \theta_1)] \, dx .$$

$$= \frac{1}{2} \int_0^{\ell} [\dot{\underset{\sim}{Y}}_{(1)} \cdot \underset{\sim}{\phi}_{(2)} \underset{\sim}{u}_{(2)} + \dot{\underset{\sim}{Y}}_{(2)} \cdot \underset{\sim}{\phi}_{(1)} \underset{\sim}{u}_{(1)}] \, dx. \quad (1.28)$$

(The dots stood for differentiation with respect to time).

We observe that the forces of the first system act on displacements of the second system and vice-versa.

Written out in full the equation (1.28) is :

$$\frac{d}{dt} < \underset{\sim}{Y}_{(1)}, \underset{\sim}{Y}_{(2)} > = \frac{1}{2} \int_0^{\ell} [\dot{Y}_1(x,t) \phi_{1(2)}(x) u_{1(2)}(x)$$

$$+ \dot{\theta}_1(x,t) \phi_{2(2)}(x) u_{2(2)}(t) +$$

$$\dot{y}_2(x,t) \phi_{1(1)}(x) u_{1(1)}(t) + \dot{\theta}_2(x,t) \phi_{2(1)}(x) \cdot$$

$$\cdot u_{2(1)}(t)] \, dx, \quad (8^{\underline{a}}).$$

The bracketed subscripts referred to the corresponding vectors $Y_{(1)}$ or $Y_{(2)}$ respectively. Now let $y_{(2)} = y_H$ be any solution of the homogenous equation corresponding to inhomogenious term, i.e. the control vector $\underset{\sim}{u}_{(2)} \equiv 0$. Then

$$\frac{d}{dt} < \underset{\sim}{Y}_{(1)}, \underset{\sim}{Y}_H > = \frac{1}{2} \int_0^{\ell} [\dot{Y}_H \phi_{1(1)} u_{1(1)} +$$

$$\theta_H \phi_{2(1)} u_{2(1)}]\, dx,$$

and

$$< \underset{\sim}{Y}_{(1)}, \underset{\sim}{Y}_H >_{t = \tau} = < \underset{\sim}{Y}_{(1)}, \underset{\sim}{Y}_H >_{t = 0} +$$

$$\frac{1}{2} \int_0^{\tau} \int_0^{\ell} [\dot{Y}_H \phi_{1(1)} u_{1(1)} + \theta_H \phi_{2(1)} u_{2(1)}]\, dx\, dt.$$

Example The brachistochrone problem.

No introduction to variational problems and principles can be complete without at least mentioning the historical origins of calculus of variations and the "brachistochrone" problem of John Bernoulli.

Its formulation and development had decisive influence on subsequent development of the subject. Let us restate this problem.

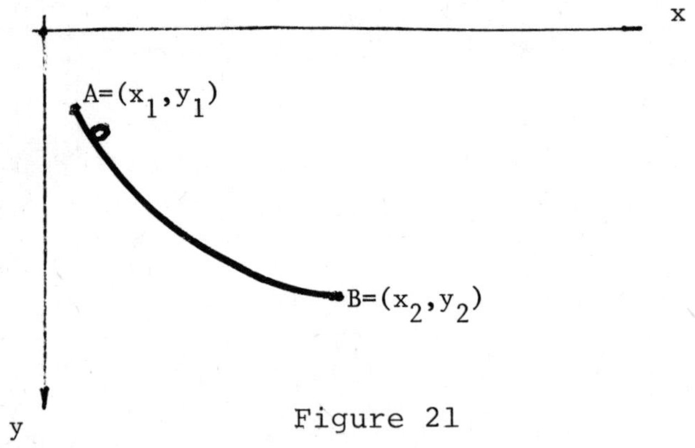

Figure 21

A historical problem of calculus and of mechanics.

At the end of the 17th century (specifically in 1697) John Bernoulli posed the following problem, known as the brachistochrone problem:How one must draw a curve connecting two points A and B, not lying in the same vertical plane, so that a particle sliding without friction along the curve from A to B will reach point B in the least time? Only continuously differentiable curves will be considered in this section.

The mathematical expression for the time taken is:

$$t = \int_A^B \frac{ds}{V} = \int_{x_0}^{x_1} \left(\frac{1}{\sqrt{2g}} \cdot \frac{\sqrt{(1 + y')^2}}{\sqrt{y - y_0}} \right) dx \quad (128^a)$$

recalling that $V = \sqrt{2gh} = \sqrt{2g(y-y_0)}$.

Of all the possible continuously differentiable functions $y(x)$ we have to find a function $\hat{y}(x)$ that assigns to the integral

$$I(y) = \int_{x_0}^{x_1} \sqrt{\left(\frac{1 + y'^2}{y - y_0} \right)}\, dx \quad (1.28^b)$$

the least possible value.

A problem which would lead to a similar expression is Fermat's principle of least time. Newton's solution of the brachistochrone problem utilized this analogy.

Other problems leading to expressions of the type (1.24) are found, for example, in attempting to find surfaces of least area, geodesic curves on 3-dimensional surfaces, etc.

Let us follow roughly Euler's ideas and examine a functional of the type

$$I(y) = \int_{x_0}^{x_1} F(x,y,y')dx \,. \quad (1.29)$$

Let us assume that the points x_0 and x_1 are fixed, but the function $y(x)$ may be varied in an arbitrary manner between the points x_0 and x_1. Let us assume that some function $y(x) = u(x)$ minimizes the functional (1.25). Then addition of any arbitrary function $\eta(x)$ to $u(x)$ must increase the value of (1.25). Let t be a "small" constant, so that the altered values of u(x) lie in a suitably small neighborhood of u(x):

$$\tilde{y}(x,t) = u(x) + t\,\eta(x)$$

Because of the assumed smoothness of y(x,t),and η(x,t) this will not cause any difficulties.

That is,only small variation will take place in the value of (1.25) if t is sufficiently small, and η(x) is bounded.

We regard I(u+tη) as a function of the parameter t.

$$\Phi(t) = I(u+t\eta) = \int_{x_0}^{x_1} F(x,\ u + t\eta,\ u' + t\eta')$$

The quantity tη(x) = δu is called the variation of u(x). From here on let us repeat the classical argument of Euler.

Since we assume that the functional has a minimum for t=0, we must have

$$\left.\frac{\partial\phi}{\partial t}\right|_{t=0} = \phi'(0) = 0 \text{ and } \delta I = 0,$$

or assuming the existence of all appropriate derivatives we have

$$\phi(0) = \int_{x_0}^{x_1} \left(\eta\frac{\partial F}{\partial u} + \eta'\frac{\partial F}{\partial u'}\right) dx = 0$$

Integrating by parts, we have the identity

$$\int_{x_0}^{x_1} \eta'\frac{\partial F}{\partial u'}\,dx = \eta\frac{\partial F}{\partial u'}\Bigg|_{x_0}^{x_1} - \int_{x_0}^{x_1} \frac{d}{dx}\left(\frac{\partial F}{\partial u'}\right)\eta\,dx =$$

$$- \int_{x_0}^{x_1} \eta\frac{d}{dx}\left(\frac{\partial F}{\partial u'}\right)dx,$$

since

$$\eta(x_0) = \eta(x_1) = 0,$$

according to our assumptions.

This result implies the following equality:

$$\int_{x_0}^{x_1} \eta\left(\frac{\partial F}{\partial u} - \frac{d}{dx} \frac{\partial F}{\partial u'} \right) dx = 0.$$

Since η is a completely arbitrary continuous function, this can be true only if

$$\frac{\partial F}{\partial u} - \frac{d}{dx} \left(\frac{\partial F}{\partial u'} \right) = 0. \qquad (1.29^a)$$

This statement is not trivial but it will not be proved here.

We note that if F is not an explicit function of t then the equation (1.27) may be rewritten in the form:

$$F - u'F_{u'} = \text{constant}. \qquad (1.29^b)$$

(subscripts denote partial differentiation) It is easy to check that (1.27^a) implies (1.29^b) by differentiating both sides of (1.29^a) with respect to t and factoring out u' . We make the assumption that the solution u = constant is of no interest.

Equation (1.29^a, or 1.29^b) is the Euler, or Euler-Lagrange equation stating a necessary condition for extremal value of the functional (1.29).

It may be written in the form :

$$\delta I = 0. \qquad (1.29^c)$$

This completes the review of a particularly simple problem of the calculus of variations and of the Euler's equation (1.27) which is a necessary condition for the minimum or maximum of the functional I(u). There seems to be a direct analogy between the condition $\delta(I(u)) = 0$ for the functional I(u) and the condition $\frac{dy}{dx}=0$ for a function $y = y(x)$ in elementary calculus. Both represent a necessary condition for the existence of a smooth extremum.

Let pause for a brief time and consider the class of problems of the brachistochrone type and answer a few simple questions. What kind of problems constitute the scope of the calculus of variations? What restrictions did we impose on the solutions of our problems before we arrived at the equation (1.29),or (1.27^a)?

Let first recall the definition of a functional. $I(u(x_1, x_2, \ldots, x_n))$ is called a functional if it represents a one-to-one correspondence between some topological space X and the field R (R is generally the field of real numbers or of complex numbers but it does not have to be so). In this case we consider the functions of n variables which are required to obey some additional conditions. Under the map $I: X \to \mathbb{R}$ to each such function there corresponds a unique real number (in the case when R is the field of real numbers)

$$f_i(x_1, x_2, \ldots, x_n) \overset{I}{\to} \mathbb{R}\ .$$

For example,consider the definite integral

$$I(y) = \int_{x_0}^{x_1} F(x, y(x), y'(x))dx. \qquad (1.29^d)$$

For any function y(x), such that F(x,y(x),y'(x)) is Riemann integrable, we have a unique real number corresponding to the value of this Riemann integral.

The calculus of variations deals with the problem of finding the minima, the maxima, or the stationary points of such functionals.

We have concluded that Euler's equation(1.29^a) is a necessary condition for the existence of a minimum (or of a maximum) of the functional of the type:

$$I(y(x)) = \int_{x=x_o}^{x=x_1} \phi(y, y', x)dx$$

Let us now investigate what kind of a minimum we can expect - an absolute minimum, or only a local minimum.

Figure 22 shows a graph of some function y(x) (not necessarily continuous) with two minima in the interval

$$x_o \leq x \leq x_1$$

One is a local minimum, the other an absolute minimum in that interval.

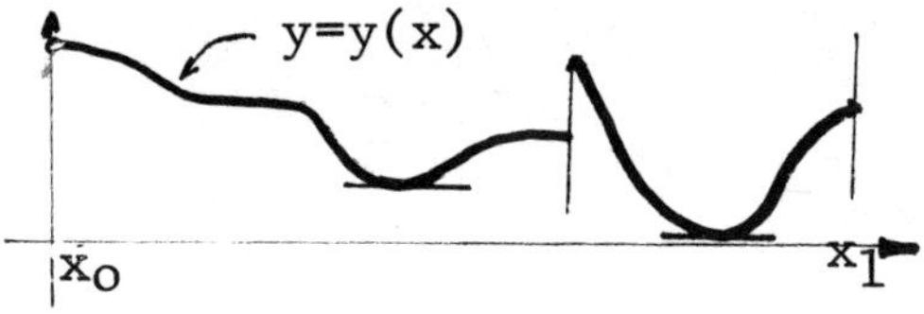

figure 22

A similar condition may exist for a functional I(y(x)). Apart from an absolute minimum, it may possess local minima in the class of admissible functions.

Before we make a decision concerning this question, we should first clarify our intuitive concepts. In case of a real function of a real variable y(x), we have a simple mapping: $x = x_i$ (a real number) corresponds to $y = y_i$ (also a real number). We can talk about a local minimum at $x = x_i$ by enclosing a sufficiently small neighborhood of x_i: $x_i - \varepsilon < x < x_i + \varepsilon$, and pointing out that in the interval $x_i - \varepsilon < x < x_i + \varepsilon$ all corresponding values of y are either greater than or equal to y_i. In the case of a functional, we have to define what we mean by admissible functions which are close to some function y(x), or are in some ε neighborhood of y(x) by assigning a suitable metric, and therefore topologizing the space of admissible functions. Here the answer is not unique, and the metric we assign in such space of function may vary with the type of problem which we have to handle.

If a simple comparison is only required, we may define on a finite interval $x_0 \leq x \leq x_1$, the distance between two functions $y_1(x)$ and $y_2(x)$ bounded in that interval, (such that $y_1(x) = y_2(x)$

at $x = x_0$ and at $x = x_1$) to be the usual sup. norm distance:

$$\rho(y_1(x), y_2(x)) = \max_{x_0 \leq x \leq x_1} |y_1(x) - y_2(x)|,$$

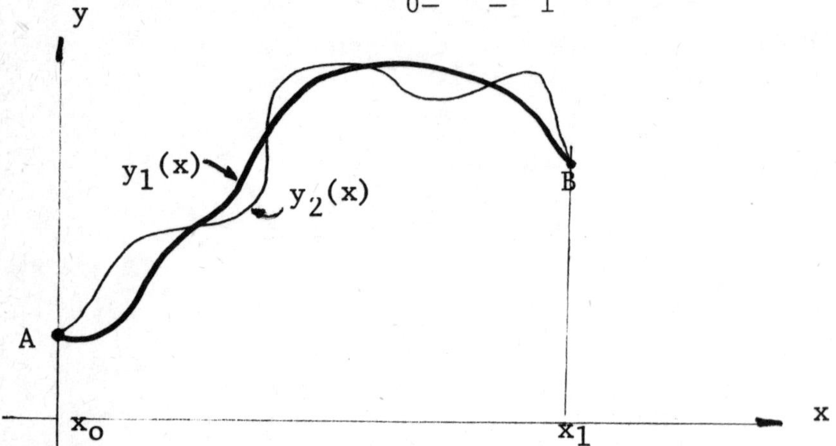

Figure 23

This is frequently called "closeness of order zero"by the engineers and "strong" closeness by some mathematicians.The corresponding convergence is also called "strong".I prefer not to use this terminology."Strong" convergence generally means convergence in the norm,and this terminology will be used here.

In some problems this or similar definition of closeness is positively useless.We want definitions that parallel our ideas concerning the minima or maxima of real functions of real variables. If we demand that the value of the functional

$$I(y) = \int_{x_0}^{x_1} \phi(y,y',x)dx$$

should change by less than some arbitrarily chosen quantity ε, after we substitute for $y(x)$ some $y^*(x)$,such that $y^*(x)$ is sufficiently close to $y(x)$ in our metric, the closeness of order zero is obviously not sufficient. The physical problem may demand that for every $\varepsilon > 0$, there should exist a $\delta > 0$ such that if the distance

$$\rho(y_1(x), y_2(x)) < \delta,$$

then

$$|I(y_1) - I(y_2)| < \varepsilon.$$

As a counterexample, we consider the functional

$$I(y) = \int_0^1 (1 + y'^2)^{1/2} dx$$

and define the metric $\rho(y_1, y_2) = \max_{x \in [0,1]} |y_1 - y_2|$

and build the following curve :

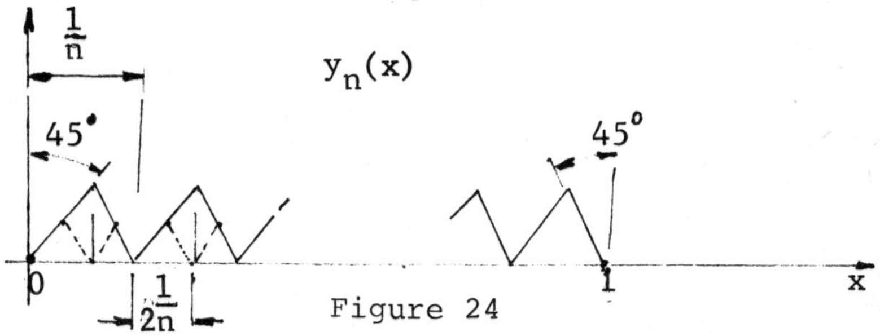

Figure 24

By choosing n, (the number of segments) to be sufficiently large, we can make this curve to be arbitrarily close to the curve $y \equiv 0$, but for any such curve the value of I(y) is always the same:

$$I(y_n) = \sqrt{2} \quad ;$$

but in the limit

$$I(y \equiv 0) = 1 \quad .$$

Therefore while the two functions are very close, the corresponding functionals are not.

We can remedy the situation by demanding a definition of $\rho(y(x), y^*(x))$ to include terms in $y'(x)$ and $y^{*\prime}(x)$. For example let us define:

$$\rho: \; X \times X \to \mathbb{R}$$

$\rho(y(x), y^*(x)) = \max |y(x) - y^*(x)| + \max |y'(x) - y^{*\prime}(x)|$, i.e. ρ is the Sobolev norm defined in subsequent chapters and Appendix B.

<u>Critique of the formulas (1.29^a)-1.29^c).</u> The first obvious criticism concerns the differentiability requirement of the formulas. There is no universal physical law concerning all variational problems of the class discussed here which implies continuous differentiability of solutions, or differentiability for example of the terms in (1.9). In particular, we have absolutely no right to expect that the functional (1.22) is minimized by a function which possess all the derivative appearing in the Euler-Lagrange equation(1.27)

An example of nondifferentiable solutions to the variational principle of Fermat is easily observed by studying light propagation in a medium with a density discontinuously changing across some surface. A stick looks broken to an observer if one end of it is immersed in a pool of water. (Fermat's principle states that a light ray emanating from a point A reaches a point B - say the eye of the the observer, along a path Γ_{AB} such that the interval of time for the travel taken along that path is the smallest possible. Of course, we have to note that the speeds of light propagation in air and in water are different.

Things can be patched up only if the corners are permitted to exist on the optimal path Γ_{AB}.

If the path represents a true path of particle mass, one physical requirement that must be obeyed at such a corner point is the law of conservation of the linear momentum. That is, in a mechanical problem involving a movement of some particle or rigid body with mass m, we have $L_{\dot{y}_i}(t = t_{0-}) = L_{\dot{y}_i}(t = t_{0+})$. This condition expressing preservation of the generalized momentum $L_{\dot{y}_i}$ associated with a generalized displacement coordinate y_i is known as the Erdman-Weierstrass corner condition.

However, an optimal path with only a finite number of corners may not exist, and we have no way of even defining such "paths" without abandoning classical analysis. Consider, as an example, the minimization problem for the functional

$$I(y) = \int_0^1 (y^2(x) + (1 - y_x^2)^2)\,dx,$$ with end point constraints $y(0) = y(1) = 0$. We could attempt to solve this problem by searching for appropriate solutions of the Euler-Lagrange equation

$$2y - 4\frac{d}{dx}[(1 - y_x^2)y_x] = 0.$$

However, a common sense argument based on geometric intuition should tell us that no such solution exists. The integrand is a sum of two non-negative terms. Hence, it is minimized if each term is minimized separately. We want $y(x)$ to have small absolute value and y_x to have the value of +1 or -1 along an optimal curve. Again a curve of the type shown on fig. 24 will make $I(y)$ as small as we like provided n is large. As $n \to \infty$ $\lim I(y_i) = 0$. Pointwise y_i converge to $y \equiv 0$. But $I(y \equiv 0) = 1$. Hence, it is easy to see that no continuous path $y = y(x)$, $y(0) = y(1) = 0$ will provide a minimum for the value of the functional $I(y)$.

The reason for the nonexistence of a "reasonable" solution to our minimization problem was the lack of closure properties of our family of approximate solutions of the extremal problem.

An entirely different reason for nonexistence of solutions to an extremal problem was given in the following example by Caratheodory. The phenomenon described here is called "rigidity". An optimization problem leads to the following Euler-Lagrange equation as the necessary condition for the existence of an extremum:

$$(*) \quad y_t = (1 + x_t^2)^{\frac{1}{2}}$$

The boundary conditions assigned to the solution are

$$x(0) = y(0) = 0$$

$$x(1) = y(1) = 1.$$

A geometric interpretation of the equality (*) shows that no differentiable solution can possibly satisfy the given two-point boundary conditions. Let us denote by ds the infinitesimal increment in the length of the projection of the solution arc into the x-t plane and by dy the increment in the length of the projection of the solution curve into the y-direction. We see that the equation (*) can be restated as a system of two equalities: dy = ds , s(0) = y(0) =0,

or

$$\int_0^1 y_t \, dt = \int_0^1 s_t \, dt \ , s(0)=y(0)=0.$$

But the minimal smooth solution to this problem is given by a straight line interval. This results in s(1) = $\sqrt{2}$, while y(1) = 1 is the imposed boundary condition. This contradiction shows that we cannot impose arbitrary boundary conditions on the end points of minimizing arcs, or even prescribe arbitrarily some surfaces on which such points must be located.

The problem of rigidity enters into many seemingly unrelated mathematical problems, and is closely related to the question: When are the initial or boundary conditions well posed for a certain class of differential equations? Clearly, in Caratheodory's example the boundary value problem was not well-posed. Hence, the corresponding minimization problem was also not well-posed.

For examples of problems in continuum mechanics and particularly in classical elasticity that are not well-posed, and for the definition of well-posedness in the sense of Hadamard see a short discussion in Ch. 2, or the referenced bibliography.

A problem similar to "rigidity" arises if the physical setting requires the end points of a trajectory to lie on some surfaces in $\mathbb{R}^n$, while the mathematical setting implies that the trajectories may not vary over a sufficiently "rich" family of curves in $\mathbb{R}^n$.

The problem is usually caused by poor mathematical modelling of a "real-life" problem.

The non-existence of solutions in a space of functions assigned to the problem puts a big question mark on any necessary conditions we may derive for the optimality of solutions.

1.23 A restatement of Hamilton's principle

Let us consider a dynamic system that is completely described by two energy functions - the kinetic energy and the potential energy. Kinetic energy could be a function of both the position coordinates and on their time derivatives. That is $T = T(q_1, q_2, .. q_n, \dot{q}_1, ... \dot{q}_n, t)$, while the potential energy depends only on the position coordinates. That is $V = V(q_1, q_2, ... q_n)$.

The position coordinates are generally parametrized with respect to time variable t.

Let us assume that all forces are derivable from the potential V. The Hamilton principle asserts that the actual motion of such conservative system is such that

$$\delta \int_{\tau_0}^{\tau_1} (T - V)\, dt = 0, \qquad (1.29^e)$$

or that the action integral

$$I = \int_{\tau_0}^{\tau_1} \mathcal{L}\, dt$$

is stationary. Here $\mathcal{L}$ denotes T-V.

1.23 A heuristic derivation of Hamilton's principle from d'Alambert's principle.

Let us consider a dynamic system consisting of n-particles with corresponding masses m_i, i=1,2,...,n. We write the d'Alambert form of Newton's second law:

$$\sum_{i=1}^{n} (F_{i_x} - m_i \frac{d^2x}{dt^2}) \cdot \delta x + (F_{i_y} - m_i \frac{d^2y}{dt^2}) \cdot \delta y + (F_{i_z} - m_i \frac{d^2z}{dt^2}) \cdot \delta z = 0.$$

This can be rewritten as a vector equation

$$\sum_{i=1}^{n} (F_i - m_i \frac{d^2 R_i}{dt^2}) \cdot \delta R_i = 0.$$

F_i denotes the force acting on the i-th particle, R_i the position vector describing the location of the i-th particle.

The term $F_i \delta R_i$ expresses the work performed due to the virtual displacement δR_i.

We identify the work $\Sigma F_i \cdot \delta R_i$ with twice the increase in the potential energy, and

$\Sigma m_i \dfrac{d^2 R}{dt^2} \cdot \delta R_i$ as the variation of kinetic energy. Then we reverse all signs and claim that this is equivalent to (1.29a)

Thus, introducing a generalized coordinate system $\{q_i\}$ such that the position or state vector $\vec{R}$ may be expressed in terms of q_i's we may convert

(1.29a) to the form $[\frac{d}{dt}(\frac{\partial L}{\partial \dot{q}_i}) - \frac{\partial L}{\partial q_i} - Q_i]\cdot \delta q_i = 0.$

or (1.30)

$$\left[\frac{d}{dt}\left(\frac{\partial L}{\partial \dot{q}_i} \cdot \delta q_i\right) - \frac{\partial L}{\partial \dot{q}_i}\right] \cdot \delta \dot{q}_i - \frac{\partial L}{\partial q_i} \cdot \delta q_i - Q_i \cdot \delta q_i = 0 .$$

1.24 A more general form of the Euler-Lagrange equations.

The author considered in some unpublished notes (soon to appear in a monograph form) a pre-distributional version of both the control theory and the classical variational results concerning the virtual work formalism. This monograph is not suitable for presentation of this material. A restricted version may be restated here using only the theory of distribution of L. Schwartz.

A distribution over a class of functions D is a continuous linear map from D into the reals or into the complex numbers. In the original formulation of Schwartz theory the class of test functions $\mathcal{D}$ consists of C^∞ functions with compact support. (The support of f(x) is the closure of the set Ω :$\{x \in \Omega \to f(x) \neq 0\}$ to avoid references to appendices we comment that in a finite dimensional space the compact sets are exactly the closed and bounded sets.) Let us rewrite a weak form (in the Sobolëv sense) of the Euler-Lagrange equations regarded as a necessary condition for the stationary behavior of the functional

$$\int_0^{\ell} f(x,y,y')dx.$$

$$< \frac{d}{dx}(f_{y'}),\ \phi(x) > - < f_y,\ \phi(x) >= 0$$

where $\phi(x) \in \mathcal{D}$, and $(\frac{d}{dx} f_{y'} - f_y)$ is an element of $\mathcal{D}^*$(the topological dual of $\mathcal{D}$).

Let us assume that f_y is continuous on $[0,\ell]$. We may allow the function $f_y'(x)$ to be continuous on $[0,\ell]$ except on a finite set of points $\{p_1,p_2...p_m\}$ of the optimizing curve, where it has simple jump discontinuities.

We assume that f_y' is of the form

$$f_y'(x) = \alpha(x) + \Sigma a_i u(x-p_i)$$

where $\alpha(x)$ is a continuously differentiable function, while $u(x)$ is the unit step function. Thus, $\frac{d}{dx}(f_{y'})$ is a continuous function except at $\{p_1 p_2...p_m\}$, and $\frac{d}{dx}(f_{y'}) = \alpha'(x) + \sum_{i=1}^{m} \delta(x-p_i)$, $\delta(x)$ denoting the Dirac delta function. Since $\delta(x-p_i) = 0$ for all $x \neq p_i$, $i = 1,2,...m$, while $\alpha'(x)$ and $f_y(x)$ are continuous we have the following result

$$\lim_{x\to p_i+} [\frac{d}{dx}(f_y') - f_y] = \lim_{x\to p_i-} [\frac{d}{dx}(f_y') - f_y].$$

If the Euler-Lagrange equations are in "the integrated form" $(f-y'f_{y'})$ = constant, the weak form leads after a similar discussion to the equality

$$\lim_{x\to p_i+} (f-y'f_{y'}) = \lim_{x\to p_i-} (f-y'f_{y'})\ ,\ i=1,2,...m. \tag{1.31}$$

The equality (1.31) is the well known Weierstrass-Erdmann corner condition.

The observation that the Weierstrass-Erdmann condition is an almost trivial consequence of the weak form of Euler-Lagrange equations was made by L. Bahar in a talk given to a meeting of Engineering Societies at Georgia Tech. in 1980. Also see [21].

An assumption that the end points of the trajectory $\hat{y} = y(x)$ are fixed is unnecessary in this discussion. Allowing the end points to vary along some curve only adds another continuous function to each side of the limit expression. We have to make some technical statements concerning the admissible choice of such curves (the transversality conditions). Here it is better to by-pass this aspect of the corner condition.

Euler-Lagrange equations for multiple integrals.
An equation analogous to (1.29^a) can be obtained by a repetition of Euler's arguments. If we ignore complications concerning the existence of strong solutions, or even of weak solutions, and do not ask questions regarding completeness, appropriate topology, the meaning of integrals that appear in the variational statement, then the arguments we offered for extremizing the value of $\int_\Omega f(y,y',x)dx$ extend in an obvious manner to integrals containing partial derivatives in $\mathbb{R}^n$. For the sake of simplicity let us take n=2 and consider only the case when partial derivatives are of order two, or less and the function f is at least three times continuously differentiable with respect to all arguments.

We seek an extremum of the functional

$$J = \iint_\Omega f(x,y,u,u_x, u_y, u_{xx},u_{yy},u_{xy})\ dxdy.$$

u_x denotes $\frac{\partial u}{\partial x}$, u_y denotes $\frac{\partial u}{\partial y}$, u_{xx} denotes $\frac{\partial^2 u}{\partial x^2}$ etc...

The Euler-Lagrange equation, that u(x,y) must satisfy if J(u) attains an extremum, is given by:

$$f_u - \frac{\partial}{\partial x}\left(\frac{\partial f}{\partial u_x}\right) - \frac{\partial}{\partial y}\left(\frac{\partial f}{\partial u_y}\right) + \frac{\partial^2}{\partial x^2}\frac{\partial f}{\partial u_{xx}} + \frac{\partial^2}{\partial y^2}\frac{\partial f}{\partial u_{xy}} + \frac{\partial^2}{\partial x \partial y}\left(\frac{\partial f}{\partial u_{xy}}\right) = 0,$$

The + and - signs originate in the integration by parts of the formal variational formula:

$$\varepsilon \iint_{\Omega} \left(\frac{\partial f}{\partial u}\eta + \frac{\partial f}{\partial u_x}\eta_x + \frac{\partial f}{\partial u_y}\eta_y + \frac{\partial f}{\partial u_{xx}}\eta_{xx} + \frac{\partial f}{\partial u_{xy}}\eta_{xy} + \frac{\partial f}{\partial u_{yy}}\eta_{yy}\right.$$

$$\left. + \ldots\right) dxdy = 0,$$ after replacing $\bar{f}$ by $f = \bar{f} + \varepsilon\eta$.

We assume that $u(x,y)$ is a four time continuously differentiable function in $\Omega \subset R^2$. Extension to higher order derivatives or to a higher number of dimensions is trivial.

Note: Some Russian authors credit Mikhail Ostrogradskiĭ with generalizing Euler's equation to higher dimensions.

An example:

$$\frac{\partial^2 u}{\partial x^2} + \frac{\partial^2 u}{\partial y^2} = 0$$

is the Euler equation that is a necessary condition for a smooth (C^1) extremum of $J(u) = \iint_{\Omega} \left[\left(\frac{\partial u}{\partial x}\right)^2 + \left(\frac{\partial u}{\partial y}\right)^2\right] dxdy.$

Solution of the brachistochrone problem.

We write the Euler-Lagrange equation for the action integral

$$J(y) = \int_0^{\ell} \{[1+(y')^2]^{\frac{1}{2}} \cdot (y)^{-\frac{1}{2}}\}dx = \int_0^{\ell} f(y,y')dx$$

$f - y'f_{y'} = \text{constant}.$

(This is the so called "integrated form" of the Euler-Lagrange equation for a Lagrangian of the form $\alpha = f(y,y')$ ·)

The first integral of this equation is

$$y\left(\sqrt{1 + (y')^2}\right) = \text{constant} = C.$$

Therefore $y' = \left(\frac{C-y}{y}\right)^{\frac{1}{2}}$.

Introducing a parameter t, with $y = C \sin^2(t/2)$, we derive the parametrized equations of the solution curve

$$\begin{cases} x = r\,(t - \sin t) + k \\ y = r\,(1 - \cos t), \end{cases}$$

where $r = \frac{C}{2}$ and k is an arbitrary constant.

The solution curve is described by this set of equations is a cycloid, as predicted by Isaac Newton.

More general brachistochrone problems.

There are several variations and generalizations of the brachistochrone problem. Kleinschmidt and Schulze [8] replaced the gravitational attraction by a central force field. Russalovskaya, Ivanov and Ivanov allowed friction and variable mass [9]. Ashby et al. solved the classical brachistochrone problem with Coulomb friction[10]. Djukič published at least two papers [11] on the brachistochrone problem with non-conservative forces and briefly considered a related optimal control problem.

Drummond and Downes solved the brachistochrone problem in an accelerated coordinate system (a moving track) in [12]. The clue to the ideas behind some of these generalizations is replacement of the equations of motion (in this case of the simple relation $v = 2gh$) by the Euler-Lagrange equations $\frac{d}{dt}\left(\frac{\partial T}{\partial \dot{q}_i}\right) - \frac{\partial T}{\partial q_i} - \frac{\partial V}{\partial q_i} = Q_i$,

where Q_i are the non-conservative forces. For example, if friction is conssidered, Q_i corresponds to the friction force acting in the direction of the coordinate q_i.

We also introduce a physically motivated constraint equating the change in total energy to the power generated by friction.

We compare the two rates of energy changes

$$\text{(1.32)} \quad \frac{d}{dt}(T + V) = \sum_i Q_i \dot{q}_i \,.$$

A historical note. While the brachistochrone problem was posed earlier by John Bernoulli (apparently around 1694) he set it in writing in 1697. Newton read it, solved it the same day and permitted it to be published anonymously.

Bernoulli immediately recognized Newton's writing by the power and ingenuity of this anonymous note and remarked "a lion is recognized by its paw."

Summary of some Classical Variational Principles

1) The principle of least action due to de Maupertuis.

Let the Lagrangian be of the form

$$(1.33)\quad \mathcal{L} = \frac{1}{2}\sum_{i,j=1}^{n} g_{ij}\dot{q}_i\dot{q}_j - V(q_1, q_2 \ldots q_n) = T - V,$$

with the action integral defined by $\int_{\partial}^{T} \mathcal{L}\, dt$.

We define $p_i = \dfrac{\partial \mathcal{L}}{\partial \dot{q}_i}$.

The Hamiltonian H is identified with the total energy of the system

$$H = \sum_{i=1}^{n} p_i\dot{q}_i - L = T + V = \frac{1}{2}\sum g^{k\ell} p_k p_\ell .$$

If $H = \text{constant} = C_H$ along the trajectory of motion, then the Maupertuis principle asserts that

$$(1.34)\quad \delta \int_0^T \sum_i \left(\frac{\partial L}{\partial \dot{q}_i}\cdot \dot{q}_i\right) dt = 0.$$

This is easily shown to be equivalent to

$$\delta \int_0^T 2T\, dt = 0.$$

Let us define a kinematic metric

$$(1.35)\quad ds = \left(\sum_{i,j} (g_{ij}\, \dot{q}_i\dot{q}_j)\right)^{\frac{1}{2}} dt = (2T)^{\frac{1}{2}} dt,$$

that is

$$\left(\frac{ds}{dt}\right)^2 = \sum g_{ij}\, \dot{q}_i\dot{q}_j = 2T.$$

Then the Maupertuis principle asserts that the action integral attains a minimum, and consequently

$$\delta \int_{\sigma_0}^{\sigma_1} (T)^{\frac{1}{2}}\, ds = 0.$$

We can redefine the metric $d\tilde{s} = (T)^{\frac{1}{2}} ds$.

In the topology defined by this metric the motion of the system corresponds to a curve of shortest distance, that is to a geodesic curve.

Henri Poincaré and later George Birkhoff gave a topological interpretation of this principle. Specifically, the metric tensor g_{ij} must satisfy the relation (1.35) for the motion of the system to be represented by a geodesic.

The least action principle of Maupertuis also indicates that the motion of a conservative system is completely determined by the kinetic energy function. Given the dependence of the kinetic energy on time one can predict the behaviour of the system.

1.25 Hamilton's principle.

Newton's second law of mechanics may be restated as follows: the rate of change of the generalized momentum is equal to the applied generalized force. If the force is derived from a potential, that is if

$$F_i = \partial V/\partial q_i$$

then Newton's law becomes an equality :

$$\frac{d}{dt}\left(\frac{\partial T}{\partial \dot{q}_i}\right) = \frac{\partial V}{\partial q_i}$$

or, after some obvious assumptions are made:

$$\frac{d}{dt}\left(\frac{\partial \mathcal{L}}{\partial \dot{q}_i}\right) - \frac{\partial \mathcal{L}}{\partial q_i} = 0 \quad .$$

This is exactly the Euler-Lagrange equation for the Lagrangian $\mathcal{L} = T - V$.

The statement of Hamilton's principle may be written symbolically

$$(1.36) \qquad \delta \int \mathcal{L}\, dt = 0.$$

1.26 The conservation of energy

Let us suppose that in a mechanical system the Lagrangian $\mathcal{L}(q_i, \dot{q}_i, t)$ and the Hamiltonian $H = \langle p, \dot{q} \rangle - \mathcal{L}$ are not explicit functions of time. That is $\mathcal{L} = \mathcal{L}(q, \dot{q})$.

Therefore $\frac{\partial \mathcal{L}}{\partial t} = 0$, and

$$\frac{d\mathcal{L}}{dt} = \Sigma\left(\frac{\partial \mathcal{L}}{\partial q_i}\dot{q}_i + \frac{\partial \mathcal{L}}{\partial \dot{q}_i}\ddot{q}_i\right) = \frac{d}{dt}\sum_i \left(\frac{\partial \mathcal{L}}{\partial \dot{q}}\dot{q}_i\right) .$$

Thus $\frac{d}{dt}\left[\Sigma\left(\dot{q}_i \frac{\partial \mathcal{L}}{\partial \dot{q}_i}\right) - \mathcal{L}\right] = 0$, or

(1.36) $$\frac{d}{dt}[\Sigma \dot{q}_i p_i - \mathcal{L}] = \frac{d}{dt} H = 0.$$

Therefore, the Hamiltonian H is conserved (is constant) along any trajectory of motion for the system. (A general theory of conservation laws is delayed till Volume 2 of this monograph.)

1.27 Hamilton's canonical equations.

Let us not assume that the system is conservative. We formally differentiate both sides of the equation.

(1.37) $$H = \Sigma p_i \dot{q}_i - \mathcal{L}(q_i, \dot{q}_i, t)$$

obtaining a "Pfafian" form $dH = \Sigma \dot{q}_i dp_i - dL$, (observe that H does not depend of $\dot{q}_i$, i.e. $H = H(\underset{\sim}{p}, \underset{\sim}{q}, t)$).
Again, we write formally

$$dL = \frac{\partial \mathcal{L}}{\partial t} dt + \Sigma \frac{\partial \mathcal{L}}{\partial q_i} dq_i + \Sigma \frac{\partial \mathcal{L}}{\partial \dot{q}_i} d\dot{q}_i$$

$$= \frac{\partial \mathcal{L}}{\partial t} dt + \left(\Sigma\left(\frac{d}{dt}\left(\frac{\partial \mathcal{L}}{\partial \dot{q}_i}\right)\right) dq_i + \Sigma \frac{\partial \mathcal{L}}{\partial \dot{q}_i} d\dot{q}_i\right.$$

$$= \frac{\partial \mathcal{L}}{\partial t} dt + \Sigma(\dot{p}_i \cdot dq_i) + \Sigma p_i \cdot d\dot{q}_i).$$

We use the Euler-Lagrange equation to replace $\frac{\partial L}{\partial q_i}$ by $\frac{d}{dt}\left(\frac{\partial L}{\partial \dot{q}_i}\right) = \dot{p}_i$.

Hence,

(1.38) $$dH = -\Sigma(\dot{q}_i dp_i - \dot{p}_i dq_i) - \frac{\partial \mathcal{L}}{\partial t} dt .$$

Directly differentiating $H(q_i, p_i, t)$ we have

(1.39) $$dH = \sum_i \left(\frac{\partial H}{\partial p_i} dp_i + \frac{\partial H}{\partial q_i} dq_i\right) + \frac{\partial H}{\partial t} dt$$

Subtracting from each other left and right hand sides of equations (1.38) and (1.39), respectively, we derive

$$0 = \left(\frac{\partial H}{\partial t} + \frac{\partial \mathcal{L}}{\partial t}\right) dt + \Sigma\left(\frac{\partial H}{\partial q_i} + \dot{p}_i\right)dq_i + \left(\frac{\partial H}{\partial p_i} - \dot{q}_i\right)dp_i$$

The increments dt, $d\tilde{q}$, $d\tilde{p}$ may be chosen independently. Thus

$$\frac{\partial H}{\partial t} + \frac{\partial \mathcal{L}}{\partial t} = 0, \quad \frac{\partial H}{\partial p_i} - \dot{q}_i = 0, \quad \frac{\partial H}{\partial q_i} + \dot{p}_i = 0,$$

or

$$(1.40) \quad \frac{\partial \mathcal{L}}{\partial t} = -\frac{\partial H}{\partial t}$$

$$(1.41) \quad \frac{\partial H}{\partial p_i} = \dot{q}_i, \qquad i = 1,2,\ldots n.$$

$$(1.42) \quad \frac{\partial H}{\partial q_i} = -\dot{p}_i$$

The equations (1.41), (1.42) are again Hamilton's canonical equations, identical with the equations (1.20).

1.28 Hertz's principle of least curvature.

A particle (or system) not subjected to external forces will move along a trajectory consistent with constraints such that its path has the least mean curvature.

1.29 Some examples.

An example of a conservative motion.

Consider the motion of a mass m attracted to two points in the plane by linear springs and subjected to gravity. This system is illustrated on Figure 25. Let k_1, k_2 be spring attraction constants, respectively. x,y are Cartesian coordinates.

$\{\bar{x}, \bar{y}\}$ denotes the position of the mass m.

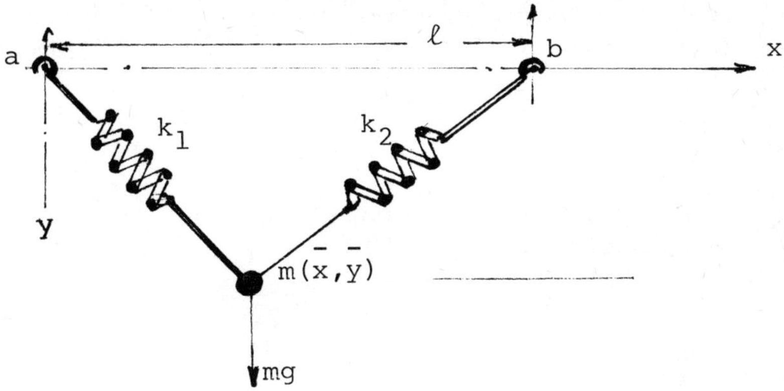

Figure 25

Let the x-axis connect the attraction points a,b. Their distance ℓ is given. The potential energy is

$$V = \frac{1}{2}\{K_1(\bar{x}^2 + \bar{y}^2) + K_2[(\bar{x} - \ell)^2 + \bar{y}^2]\} + mg\bar{y}$$

and the kinetic energy is

$$T = \frac{m}{2}(\dot{\bar{x}}^2 + \dot{\bar{y}}^2) \cdot \frac{M}{2}(\dot{\bar{x}}^2 + \dot{\bar{y}}^2)$$

Hence, the stationary property of

$$\int_0^t (\frac{m}{2}(\dot{\bar{x}}^2 + \dot{\bar{y}}^2) - \frac{1}{2}\{K_1(\bar{x}^2+\bar{y}^2)+K_2[\bar{x}-\ell)^2+\bar{y}^2]\}+mg\bar{y})\ dt$$

will coincide with the dynamic behavior of this system.

In this case the actual motion of this system is given by the Euler-Lagrange equations for the Lagrangian integral $\int_0^t L = \int_0^t (T-V)dt$, which are

$$m\ddot{x} + k_1 x + k_2(x-\ell) = 0 ,$$
$$m\ddot{y} + k_1 y + k_2 y + mg = 0 .$$

Exercise. Show that equating to zero the inertia terms ($\ddot{x} = \ddot{y} = 0$) results in the equations of static equilibrium ,corresponding to an extremum value of the potential energy. (Why?)

A slightly more complex conservative system is illustrated on figure 26. The analysis is more tedious,but in principle it is the same as in the preceeding example.

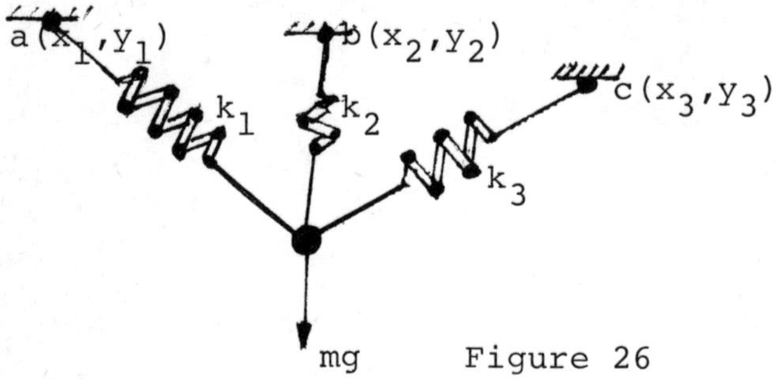

Figure 26

An example of a non-conservative system.

<u>The follower problem</u>

We consider a rod (or a column) subjected to a force of constant magnitude applied at the free end of the rod that is always directed along the axis of the vibrating rod.

Figure 27

The differential equation of motion can be derived either by summation of all forces,including the inertia forces acting on the rod,or else by following the Euler-Lagrange formalism.

The differential equation of motion is

(1.44) $a^2 u_{xxxx} + Pu_{xx} + mu_{tt} = 0,$

where $a^2 = EI(x)$.

The boundary conditions assigned to this problem are :

(*) $u(0,t) = u_x(0,t) = u_{xx}(\ell,t) = u_{xxx}(\ell,t) = 0.$

The initial conditions are :

$$u(x,0) = u_o(x)$$
$$u_t(x,0) = v_o(x) \ .$$

We look for solutions to this problem in the Sobolëv space $H^3(\Omega \times [0,T])$, Ω denoting the spatial domain $0 \le x \le \ell$.

Since mu_t^2, $(u_{xx})^2$, $(u_x)^2$ are all integrable funct-tions on $\Omega \times [0,T]$, the integral

$$\tfrac{1}{2} \int_\Omega (mu_t^2 + a^2 u_{xx}^2 - Pu_x^2)\,dx = \mathcal{E}(t)$$

is defined, and the system has a finite total energy.

Note: Observe that P cannot be regarded as a constant in "integration by parts".

Hence, this system (given by equation 1.44) is non-conservative if $P \neq 0$.

This example is particularly interesting, since the operator $Lu = a^2 u_{xxxx} + Pu_{xx} + mu_{tt}$ "looks" like a self-adjoint operator.

Problem. A certain amount of confusion was in evidence in engineering discussion of the static version of the follower problem obtained by setting $mu_t \equiv 0$, and considering the correspond-ing "static equation" derived by ignoring the acceleration term mu_{tt}.

$$\begin{cases} a^2 u_{xxxx} + Pu_{xx} = 0 \\ a^2 > 0, \\ P > 0 \ . \end{cases}$$

(The usual "practical argument" if m is "small"!) It is easy to show that the only solution satisfying the prescribed boundary conditions is $u \equiv 0$. (Hint: define $w = u''$ and consider the corresponding second order equation with $w(\ell) = w'(\ell) = 0$).

Therefore, no information about the dynamics of this system may be gained by examining first the "simpler" static case.

Problem. "What "goes wrong" when the inertia term is ignored?

Problem. Compare the adjoint of the operator

$$L = a^2 \frac{\partial^4}{\partial x^4} + P \frac{\partial^2}{\partial x^2} + m \frac{\partial^2}{\partial t^2}, \quad (L: H^3 \to L_2(\Omega\times T))$$

with boundary conditions (*)-with the adjoint of L

$$M = a^2 \frac{\partial^4}{\partial x^4} + P \frac{\partial^2}{\partial x^2}, \quad M: \; H^3 \to L_2(\Omega).$$

A case of a varying and possibly non-positive mass

r, θ will denote the usual polar coordinate system. Suppose, a particle of mass $m(t)$ is forced to follow a path $r = \psi(\theta)$ in a planar motion. Its angular momentum is

$$p_\theta = m(t)\cdot r^2 \cdot \theta' = m(t)\cdot \psi^2(\theta)\cdot\theta' \qquad (' = \frac{d}{dt}).$$

The moment about the origin exerted by the field of force is given by a time dependent relation

$$\mathcal{M} = c(t)f(r,\theta) = c(t)f(\psi(\theta),\theta) = c(t)\phi(\theta).$$

The generalized Newton's second law states that

$$\mathcal{M} = -(p_\theta)', \text{ or}$$

$$(1.45) \qquad (m(t)\psi^2(\theta)\theta')' + c(t)\phi(\theta) = 0,$$

is the equation of motion along the constraint path $r(t) = \psi(\theta(t))$, ($r \geq 0$). We assume that $\phi(\theta)\psi(\theta) > 0$ a.e. and that mass $m(t)$ is always positive. We shall assume however that $c(t)$ is of arbitrary sign, and in particular we shall consider the possibility of $m(t)c(t)$ being negative. Note: It would make no difference if we assumed the "usual" physical behavior of inertia opposing the generalized force applied to the system, but inserted a negative mass $m(t)$ into the equation.

If at any time $t = t_0$ the product $a(t_0)c(t_0)$ is negative, then if the product $\phi(\theta)\psi(\theta)$ is in some sense sufficiently large on some interval (t,T) then a phenomenon of non-continuability occurs. A precise statement is given below.

The following theorem was proved by Komkov in [19]. Suppose that $a(t_0)c(t_0) < 0$ for some t_0. Then a necessary and sufficient condition for the existence of a non-continuable solution is given by the inequality

$$\int_{\xi_0}^{\infty} \frac{\psi(\xi)}{\left[1+\int_{t_0}^{\xi} (\phi(s)\psi(s))ds\right]^{\frac{1}{2}}} d\xi < \infty \quad , \text{(i)}$$

valid for some $\xi_0 \in \mathbb{R}$.

The proof is non-constructive and there is no need to repeat it here. For details see [19]. See also the paper of Burton and Grimmer [20].

The phenomenon of non-continuability is in itself of great interest. Roughly speaking, the solution becomes infinite in the middle of some compact interval not because of resonance (there is no vibration and no natural frequencies have been introduced into the problem!) but because of the structural property of the nonlinearities occuring in the equation. The system acquires infinite energy in a finite amount of time! Clearly this is impossible for a linear system even in the presence of a negative mass. Putting $\phi(\theta) = \theta$, $\psi(\theta) \equiv 1$, we see that the integrability condition (i) can not be satisfied. Hence all solutions of the linear system are continuable no matter what signs are attached to the coefficient functions.

Physically this phenomenon is important in electronic and electromagnetic applications, where coupled mechanical, electromagnetic and electronic circuits simulate negative friction, negative mass, or negative inertia. In electronic circuitry a 180° phase shift accomplishes the changes of sign required in the appropriate differential equation.

An example of a negative resistance circuit is given below.

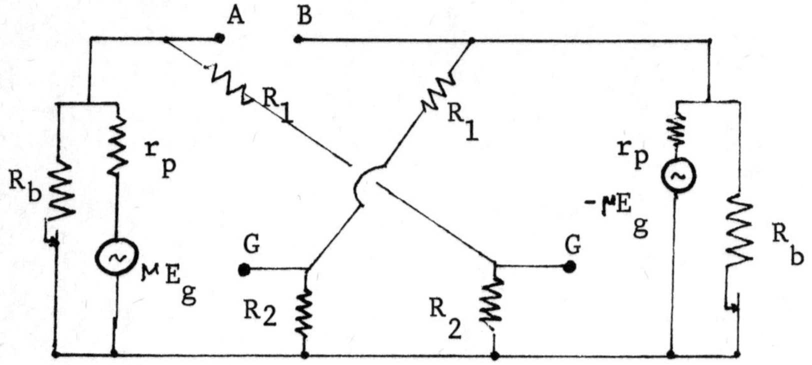

Figure 27

Assuming $\frac{r_p}{R_b}$ to be small, the effective resistance is computed to be

$$R_{AB} = \frac{2r_p}{\left\{1-\frac{\mu R_2}{R_1+R_2}\right\}}$$

which is negative if

$$\frac{R_1+R_2}{R_2} < \mu .$$

The importance of the electronic circuit exhibited above lies not in its particular design, but in the fact that it is possible to design it and to achieve effective negative resistance. Combining electromagnetic and electronic circuitry it is also possible to simulate the effects of a negative inertia, or of negative mass. One can imitate negative gravity by vibrating with appropriate frequency the support of an inverted (upside down) pendulum as shown on figure 28.

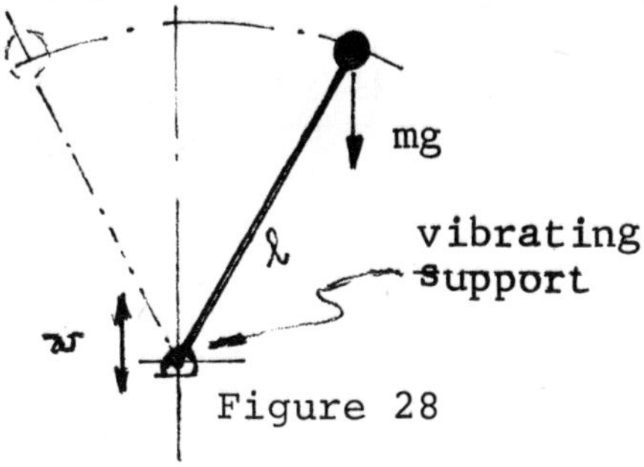

Figure 28

1.30 Constrained problems and Lagrangian multipliers.

Let us introduce in a very simple manner the Lagrange approach to constrained optimization. Let us consider a nonlinear programming.

Problem: minimize the value of the function $f(x,y)$ ($\in \mathbb{R}^2$) with "a geometric constraint" condition

$g(x,y) = 0.$

We assume that $g(x,y) = 0$ is not a natural condition, that is the unconstrained minimum of $f(x,y)$ occurs at some point $(\bar{x},\bar{y})$ where $g(\bar{x},\bar{y}) \neq 0$, or that the "absolute" minimum (with no constraints) does not coincide with the constrained minimum of $f(x,y)$. Moreover, we assume that the problem is not trivial, that is, $f(x,y)$ is not constant on the curve $g(x,y) = 0$. The "level" curves $f(x,y)$ = constant and the curve $g(x,y)=0$ are assumed to be smooth. That is, grad$(f(x,y))$ is a continuous function in some neighborhood of the constraint curve $g(x,y) = 0$, and $g(x,y)$ is a differentiable function. (All of these assumptions can be weakened later.)

We claim that $f(x,y)$ assumes its minimum (or maximum) on the curve $g(x,y) = 0$ at the point $(\bar{x},\bar{y})$ such that $\langle \tau, \text{grad } f \rangle = 0$, where τ denotes the unit vector $\tilde{\cdot}$ tangential to $g(x,y) = 0$.

(That is the curves $f(x,y)$ = constant and $g(x,y) = 0$ have a common normal at the point $(\bar{x},\bar{y})$. Let us offer an intuitive proof. Suppose that $f(x,y)$ assumes its minimum at the point $(\bar{x},\bar{y})$ on the curve $g(x,y) = 0$, i.e. $g(\bar{x},\bar{y}) = 0$. $f(\bar{x},\bar{y}) = c_o - \min\{f(x,y) \mid x,y \in g(x,y) = 0\}$, but grad $f|(x=\bar{x}, y=\bar{y})$ is not orthogonal to $\tau(\bar{x},\bar{y})$. Therefore, $f(x,y)$ decreases along $g(x,y) = 0$ in the direction in which the projection of (-grad f) onto the line tangent to $g(x,y) = 0$ (at $\bar{x},\bar{y}$) is positive. This contradicts the assumption that $f(\bar{x},\bar{y})$ is the minimum of the function $f(x,y)$ on the curve $g(x,y) = 0$.

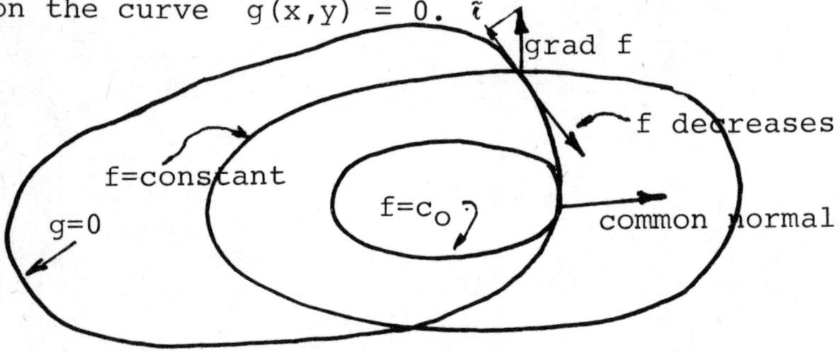

Figure 29

The necessary condition of common normal at $(\bar{x},\bar{y})$ for $g(x,y) = 0$ and $f(x,y)$ = const. is

$$\frac{\left(\frac{\partial f}{\partial y}\right)}{\left(\frac{\partial g}{\partial y}\right)} = \frac{\frac{\partial f}{\partial x}}{\frac{\partial g}{\partial x}} = -\lambda,$$

where $-\lambda$ is the value of this ratio, with the minus sign placed for convenience.

Thus

$$\frac{\partial f}{\partial x} + \lambda \frac{\partial g}{\partial x} = 0$$

and

$$\frac{\partial f}{\partial y} + \lambda \frac{\partial g}{\partial y} = 0 .$$

These are necessary conditions for a stationary behaviour of the function $F(x,y,\lambda) = f(x,y) + \lambda g(x,y)$. The constant λ is the "Lagrangian multiplier".

A simple example: Minimize the function $f(x,y) = (x-2)^2+(y-2)^2$, with an imposed constrain :

$g(x,y) = x+y-2 = 0$.

The common normal, or the orthogonality condition is $\langle \text{grad } f, \tau \rangle = 0$. Using $\mathbf{i}$, $\mathbf{j}$ as basis of our vector space, we can express this orthogonality relation in the form :

$$\langle \mathbf{i}\frac{\partial f}{\partial x} + \mathbf{j}\frac{\partial f}{\partial y} , \mathbf{i}-\mathbf{j} \rangle = 0 .$$

Thus $2x-2y = 0$ is the orthogonality condition, while $x + y = 2$ is the constrain.

The extremal point is : $x=1$, $y=1$ .It turns out to be a minimum of $f(x,y)$,but this cannot be deduced from our analysis.

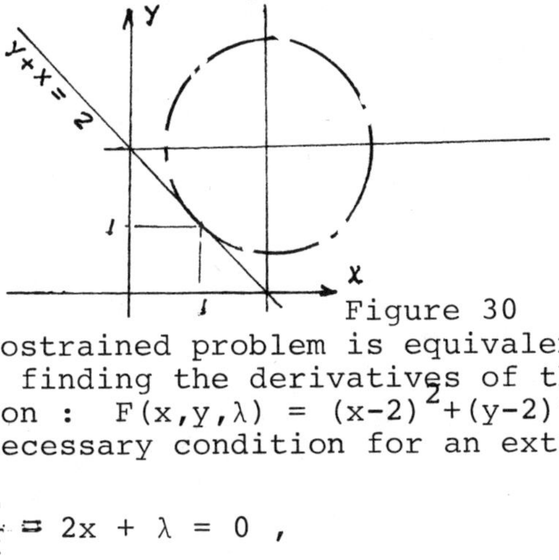

Figure 30

The costrained problem is equivalent to the problem of finding the derivatives of the Lagrange function : $F(x,y,\lambda) = (x-2)^2+(y-2)^2 + \lambda(x+y-2)$.

The necessary condition for an extremum is :

$$\frac{\partial F}{\partial x} = 2x + \lambda = 0 ,$$

$$\frac{\partial F}{\partial y} = 2y + \lambda = 0 ,$$

$$\frac{\partial F}{\partial \lambda} = x + y - 2 = 0 .$$

Solving for x , y , λ we obtain $\bar{x}=1, \bar{y}=1, \bar{\lambda}=-2$.

That was the general idea of Lagrange, who first suggested that the minimization problem:

$f(\underset{\sim}{x}) \to \min$, subject to the constrain $g(\underset{\sim}{x})=0$,

$\underset{\sim}{x} \in R^n$, may be replaced by the non-constrained problem of minimizing $F = f(x) + \lambda g(x)$.

We also observe that in our simple problem the minimum of $f(x,y) = (x-2)^2 + (y-2)^2$, subject to the constrain $g(x,y) = 0$ is exactly the same as the minimum of $f(x,y)$ with the inequality constraint : $g(x,y) = x+y-2 \leq 0$.
Then the admissible region is the half-plane as shown shaded on figure 31.

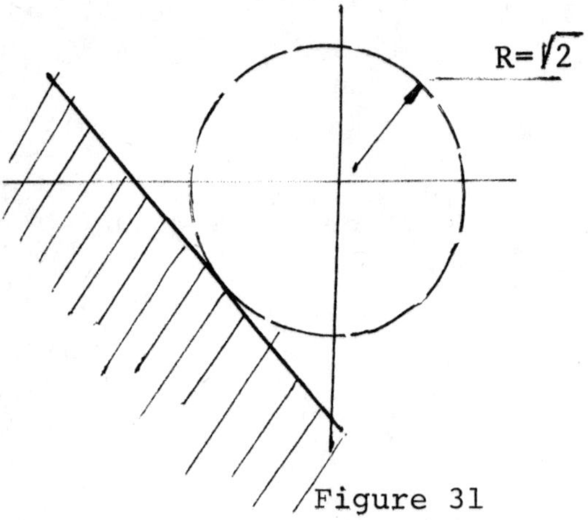

Figure 31

If the constraint conditions $y \geq -1/4$, $x \geq -1$ are also imposed on the problem,

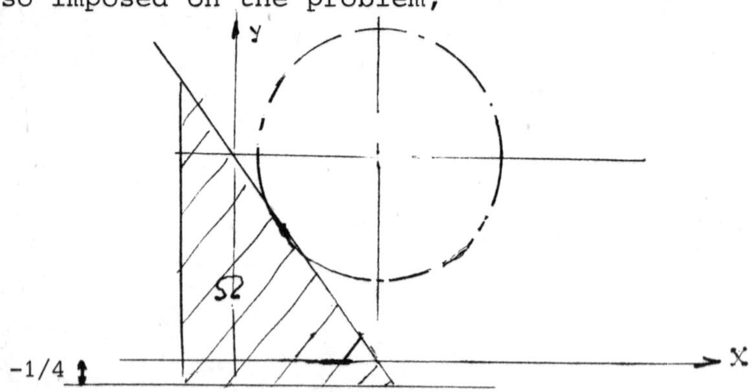

Figure 32

the solution of the minimization problem is not changed. Such constraint is called inessential. The admissible region Ω is shown shaded on figure 32.

We observe that the minimum of f(x,y) does not occur at one of the verteces of the admissible region, as one would expect in a linear programming problem.

Example: Minimize $f(\underset{\sim}{x}) = \sum_{i=1}^{n} (x_i - a_i)^2$

subject to the constraint

$\sum_{i=1}^{n} x_i^2 \leq 1$, i = 1,2,...n, where a_i are given positive real numbers.

Solution:

If $\sum_{i=1}^{n} a_i^2 \leq 1$ then the point $\underset{\sim}{x} = \underset{\sim}{a}$ is the solution. (Note that $f(\underset{\sim}{x}) \geq 0$ and $f(\underset{\sim}{a}) = 0$, $\underset{\sim}{a} = \{a_1, a_2 \ldots a_n\}$.

If $\sum a_i^2 > 1$ then $\underset{\sim}{x} = \underset{\sim}{a}$ violates the constraint $\sum x_i^2 \leq 1$, i.e. $\underset{\sim}{x} = \underset{\sim}{a}$ does not lie in the unit sphere, and the constrained minimum differs from the unconstrained minimum given by $\underset{\sim}{x} = \underset{\sim}{a}$.

We introduce a Lagrangian multiplier λ, imbed the n-dimensional problem in a (n+1)- dimensional space with coordinates $\{x_1, x_2, \ldots, x_n, \lambda\}$ and minimize

$$\tilde{L} = f(\underset{\sim}{x}) + \lambda\, g(x) = \sum_{i=1}^{n} [(x_i - a_i)^2 + \lambda[(\sum_{i=1}^{n} x_i^2) - 1\,]\,; \quad \text{grad } \tilde{L} = \sum(2(x_i - a_i) + 2\lambda x_i) \cdot \mathbf{e}_i,$$

where $\mathbf{e}_i$ are unit vectors .

$$\frac{\partial \tilde{L}}{\partial \lambda} = (\sum_i x_i^2 - 1) \; .$$

Then grad $\tilde{L} = 0$,

$$\frac{\partial \tilde{L}}{\partial \lambda} = 0 \, ,$$

are the necessary conditions for a constrained minimum of $f(\underset{\sim}{x})$.

Thus $(x_i + \lambda) - a_i = 0$, $i = 1,2\ldots n$,

$$\sum_{i=1}^{n} x_i^2 - 1 = 0 ,$$

or

$$x_i = a_i - \lambda,\ \Sigma(\lambda - a_i)^2 = 1 ,$$

$$\lambda = \frac{\Sigma a_i}{n} + \sqrt{\frac{(\Sigma a_i)^2}{n^2} - (\Sigma a_i^{\,2}) + 1}$$

is the computed value of the Lagrangian multiplier λ. $\bar{x}_i = a_i - \lambda$ are the coordinates of the extremal point.

Problem. Derive the equations of motion of a bead sliding on a wire bent into a shape $z = \frac{1}{x}$ (hyperbola) that is rotating around the z axis

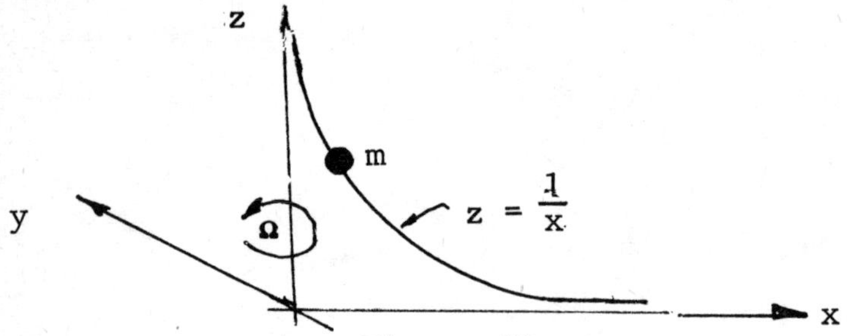

Figure 33

with a constant angular velocity ω.

Solution. The constraints can be replaced by

$z = \frac{1}{r}$, $\dot{\theta} = \omega =$ constant after a change to the $\{r,\theta,z\}$ cylindrical polar coordinate system.

The potential energy is $V = mgz$.
The kinetic energy is

$$T = \tfrac{1}{2} m\cdot(\dot{r}^2 + (r\dot{\theta})^2 + \dot{z}^2)$$

We can introduce the action integral

$$\int L = \int_{t_o}^{t_1} T-V = \tfrac{1}{2}\int_{t_o}^{t_1} m[(\dot{r}^2 + r^2\,\dot{\theta}^2 + \dot{z}^2) - gz]\,dt$$

and the modified Lagrangian integral

$$\int \bar{L} = \int L + \lambda_1 \int_{t_o}^{t_1}[(z - \tfrac{1}{r})\,dt + \lambda_2 \int(\dot{\theta} - \omega)\,dt = \mathcal{W}.$$

The equations of motion are

$$\frac{\partial \bar{L}}{\partial z} - \frac{d}{dt}\left(\frac{\partial \bar{L}}{d\dot{z}}\right) = 0$$

$$\frac{\partial \bar{L}}{\partial r} - \frac{d}{dt}\left(\frac{\partial \bar{L}}{\partial \dot{r}}\right) = 0$$

$$\frac{\partial \bar{L}}{\partial \theta} - \frac{d}{dt}\left(\frac{\partial \bar{L}}{\partial \dot{\theta}}\right) = 0$$

$$\frac{\partial \bar{L}}{\partial \lambda_1} = 0$$

$$\frac{\partial \bar{L}}{\partial \lambda_2} = 0 .$$

However, in this case we can eliminate r, $\dot{r}$ and θ directly, obtaining a non-constrained Lagrangian. The action integral is :

$$\tfrac{1}{2}m\int_{t_o}^{t_1}\{[\frac{1}{z^4}\,\dot{z}^2 + \frac{\omega^2}{z^2} + (\dot{z})^2] - 2gz\}dt.$$

The equations of (constrained) motion are

$$\frac{2\dot{z}^2}{z^5} + \frac{\omega^2}{z^3} + g + \frac{d}{dt}\,(\frac{\dot{z}}{z^4} + \dot{z}) = 0 ,$$

or

$$\ddot{z} + \frac{\ddot{z} - 2\dot{z}^2}{z^5} + \frac{\omega^2}{z^3} + g = 0,$$ while the unconstrained equation of motion is trivial:

$$-\ddot{z} = g$$

(positive z axis points up!)

Obviously, the unconstrained bead would fall with a constant acceleration equal to g.

As an Exercise we could prove that the geodesics (lines of shortest distance) on a sphere are great circles. One possible approach is to minimize the length of a curve connecting two points in space with a constrain that the curve Γ lies on a sphere. It does not cause any loss of generality if we assume that the sphere has a unit radius. We assume that the curves are shorter than half the circumference of the sphere.

We introduce a parametrized curve $x=x(t), y=y(t), z=z(t)$ and compute

$$ds = \sqrt{\dot{x}^2+\dot{y}^2+\dot{z}^2}\, dt.$$

Our minimization problem is: $\ell = \int_{t_1}^{t_2} ds \to \min.$

with constraint $x^2 + y^2 + z^2 = 1.$

Thus, we could minimize

$$\int_{t_1}^{t_2} [\dot{x}^2 + \dot{y}^2 + \dot{z}^2)^{\frac{1}{2}} + \lambda(x^2+y^2+z^2-1)]\, dt.$$

It is easier to introduce (r,θ,ϕ) spherical coordinates.

$$z = r \cos\theta$$
$$y = r \sin\theta \cos\phi$$
$$x = r \sin\theta \sin\phi$$

$$\dot{z} = (-r \sin\theta)\dot{\theta} \qquad (\dot{r} \equiv 0)$$

$$\dot{y} = (r \cdot \cos\theta \cdot \cos\phi)\dot{\theta} - r(\sin\theta \sin\phi)\dot{\phi}$$

$$\dot{x} = (r \cos\theta \sin\phi)\dot{\theta} + (r \sin\theta \cos\phi)\dot{\phi} .$$

The minimization problem consists of finding the minimum value of the integral:

$$(1.46) \quad \int_{t_o}^{t_1} [\sin^2\theta (\dot{\theta})^2 + \cos^2\theta \cos^2\phi \; (\dot{\theta})^2 +$$

$$\sin^2\theta \sin^2\phi (\dot{\phi})^2 - 2(\sin\theta \cos\theta \sin\phi \cos\phi) \dot{\theta} \dot{\phi}$$

$$+ \cos^2\theta \sin^2\phi (\dot{\theta})^2 + \sin^2\theta \cos^2\phi \cdot (\dot{\phi})^2$$

$$+ 2(\sin\theta \cdot \cos\theta \cdot \sin\phi \cdot \cos\phi) \dot{\theta} \dot{\phi}]^{\frac{1}{2}} dt$$

$$= \int_{t_o}^{t_1} [\sin^2 \theta \cdot (\dot{\phi})^2 + \cos^2 \theta \cdot (\dot{\theta})^2 + \sin^2 \theta \cdot (\dot{\theta})^2]^{\frac{1}{2}} dt$$

$$= \int_{t_0}^{t_1} [\sin^2 \theta \cdot (\dot{\phi})^2 + (\dot{\theta})^2]^{\frac{1}{2}} dt \quad .$$

We can arrange, without any loss of generality, $\{\theta,\phi\}$ coordinates so that $\theta(t_o) = \theta(t_1)=0$. Then $\theta \equiv 0$ along the arc Γ is an obvious minimum. Thus the integral (1.46) is minimized along the great circle arc $\theta \equiv 0$. We clarify "without any loss of generality" remark by observing that the integral (1.46) is invariant with respect to rotations of the sphere, i.e. changes of coordinate $\{\theta,\phi\} \rightarrow \{\theta',\phi'\}$.

An example of a constrained system. We consider a planetary gear system which may be replaced by an ideal system of two cylinders rolling over each other without slippage. Let gear A be represented by a cylinder of radius r_1, gear B by a cylinder r_2.

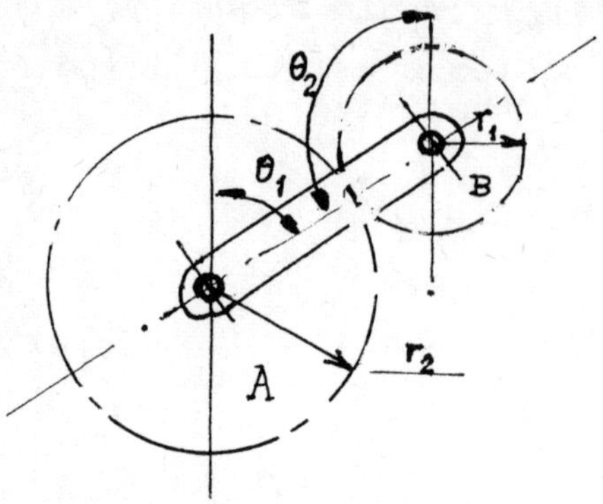

Figure 34

The moments of inertia of the masses of the two gears (relative to their centers of mass) are respectively I_A and I_B.
The mass of the gear B is m_B. The center 0_1 of the gear A is fixed, while the center 0_2 of the gear B is at a constant distance of $r_1 + r_2$ from the center 0_1 of A. That is, the gear B is behaving as the planet in this planetary gear arrangement, while the gear A is the (fixed) sun. The constrained system has two degrees of freedom.Yhe unconstrained system has four degrees of freedom and two constraints.(the center 0_1 being regarded as fixed is not an additional constrain. The coordinates fully determining the state of the system are $x_B, y_B, \theta_1, \theta_2$.The constraints are $x_B^2 + y_B^2 = (r_1 + r_2)^2$, and $r_1\dot{\theta}_1 = r_2\dot{\theta}_2$, or in the virtual displacement form: $r_1\delta\theta_1 = r_2\delta\theta_2$. With a suitable choice of coordinates θ_1, θ_2 we can write $r_1 \cdot \theta_1 = r_2 \cdot \theta_2$.

The potential energy is : $V = m_B g(r_1 + r_2) \cos \theta_1$.

The kinetic energy is :

$$T = \tfrac{1}{2} m_B (r_1 + r_2)(\dot{\theta}_1)^2 + \tfrac{1}{2} I_A (\dot{\theta}_1)^2 + \tfrac{1}{2} I_B (\dot{\theta}_2)^2 .$$

This system is conservative and (T - V) is the unconstrained Lagrangian.

The equations of motion can be derived from the constrained action integral:

$$\int_0^T \bar{L}\,dt = \int [\,(T(\dot{\theta}_1,\dot{\theta}_2) - V(\theta_1) + \lambda(r_1\theta_1 - r_2\theta_2)\,]\,dt,$$

i.e.

$$-\frac{d}{dt}\left(\frac{\partial T}{\partial \dot{\theta}}\right) + \frac{\partial V}{\partial \theta} + \lambda r_1 = 0;\; -\frac{d}{dt}\left(\frac{\partial T}{\partial \dot{\theta}_2}\right) + \frac{\partial V}{\partial \theta_2} - \lambda r_2 = 0.$$

That is

$$\begin{cases} m_B(r_1+r_2)\,\ddot{\theta}_1 + I_A\ddot{\theta}_1 + m_B g(r_1+r_2)\sin\theta + \lambda r_1 = 0 \\ I_B\,\ddot{\theta}_2 - \lambda r_2 = 0, \end{cases}$$

are the equations of motion, where λ is the Lagrangian multiplier.

Eliminating λ (for example substituting $\lambda = \frac{I_B\dot{\theta}_2}{r_2}$ into the first equation) and eliminating θ_2 from this equation by using the constraint $\theta_2 = \frac{r_1}{r_2}\theta_1$ we finally derive a differential equation of motion for the variable θ_1.

Problem.

Compare the constrained and unconstrained behavior of an elastic pendulum

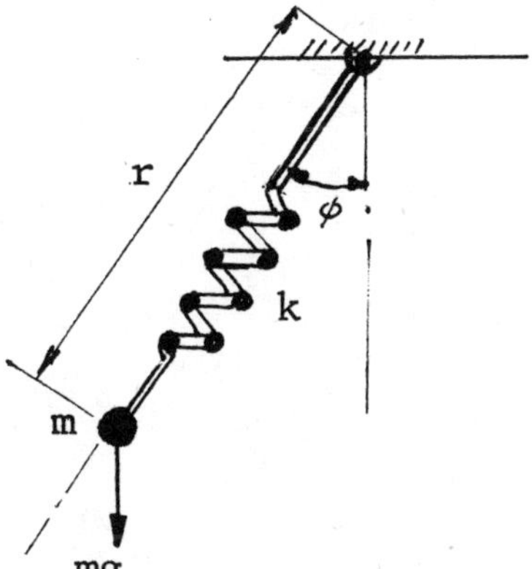

Figure 35

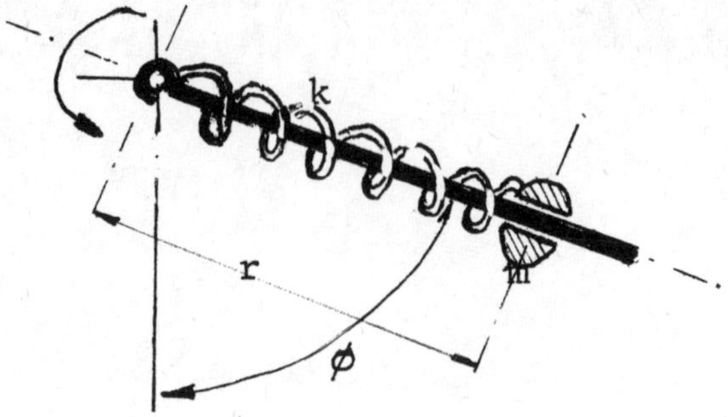

Figure 36

The mass suspended from an elastic rod as shown on figure 35 swings freely, while the mass on figure 36 is sliding without friction along a weightless rod which rotates at a constant angular velocity ω about the suspension point. Interpret the physical meaning of the Lagrangian multiplier λ appearing in the term $\lambda(\phi-\omega t)$ in the constrained Lagrangian and of the terms in the Euler-Lagrange equations.

We assume that in both cases the motion is restricted to a vertical plane.

The solution.

We use the cylindrical polar coordinates $\{r,\phi\}$.

The unconstrained Lagrangian is

$$L = \frac{m}{2}(r^2\dot{\phi}^2 + \dot{r}^2) - mgr\cos\phi = \frac{Kr^2}{2} .$$

The Euler-Lagrange equations are

$$m\ddot{r} - mr(\dot{\phi})^2 + Kr - mg\cos\phi = 0 ,$$

$$m\frac{d}{dt}(r^2\dot{\phi}) + mgr\sin\phi = 0 .$$

The terms in the first equation are identified as Newton's radial forces (mass times radial acceleration), centrifical force, the force of the spring (or elastic force) and the radial component of the gravity force. The terms in the second equation are: the rate of change of the angular momentum, and the angular component of the gravity force. The constrained Lagrangian is

$$\bar{L} = \frac{m}{2}\,(r^2(\dot{\phi})^2 + \dot{r}^2) - Kr^2 - mgr\cos\phi$$

$$+ \lambda(\phi - \omega t).$$

Observe that the constraint $\phi = \omega t$ is holonomic (instead of the more obvious constraint $\dot{\phi} = \omega$). In each case $\lambda = \lambda(t)$, but the constraint $\dot{\phi} - \omega$ gives an expression for $\dot{\lambda}$, while using the constraint $\phi - \omega t = 0$ we can eliminate λ by using both the Euler-Lagrange equations and the constraint condition.

The Euler-Lagrange equations are

$$\begin{cases} m\ddot{r} - mr(\dot{\phi})^2 + mg \cdot \sin\phi + Kr = 0 \\ m\dfrac{d}{dt}\,(r^2\dot{\phi}) + mgr(\sin\phi)\lambda = 0\ , \quad \phi = \omega t\ . \end{cases}$$

Solving the second equation for λ and making use of the constraint $\phi = \omega t$, we derive

$$\lambda = 2\,m\,r\,\dot{r}\cdot\omega + m\,g\,r\cdot\sin\phi.$$

$\lambda(t)$ is the moment about the origin of the contact force transmitted between the wire and the bead. It is exactly the moment one would have to apply to keep the unconstrained bead rotating about the origin with the constant angular velocity ω. In functional analytic terms λ is an element of the topological dual of the space of functions assigned to ϕ. Again, we conclude that it has to have the physical dimension of the moment of a force.

1.31 The Weierstrass E-function and the mathematical control theory.

Let us again consider the brachistochrone problem, that is of finding a curve of fastest descent. Instead of following Euler and perturbing the optimal curve Weierstrass offered a new insight into this class of problems. Suppose that a position on an optimal path is given, Then it suffices to prescribe the optimal slope for the optimal solution curve at that point. After infinitesimal travel in the direction of the optimal slope it suffices to choose a new direction, etc. In fact one can direct the particle at each point of the curve steering it in the optimal direction thus generating the trajectory. From this point of view we regard the integral

$J(y) = \int_0^{\ell} f(y,y',x)\,dx$ as a function of $y'(x)$.

If two controls are adopted, Z and U then over a "short" distance the value of J(y) would have differed by

$$\int\{(f(x,y,U) - f(x,y,Z)) + (Z-U)\,f_{y'}\}dx,$$

$$= \int E\,(x,y,Z,U)\,dx$$

where

$$E = f(x,y,U) - f(x,y,Z) + (Z-U)f_{y'}(x,y,Z)$$

The second term is the linear term in the Taylor series expansion of the f(x,y,U) about point Z i.e.

$$f(x,y,U) = f(x,y,Z) + (U-Z)\,f_{y'}(x,y,Z) + \ldots$$

In a higher dimensional case we have

$E(x,y,Z,U) =$

$$f(x,\tilde{y},\tilde{U}) - f(x,\tilde{y},\tilde{Z}) - \sum_{i=1}^{n}\{u_i - z_i).f_{y'}(x,\tilde{y},\tilde{Z})\}$$

The implications of the positivity of the Weierstrass E-function (excess function) may be better understood if we relate the property of E-function to covexity of f(y,y',x).

Let us recall that a set $\Omega \subset R^n$ is convex if for every pair of points ξ, η in Ω, the straight line interval connecting ξ, η lies entirely in Ω.

That is, for any value of a parameter τ, $0 \leq \tau \leq 1$, the point $\tau\cdot\xi + (1-\tau)\cdot\eta$ lies in Ω. A function f(x), $x \in \Omega$, is convex if for any points ξ,η

$$(*)\quad f[(1-\tau)\cdot\xi + \tau\cdot\eta] \leq (1-\tau)\cdot f(\xi) + \tau\cdot f(\eta),$$

with $0 \leq \tau \leq 1$.

A function f(x) is concave if (-f(x)) is convex.

The inequality (*) implies that if we choose points $(\xi,f(\xi))$, $(\eta,f(\eta))$ on the graph of $y = f(x)$, $\xi < \eta$, then on the segment $\xi \leq x \leq \eta$ the graph of the function lies below the line segment connecting $(\xi,f(\xi))$ and $(\eta,f(\eta))$. Thus for $\tau \neq 0$ we have

$$f(\eta) - f(\xi) \geq \frac{f[\xi + \tau(\eta - \xi)] - f(\xi)}{\tau}.$$

If this inequality is strict, i.e. we replace "$\geq$" by "$>$" sign, then f(x) is called strictly convex.

For a convex function f we can introduce the excess, or E-function:

$$E_f(\xi,\eta) = f(\eta) - f(\xi) - [f'(\xi) \cdot (\eta - \xi)]$$

The following is easily proved:

<u>Theorem</u>. f(x) is convex in Ω if and only if $E_f(\xi,\eta) \geq 0$ for any choice of $x = \xi, x = \eta \in \Omega$.

$(\underset{\sim}{\eta} - \underset{\sim}{\xi}) \cdot f'(\underset{\sim}{\xi})$ is in fact the directional derivative

$$\sum_{i=1}^{n} \frac{\partial f}{\partial \xi_i}(\eta_i - \xi_i) = \lim_{\tau \to 0} \frac{f(\underset{\sim}{\xi} + \tau \underset{\sim}{h}) - f(\underset{\sim}{\xi})}{\tau}$$

$= f'(\underset{\sim}{\xi}, \underset{\sim}{h})$, where $\underset{\sim}{h} = \underset{\sim}{\eta} - \underset{\sim}{\xi}$, and $f'(\underset{\sim}{\xi}, \underset{\sim}{h})$ denotes the directional derivative of $f(\underset{\sim}{x})$ in the direction of $\underset{\sim}{h}$ computed at $\underset{\sim}{x} = \underset{\sim}{\xi}$.

1.32 <u>Formulation of a control problem.</u>

The classical problem of Bolza with equality (or inequality) constraints is that of finding a class of arcs

$$x_i = x_i(t), \ t_o \leq t \leq t_1, \ i = 1,2,\ldots n,$$

that minimize a functional $I_o(\underset{\sim}{x})$ given by I_o =

$$I_o(\underset{\sim}{\chi}) = g_o(t_o, t_1, \underset{\sim}{b}) + \int_{t_o}^{t_1} L(t, \underset{\sim}{\chi}(t), \dot{\underset{\sim}{\chi}}(t))dt,$$

subject to constraints:

$$\xi_\alpha(t, \underset{\sim}{\chi}, \dot{\underset{\sim}{\chi}}) < 0, \ \alpha = 1,2,\ldots,k,$$

$$\xi_\alpha(t, \underset{\sim}{\chi}, \dot{\underset{\sim}{\chi}}) = 0, \ \alpha = (k+1),(k+2),\ldots,m,$$

$$t_o = t_o(\underset{\sim}{b}), \quad t_1 = t_1(\underset{\sim}{b}),$$

$\xi_i \in B_i$, $i = 1,2$, where B_i is an open subset of $\mathbb{R}^n$. That is, we minimize the functional I_o, with constraints given by a system of differential equations and differential inequalities.

A similar problem was formulated by Lagrange:

Minimize $\int_0^{\ell} L(\underset{\sim}{\chi},\dot{\underset{\sim}{\chi}},t)dt = I(\underset{\sim}{\chi})$, subject to conditions of the form :

$$\phi_\alpha(\underset{\sim}{\chi}, \dot{\underset{\sim}{\chi}}, t) \leq 0 , \quad \alpha = 1, 2,...k$$

$$\phi_\beta(\underset{\sim}{\chi}, \dot{\underset{\sim}{\chi}}, t) = 0, \quad \beta = K+1,...m.$$

Weierstrass made an observation that the optimal arc may be constructed by considering its construction as a sequence of decision-making steps. Fixing an intermediate point on the optimal arc, a decision is made regarding the optimal direction. A modern version of the Weierstrass approach may be restated in the control-theoretical terms.

The control vector $u(t)$ $(u=\dot{\chi})$ can be introduced. $\underset{\sim}{u}(t)$ is assumed to be piecewise continuous or at least measurable function. We seek an admissible arc $\underset{\sim}{\chi} = \underset{\sim}{\chi}(t)$ which minimizes

$$\int_a^b L(\underset{\sim}{\chi}, \underset{\sim}{u}(t), t)\, dt$$ subject to conditions

$$\phi_\alpha(\underset{\sim}{\chi}, \underset{\sim}{u},t) \leq 0$$

$$\phi_\beta\ \underset{\sim}{\chi}, \underset{\sim}{u}, t) = 0 .$$

Arc $\chi = \chi(t)$ is admissible, if it is sufficiently smooth to satisfy the physically dictated requirements on smoothness, and if $\chi(t)$ satisfies all constraints for any choice of an admissible control vector $\underset{\sim}{u}(t)$.

Pontryagin's problem is almost identical with the problem of Bolza, except in the manner in which it is formulated.

We seek a solution pair $\{\tilde{x}(t), \tilde{u}(t)\}$ of the system of differential equations

(*) $\dot{\chi} = f(t, \chi(t), u(t))$, $0 \leq t \leq \tau$, where a control $\tilde{u}(t)$ is a measurable function of t, and satisfies a condition of the type $||\tilde{u}|| \leq \mu$ for some $\mu>0$, such that the functional

$$I(\tilde{\chi}, \tilde{u}) = \int_0^\tau L(\tilde{\chi}, \tilde{u}, t)\, dt \text{ is minimized.}$$

Theorem (Pontryagin).

There exist multipliers $\lambda_o, \lambda_1 \ldots \lambda_n$ and a function $H(\tilde{\lambda}, \tilde{\chi}, \tilde{u}) = \lambda_o L + \Sigma\lambda_i f^i$ satisfying the "Hamiltonian equations"

$$H_{\lambda_i} = \dot{\chi}_i \quad , \quad -H_{\chi_i} = \dot{\lambda}_i \quad .$$

If $\{\tilde{\chi}_o(t), u_o(t)\}$ is the optimal pair then

$H(\tilde{\lambda}_o, \tilde{\chi}_o, \tilde{u}_o) \leq H(\tilde{\lambda}, \tilde{\chi}, \tilde{u})$ for any admissible pair $(\tilde{\chi}, \tilde{u})$. Moreover, if the set of admissible pairs $\{\tilde{\chi}, \tilde{u}\}$ is open, then

$$H_{u_i}(t, \tilde{\chi}_o, \tilde{u}_o, \tilde{\lambda}) = 0, \quad i = 1,2\ldots n .$$

(for all $t \in [0,T]$).

A set of inequality constraints

$$\Phi_\gamma = \int_0^T \phi_\gamma(t, \tilde{\chi}, \tilde{u}) \leq 0, \quad \gamma = 1,2,\ldots r,$$

may be stated as a part of the problem.

Then the corresponding Hamiltonian $\bar{H}$ is given by

$$\bar{H} = H + \sum_{\gamma=1}^{r} \mu_\gamma \phi_\gamma \quad ,$$

where the Lagrangian multipliers μ_γ satisfy the inequality $\mu_\gamma \geq 0$.

1.33. <u>An example of a direct application of the Weierstrass test.</u>

We shall restate classical results. All derivatives shown in this discussion <u>exist</u>. Let the Lagrangian be given by $L(x,y,y') = 2(y')^2 - (y')^4 - yy'$.

Then $E(x,y,y',p) = 2(y')^2 - (y')^4 - yy' - 2p^2 + p^4 + yp - (y'-p)(4p-4p^3-y) = 2(y'^2-p^2) - 2y'p(1-p^2) - (y'^4 + 3p^4)$.

Therefore $E(x,y,y',p) < 0$ if

$$y'^2 - p^2 < y'p(1-p^2) + y'^4 + 3p^4 .$$

More specifically, it suffices if $1 > p \geq \frac{1}{2}$ and y' is sufficiently close to p.

It should be clear that a direct application of Weierstrass criterion to determine the existence of a minimum or maximum is an exception rather than the rule. Its importance emerges in theoretical studies after realizing that many effective modern ideas can be directly traced or related to this criterion.

The Weierstrass function E can be interpreted as the quadratic approximation to the Taylor series expansion of the Lagrangian. Assumming sufficient (twice continuous) differentiability, let us write formally the first two terms of Taylor series with a quadratic remainder

$$L(x,y,y') = L(x,y,p) + \frac{y'-p}{1!} L_p(x,y,p) + \frac{(y'-p)^2}{2!} L_{pp}(x,y,q), \text{ where } p \leq q \leq y',$$

if $p \leq y'$ and $p \geq q \geq y'$ if $p \geq y'$.

Then

$$E(x,y,y',p) = L(x,y,y') - L(x,y,p) - (y'-p)L_p(x,y,p) = \frac{(y'-p)^2}{2!} L_{pp}(x,y,q).$$

Therefore

$E(x,y,y',p) > 0$ if

$L_{pp}(x,y,q) > 0$ for all q-s between y' and p. This is known as the Legendre condition.
Conversely, if $L_{pp}(x,y,q) > 0$ for all q in some neighborhood of $\hat{y}'(x)$ along a chosen trajectory $y = \hat{y}(x)$ and $\hat{y}(x)$ satisfies the Euler-Lagrange equations, then $\hat{y}$ provides a local minimum for the action integral in the class of admissible (smooth) trajectories.

As a simple exercise let us check that straight lines provide a local minimum to the distance funcion between two points in the Euclidean space. We define the distance between two points $(x_1 y_1),(x_2 y_2)$ along an arc: Γ: $y = y(x)$

$$d_\Gamma(p_1,p_2) = \int_\Gamma \{(\frac{dx}{dx})^2 + (\frac{dy}{dx})^2\}^{\frac{1}{2}} ds .$$

Since we assume that it is possible to express y as a function of x on Γ, we have

$$d(p_1,p_2) = \int_{x_1}^{x_2} [1 + (y')^2]^{\frac{1}{2}} dx.$$

We apply Legendre test to the solution y' = const.

$$F_{pp} = \frac{\partial^2}{\partial p^2} (1 + p^2)^{\frac{1}{2}} = (1 + p^2)^{-\frac{1}{2}} (1 - \frac{p^2}{1+p^2})$$

Clearly F_{pp} is positive, since $1 > \frac{p^2}{1+p^2}$ for any constant value of the real variable p.
The Euler-Lagrange equation assumes the form:

$$\frac{d}{dx}\left\{y'\cdot\left[1 + (y')^2\right]\right\}^{-\frac{1}{2}} = 0 ,$$

or

$$y'.(1+(y')^2)^{-\frac{1}{2}} = \text{constant} = C.$$

This relation implies that $(C^2-1)(y')^2$ = constant, and therefore ,if y' is continuous then $y' \equiv$ constant.The solution curve is a straight line.

We comment that a direct application of the Weierstrass test would have been more difficult even in this simple case.

Note. The condition $L_{pp} > 0$, or $L_{pp} < 0$ is called in the literature " the strong Legendre condition". This may cause some confusion if we associate the term strong condition with "the strong minimum", and the "weak condition" $L_{pp} \geq 0$ with the weak minimum in the 19-th century terminology of the calculus of variations. Even worse confusion does arise if such minima or maxima arise as weak or strong limits in the weak, i.e. Tikhonov topology, or strong, i.e. norm topology.

Exercise. Show that the cycloid provides a local extremum to the brachistochrone problem.

We minimize the functional

$$\int_{x_0}^{x_1} \{[1 + (y')^2]^{\frac{1}{2}} \cdot (y)^{-\frac{1}{2}}\} \, dx, \; y(0) = 0.$$

The Euler-Lagrange equations have a parametrized solution derived by Newton:

$$x = A(t - \sin t)$$
$$y = A(1 - \cos t)$$

forming a "pencil of extremals" passing through the origin (0,0).

$L_{y'y'} = (y)^{\frac{1}{2}}(1+(y')^2)^{-3/2}$ is positive.

Therefore Newton's solution is a local minimum in the class of twice differentiable functions.

1.33. <u>Some comments on the classical approach to a variational formulation.</u>

The Euler-Lagrange equations discussed in this chapter were nonlinear ordinary differential equations. The system's state was parametrized with respect to a single variable, generally identified with time. The existence of solutions was generally not questioned. The fact that the techniques that work for ordinary differential equations are easily extended to systems of partial differential equations was observed by many mathematicians. The Russians credit Mikhail Ostrogradski with generalizing Euler's equations to functionals containing partial derivatives of an arbitrary order. One of the simplest problems of this type, minimization of $\int_\Omega |\mathrm{grad}(w)|^2 \, d\underset{\sim}{x}$

over a region Ω in $\mathbb{R}^n$ with boundary values assigned on $\partial\Omega$, has a long and interesting history. Let us consider this problem in two dimensions. This is known as the Dirichlet problem.

Find a function $\omega(x,y)$ that minimizes the integral:

$$J(\omega) = \iint_\Omega \left(f\left(\frac{\partial\omega}{\partial x}, \frac{\partial\omega}{\partial y}, \omega, x, y\right)\right) dxdy = \iint_\Omega \left(\frac{\partial\omega}{\partial x}\right)^2 + \left(\frac{\partial\omega}{\partial y}\right)^2 dxdy$$

in an open region $\Omega \subset \mathbb{R}^2$ with $\omega(x,y) = \mu(x,y)$ assigned on the boundary $\partial\Omega$ of Ω.

We formally write the Euler equation

$$\frac{\partial}{\partial x}\left(\frac{\partial f}{\partial \omega_x}\right) + \frac{\partial}{\partial y}\left(\frac{\partial}{\partial \omega_y}\right) - \frac{\partial f}{\partial \omega} = 0$$

where $\omega_x \equiv \frac{\partial\omega}{\partial x}$, $\omega_y = \frac{\partial\omega}{\partial y}$, and ω is selected from a class of functions satisfying the boundary conditions, Thus we obtain the Laplace equation as a necessary condition for a minimum of the Dirichlet integral. That is, a minimizing function $\tilde{\omega}(x,y)$ must satisfy the Laplace equation

$$\frac{\partial^2\tilde{\omega}}{\partial x^2} + \frac{\partial^2\tilde{\omega}}{\partial y^2} = 0 \; .$$

Everything seems to be in order.

A minimal value for integrals $J(\omega)$ does exist, since $J(\omega)$ is non-negative and has zero as a lower bound; and for any collection of admissible functions $\{\omega_i\}$, $J(\omega_i)$ has a greatest lower bound.

It should be true that any such collection of admissible functions $\{\omega_i\}$ must contain a minimal element $\tilde{\omega}$, so that $J(\tilde{\omega}) = \min J(\omega_i)$, $i = 1,2,\ldots$ This statement is known as Dirichlet's principle. Dirichlet and Riemann believed in the validity of this principle. Riemann offered arguments whose truth depended on the truth of the Dirichlet's principle. Unfortunately, in the form stated above it is false. K. Weierstrass subjected the statement of the Dirichlet's principle to a rigorous scrutiny.

The following counter-example is a modification of Weierstrass' example given in [22].

Consider the class of functions $\omega(x,y)$ that are piecewise continuous in the interior of a square $\Omega \subset R^2$: $\{x,y| \le x \le +1, -1 \le y \le 1\}$, and satisfy the boundary conditions on $\partial\Omega$

$$\left.\begin{aligned}\omega(x,-1) &= -x\\ \omega(x,+1) &= x\end{aligned}\right\} -1 \le x \le +1,$$

$$\left.\begin{aligned}\omega(-1,y) &= -y\\ \omega(+1,y) &= y\end{aligned}\right\} -1 \le y \le +1 .$$

We observe that the values assigned are continuous on the boundary of the square region Ω.

Proof of the counter-example.

We claim that the functional

$$J(\omega) = \iint_{\Omega} (x^2 \cdot \omega_x^{\,2} + y^2 \cdot \omega_y^{\,2})\, dxdy \text{ does not}$$

attain a minimum over a class of admissible functions $\omega(x,y)$.

Consider the sequence of functions :

$$\omega_n(x,y) = \frac{\arctan(nxy)}{\arctan(n)}, \quad n=1,2,\ldots ,$$

then $$J(\omega_n) = \frac{2}{n^2(\arctan n)^2} \int_{-1}^{+1}\int_{-1}^{+1} \frac{x^2y^2}{(x^2y^2+n^{-2})^2}\, dxdy$$

and $\lim_{n\to\infty} J(\omega_n) = 0.$

But the only piecewise continuous function $\omega(x,y)$ that satisfies the relation:

$$\iint_{\Omega} (x^2\omega_x^{\,2} + y^2\omega_y^{\,2})\, dx\cdot dy = 0 \text{ is a function}$$

$\omega(x,y) \equiv$ constant almost everywhere in Ω, since otherwise the integrand would be positive.

Thus, such a function cannot be continuous in some neighbourhood of the boundary $\partial\Omega$.

In this example the region Ω had a smooth boun-

dary,except at the four corners of this region.
A problem that is caused by the shape of the region Ω is illustrated by considering the Dirichlet problem for a punctured disc. Ω is the interior of the unit disc in $\mathbb{R}^2$ with the center deleted, i.e.

$$\Omega = \{x,y \mid 0<x^2+y^2<1\} .$$

We assign boundary values : $\omega(x,y) = 0$ when $x^2+y^2 = 1$, and $\omega(x,y) = 1$ when $x = y = 0$.

We wish to minimize $\iint_\Omega |\text{grad } \omega|^2 \, dx\, dy$.

Again,there does not exist a piecewise continuous function $\omega(x,y)$ that is continuous in some neighbourhood of the boundary $\partial\Omega$ and minimizes this integral.

Around the year 1900 D.Hilbert reformulated correctly the statement of Dirichlet's principle and offered a rigorous proof.(This is known as the "modified Dirichlet's principle".) Some variants of Hilbert's theory may be found in the works of Krylov and Bogoliubov and ,for example, in the N.Y.U. Lecture Notes of Richard Courant.

Lower simicontinuity and/or the existence of an appropriate support functional,or some form of compactness is essential in all arguments concerning the existence of a minimizing function in a class of admissible functions.

A characterization of an optimal element (generally of an optimizing function) in the absense of an existence proof may turn out to be a disguised version of Oscar Perron's counterexample that is given next.

Let us postulate the existence of the largest positive integer N.That is the integer N satisfies $N \geq n$ for any positive integer n. Then we can proove that $N = 1$.

Proof. If $N > 1$ then $N^2 > N$. But $N \geq N^2$ by our hypothesis.Therefore $N = N^2$,and $N = 1$.

There is nothing wrong with our proof.The nonsensical characterization we obtained for our optimum was caused by our assumption that such optimum(i.e.the largest positive integer) does exist. A similar nonsense could be asserted if we derive some necessary or sufficient conditions for the extremum of a functional without first checking that such extremal value is attained in the class of admissible functions.The counterexamples to Dirichlet's principle serve as a warning to a search for sufficiency conditions.Solutions to Euler-Lagrange equations may not exist if the corresponding variational problem does not have a solution in a class of admissible functions (admissibility to be decided by the physics of the problem). Unfortunately many engineering arguments contained assumption of the existence of a solution. Then the properties of the solutions were derived. Some well known paradoxes, such as the Stokes' paradox have their origin in this type of argument. This touches on a basic difficulty in the modeling of physical phenomena. The existence of solutions in a mathematical model should realistically reflect our physical experience.

References for Chapter 1

[1] H.C. Corben and P. Stehle, Classical mechanics, J. Wiley and Sons, New York, 1960.

[2] H. Goldstein, Classical mechanics, Addison Wesley Reading, Mass., 1953.

[3] R. Abraham, Foundations of Classical mechanics, Benjamin, New York, 1967.

[4] L.D. Landau and E.M. Lifshitz, Mechanics, Addison Wesley, Reading Mass., 1960.

[5] Lord Rayleigh, The theory of sound, Dover Publications, New York.

[6] E.J. Routh, Treatise on the dynamics of rigid bodies, Macmillan, London, 1905 (also reprinted by Dover Publications, New York).

[7] Ernst Mach, The science of mechanics, 5th English edition, Open Court Publishing Co., La Salle, Ill., 1942.

[8] W. von Kleinschmidt and H.K. Schulze, Brachistochronen in einem zentral-symmetrischen Schwerefeld, ZAMM, 50, 1970, p. 234-236.

[9] A.V. Russalovskaya and G.I. Ivanov and A.I. Ivanov, On the brachistochrone for a particle with variable mass subjected to motion with friction and an exponential law of mass variation, Dopovidi Akad. Nauk Ukrain. S.S.R. 11, (1975), p. 97-112.

[10] N.W. Ashby, E. Brittain , W.F. Love and W. Wyss, Brachistochrone with Coulomb friction, American J. Physics, 43,(1975), p. 902-906.

[11] D.S. Djukič, On the brachistochronic motion of a non-conservative dynamic system, Zbornik Rad. Matem. Instit. Beograd, (N.S.), 3, #11 (1979) p. 39-46.

[12] J.E. Drummond and G.L. Downes, The brachistochrone with acceleration: A running track, J. Optimization Theory and Applications, 7, (1971), p. 444-449.

[13] Isaac Newton, Mathematical papers, 7 volumes, edited by D.T. Whiteside (1967-1976), Cambridge University Press, Cambridge.

[14] F. John, Continuous dependence on data for solutions of partial differential equations with a prescribed bound,Comm.Pure Appl.Math.,vol.13,(1960),p.551-585.

[15] L.E. Payne, On stabilization of ill-posed Cauchy problems in nonlinear elasticity Contemporary Mathematics, vol. IV, Problems in elastic stability and vibrations, Vadim Komkov editor, (1981) American Math. Society, Providence, R.I.

[16] H. Hertz, Prinzipien der Mechanik in neuen Zusammenhang dargestellt, Barth Verlag, Leipzig, 1910.

[17] G.D. Birkhoff , Dynamical systems, American Math. Society, New York, 1928.

[18] H. Poincaré, Oeuvres de Henri Poincaré, Gauthier Villars, Paris, 1952.

[19] V. Komkov, Continuability and estimates of solutions of $(a(t)\cdot\Psi(x)\cdot x')' + c(t)f(x) = 0$ Annales Pol. Math., XXX, (1974), p. 125-137.

[20] T.Burton and R.Grimmer,On solvability of solutions of second order differential equations,Proc.Amer. Math.Society, 29 ,#2,(1971),p.277-283.

[21] H.Anton and L.Y.Bahar,On the use of Schwartz distributions in the Lagrange variational problem,Hadronic Journal, 1 ,(1978),p.1215-1226.

[22] K. Weierstrass, Gesammelte Werke, Meier und Müller Verlag, Berlin, 1894 .

[23] E.Cartan,Leçons sur les invariants,Hermann et Cie., Paris,1922.

[24] C. Caratheodory, Variationsrechnung, Teubner Verlag, Leipzig, 1935. English translation by Chelsea Publ. Co., London, 1939.

[25] A. Kneser, Lehrbuch der Variationsrechnung, Vieweg, Brunswick, 1925.

[26] The Karman and M.A. Biot Mathematical methods in engineering, McGraw Hill, New York, 1940.

[27] K.C. Valanis and Vadim Komkov, Irreversible thermodynamics from the point of view of internal variable theory, Archives of Mechanics, Vol. 32, #1, (1978), p. 33.58.

Chapter 2

THE LEGENDRE TRANSFORMATION, DUALITY AND FUNCTIONAL ANALYTIC APPROACH

The Legendre transformation

2.0 Some general comments.

The growth of science is accompanied by a steady diversification of knowledge.At the same time a deeper understanding of diverse phenomena leads to discoveries of common laws of physics and of mathematics,unifying the mathematical treatment. Similarity between inertial and gravitational forces lies at the origins of Einstein's "thought experiments" and the special theory of relativity. Mach's explanation of inertial forces in Newton's bucket experiment is an example of a brilliant unification attempt. Basically, E. Mach suggested that the mass of the entire universe affects the local laws of mechanics so far discovered by mankind. (We refer to his famous statement about the distance stars causing it all.)

One cannot escape some comparisons between Mach's philosophy and the earlier outlooks on science dominated by Berkeley, Kant and Hegel. While Bishop Berkeley was primarily interested in asserting predominance of human senses and the influence of God on all the so called "laws of physics" discovered in the western world, his arguments forced the scientific community to evaluate the philosophy of science. What do we "perceive" by results of experiments? How is that related to "reality"? What do we mean by "physical reality"? Consequently, are all laws of physics "local laws"?

In this sense Ernst Mach continues the debate by asserting that we perceive laws that are universal in the broadest possible sense, unless we accept the "super-universes" cosmological theory in which our entire universe is perhaps a black hole, a quark, or some unstable particle, and thus our universal laws become again local laws on a truly grand scale.

On somewhat similarly grandiose scale some scientists, such as W. Hamilton claimed that all known laws of mechanics may be restated in the form of variational principles.

Such fundamental statements have been formulated both heuristically (Hamilton's "theological" assertion) and as rigorous mathematical principles, such as Gauss' principle of least curvature or Hertz's principle of the simplest trajectory. In the "variational notation"(that was regarded as heuristic by mathematicians)the variational statements, such as principles of d'Alambert and of Hamilton may be written as simple formulas

a) $\sum_{i=1}^{N} (F_i - m\, a_i)\cdot\delta q_i = 0,$

b) $\delta \int^{t_2} \mathcal{L}\, dt = 0,$

respectively.

Here F_i are forces a_i - corresponding accelerations m the mass of a single particle and δq_i virtual (infinitesimal) displacements. Formula a) states that the virtual work performed by all external and inertial forces is "equal" to zero.

Deliberately we examine the simplest possible version of d'Alambert's principle.

In a modern interpretation we would assert that the work performed is an infinitesimal of order smaller than $|\delta q_i|$.

Formula b) asserts that there exists a Lagrangian density function $\mathcal{L}(q_i, \dot{q}_i, t)$ such that the equations of motion of the system may be derived as a consequence of the stationary

behavior of the action integral $\int_{t=t_1}^{t=t_2} \mathcal{L}\, dt$

under infinitesimal changes (or perturbations) of the trajectory of motion. t-represents time in the classical formulation. Again, we could examine the simplest possible formulation and assume that the time period $[t_1, t_2]$ is fixed and the trajectory passes through known points at the times t_1, t_2, that is $\tilde{q}(t_1) = \tilde{q}_1$, $\tilde{q}(t_2) = \tilde{q}_2$ are assigned values for the trajectory $\tilde{q} = \tilde{q}(t)$. These restrictions are not essential but were stated to simplify the introductory presentation.

Let us restate the principle of least action emphasizing the variational nature of this principle. Among all possible trajectories of motion the laws of nature choose a trajectory that assigns an extreme value (generally a minimum) to the action integral. It is not only a mathematical statement, but a philosophical viewpoint concerning the so called "laws of nature." It reflects the history of western scientific philosophy since the 17-th century. Around 1650 Fermat announced the principle of minimum time for propagation(of light)in a non-uniform medium.Among the trajectories that are available for the path of a light ray between two given points in space,the one traversed by the ray minimizes the time of travel.In the case of transition from one uniform medium to another one Fermat's principle implies a principle of optics known as Snell's law.It is remarkable that Fermat's principle inspired Newton's solution of the brachistochrone problem.The similarity of thewave,or light propagation and the descent of a mass obeying the law of gravity inspired statements about the simplicity of the basic laws of nature. In 1774 M.deMapertuis elevated a simple form of the principle of least action to the status of a general law of nature.Euler systematized it and introduced the mathematical formalism that now bears his name.J.Lagrange generalised Euler's results and offered an easily comprehended physical interpretation relating some terms of Euler's formulas to the concepts of classical mechanics.

Hamilton and Jacobi reexamined the Euler-Lagrange theory and pointed out the applicability of the variational approach to areas of science outside classical mechanics.Hamilton relaxed some of the restrictions imposed by Lagrange,such as the constant total energy condition,and stated clearly the conditions when the Lagrangian $\mathcal{L}$ may be represented as a difference between the kinetic and the potential energy of the system,i.e. $\mathcal{L} = T - V$.

Hamilton's principle restates the principle of least action by asserting that a "real" physical system acts in such manner that the integral of the Lagrangian is minimized along the trajectory of motion.Moreover,the Hamiltonian generates the equations of motion by means of the the so called canonical equations.

Consequences of these early developments are almost impossible to catalog.Certainly they had a profound influence on the developments of all areas of modern physics.For example,a starting point for Planck's critique of the distribution of energy in the spectrum of white light was his observation that the current explanation was inconsistent with Hamilton's principle.The resulting ultraviolet catastrophe appeared to be contradicting our common sense experiences. In almost any theoretical discussion of modern quantum mechanics, statistical mechanics, thermodynamics, chemistry, relativity, quantum electrodynamics one may find the postulation of the Lagrangian or the Hamiltonian, a discussion of invariants, of symmetries and of the conservation laws. Variational theories are used to unify diverse areas of science and to provide both methodological and philosophical foundations.

For a general history of the calculus of variations see Forsyth [12]. For a detailed discussion of variational principles including the Euler-Lagrange and Hamilton's theories read N. Akhiezer[1], Bliss[7], C. Lanczos[65], L. El'sgolc[9], L.S. Polyak[66]. Applications of variational principles to modern physics may be found in L. Landau and E. Lifshitz[67], [68] particularly in the chapter entitled "Variational principles", W. Yourgrau and S. Mandelstam [150] or in Morse and Feshbach [117].

For applications to continuum mechanics read T. Oden and Reddy[41], or a more classical text of H.L. Langhaar[74]. However there is no substitute for the original sources. These may include J. Bernoulli[6], L. Euler's Opera Omnia[10], W.R. Hamilton, M. Planck, Weierstrass. Also a wealth of additional sources may be found in the references of [1], [53], [9], [72], [67], [41].

2.1 The relation between the Euler-Lagrange equations, and the Legendre transformation.

The Euler-Lagrange equation for the problem of extremizing the value of

$$\int_{t_0}^{t_1} \mathcal{L}(x,\dot{x}, t)\, dt$$

is

$$\frac{\partial \mathcal{L}}{\partial x} - \frac{d}{dt}\left(\frac{\partial \mathcal{L}}{\partial \dot{x}}\right) = 0. \qquad (2.1)$$

(See, for example, [72] for a "classical" derivation.)
The Legendre transformation introduces the generalized momentum

$$p = \frac{\partial \mathcal{L}}{\partial \dot{x}} \qquad (2.2)$$

Hence (2.1) can be written as

$$\frac{\partial \mathcal{L}}{\partial \dot{x}} = p = - \frac{\partial H}{\partial \dot{x}} \qquad (2.3^a)$$

and $\frac{\partial H}{\partial p} = \dot{x}$ where $H = \dot{x}\, p - \mathcal{L}$. (2.3^b)

We refer to appendices A and B and proceed to formulate the Hilbert space analogs of equations (2.3^a), (2.3^b), and (2.1). The basic ideas of generalization of equations (2.3^a), (2.3^b) go back to Korn who observed the existence of dual variational principles in the theory of elasticity of materials not necessarily obeying Hooke's law. Friedrichs has interpreted Korn's results in terms of modern operator theory [15].

An intuitive approach to the postulation of duality and formulation of dual variation principles originated with B. Noble[38] and has been successfully applied by numerous authors to problems of mathematical physics (see for example

[8], [30], [28] or the monograph of A.M.Arthurs. See reference [4] . In this chapter and the remainder of this work we shall use the higher dimensional version of equations (2.1-2.3^{b}) , as given in chapter 1,sections 1.25-1.27.

2.2 A simple physical example.

A deflection of a membrane (let us think of it as a soap bubble) is modelled by the Poisson equation.We visualise a soap bubble spread over a rectangular opening and subjected to a static load q(x,y) acting on one side of its surface.In the region Ω,occupied by the surface of the membrane, $\Omega = \{x,y \mid |x|<A, |y| < B\}$,the membrane deflection satisfies the differential equation :

$$(2.7) \qquad -\Delta u \equiv -\frac{\partial^2 u}{\partial x^2} - \frac{\partial^2 u}{\partial y^2} = \underset{\sim}{q}(x,y), \ ||\underset{\sim}{q}|| \le K,$$

for some $K \in \mathbb{R}$, with $||\cdot||$to be defined later, while on the boundary $\partial\Omega$ of the region Ω (i.e. if $|x|= A$ or if $|y|= B$) we have :

$$(2.8) \quad u(x,y) \equiv 0.$$

Symbolically we write the equation (2.7) in an arbitrary region $\mathfrak{D}$ as

$$(2.4) \quad Au - q = 0 ,$$

that is A stands for $-\Delta$. The total work performed by the distributed load $\underset{\sim}{q}(x,y)$ is given by $<\frac{1}{2} u, q>$ which is given by

$$(2.9) \tfrac{1}{2} \iint_{\mathfrak{D}} [u(x,y)\cdot q(x,y)]dxdy$$ if both u and $\underset{\sim}{q}$ are

square integrable functions in $\mathfrak{D} \subset \mathbb{R}^2$. That is, we identify the bilinear product <,> with the $L_2(\mathfrak{D})$ product. (We can identify $\mathfrak{D}$ with Ω.)

The strain energy of the membrane is equated with the work of deformation

$$(2.10) \quad V = \tfrac{1}{2} \iint_D u(x,y)\, q(x,y)\, dxdy = \tfrac{1}{2} \iint_D \{(-\Delta u)\cdot$$

$$\cdot u(x,y)\} \,dxdy = \tfrac{1}{2}\iint_{\mathfrak{D}} [(\frac{\partial u}{\partial x})^2 + (\frac{\partial u}{\partial y})^2]\, dxdy ,$$

after integration by parts.

We have not assigned the space of functions in which we locate the solution u(x,y) of the problem.

Looking at the equation (2.7) one would expect that the solution should be twice differentiable.

A physical interpretation of the solutions (properties of a soap bubble) does not uphold this conclusion. Certainly the solution should be continuous (we do not want burst soap bubble shapes) but twice differentiability is an excessive and unnecessary restriction.

The shape of the membrane corresponds to the minimum value of the potential energy:

$$(2.11) \quad U = \iint_D \{\tfrac{1}{2}[(\frac{\partial u}{\partial x})^2 + (\frac{\partial u}{\partial y})^2] - u \cdot q\} dx \cdot dy$$

To compute the value of U with $q(x,y) \in L_2[\mathfrak{D}]$ we need $u(x,y)$ to be square integrable and its first derivatives $\frac{\partial u}{\partial x}$, $\frac{\partial u}{\partial y}$ to be square integrable in $\mathfrak{D}$. Hence, we specify $\underset{\sim}{u} \in H_0^1(D)$ (or in Russian notation following Sobolëv's work [48], $\underset{\sim}{u} \in W_o^{1,2}(\mathfrak{D})$).

We observe that the value of the functional $U(u)$, as given in (2.11), is defined if $u \in H_o^1(D)$ and $u \cdot q$ is integrable in $\mathfrak{D}$. We do not need twice continuous differentiability that is tacitly implied in the original equation (2.7) describing the physical problem - that is the behaviour of a soap bubble subjected to a non-uniform pressure.

We recognise that U in (2.11) has the value equal to one half of the functional $F = \langle u,q \rangle$ (2.5). This type of minimization problem is easily solved numerically. For example, let us take $q(x,y) = \text{const.} \equiv 1$, and $A=B=1$. We expect the solution to be symmetric about the x-, and the y - axis. We need to tabulate the values of $u(x,y)$ only in the first quadrant. The values on the x-axis are given in the table 2.1. The values derived by a direct minimization algorithm are compared with the values computed by summing the first few terms in the Fourier series expansion of the solution, with the summation terminating if the next term does not contribute to the 4-th decimal place of the value.

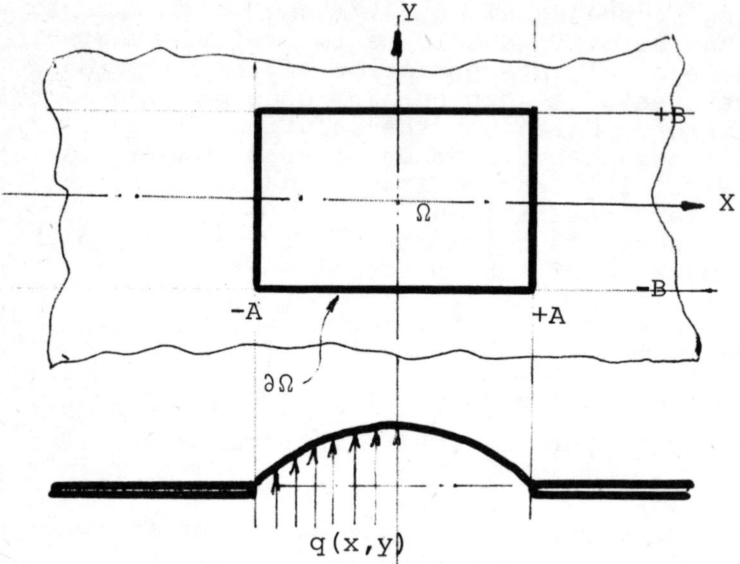

Figure 2.1

coordinates x,y	direct minimization technique	Fourier series(prediction) values of u(x,y)
0,0	.585	.581
.25,0	.543	.541
.5 ,0	.444	.442
.75,0	.301	.279
1.00,0	.000	.000

Table 2.1

2.3 Conjugate convex functionals.

Let Φ be a convex functional (see appedix B for definition). $\Phi : \Omega \subset B \to R$ is defined on a convex set Ω B,where B is a Banach space and B* is its dual. The set Ω^* conjugate to Ω relative to Φ ($\Omega^* \subset B^*$) is defined as:

(2.12) $\Omega^* = \{p \in B^*: \sup_{x \in \Omega} \langle p,x\rangle - \Phi(x) < M$ for some $M \in \mathbb{R}\}$ The conjugate functional $\phi^*(p)$ is defined by the relation

$$(2.13) \qquad \phi^*(p) = \sup_{x \in \Omega} \{\langle x,p\rangle - \Phi(x)\} .$$

Proposition. If Φ is convex, then so is Φ^*.

Proof. For any number $\Lambda, 0 \leq \Lambda \leq 1$, check that

$$\sup_{x \in \Omega} \{\langle x, \Lambda p_1 + (1-\Lambda)p_2\rangle - \Phi(x)\} \leq$$

$$\Lambda \sup_{x \in \Omega} \{\langle x, p_1\rangle - \Phi(x)\} + (1-\Lambda) \sup_{x \in \Omega} \{\langle x,p_2\rangle$$

$$- \Phi(x)\} .$$

A geometric interpretation of the convex functional Φ^* can be described intuitively. Consider a convex set $\Omega \subset \mathbb{R} \times \mathbb{B}$, where $\mathbb{B}$ is a Banach space. A family of hyperplanes in $\mathbb{R} \times \mathbb{B}$ is given by an element of $\mathbb{R} \times \mathbb{B}^*$, i.e. an ordered pair (s, p), $s \in \mathbb{R}$, $p \in \mathbb{B}^*$. A particular hyperplane is obtained by setting $\langle (r,x), (s,p)\rangle = C$, $C, r \in \mathbb{R}$, $x \in \mathbb{B}$, or $rs + \langle x,p\rangle = C$, where C is some real number. By rescaling we can always choose $s = -1$, provided $s \neq 0$. Then $C = \langle x,p\rangle - r$ defines a family of hyperplanes of $\mathbb{R} \times \mathbb{B}$. $\Phi^*(p) = \sup_{(x \in \Omega)}(C)$ is the sup. of values for which $\Phi^*(p)$ is a support hyperplane of $[\Phi, C]$.

Convexity, and concavity of certain functionals are closely tied to the existence of maxima and minima of functionals.

2.4 The equation $Ax = f$, $A = T^*T$.

Equations of this type are very common in physics and engineering mechanics. The Laplace equation, the biharmonic equation, the Lagrange-Germain equation describing a static deflection of a thin plate are all of this type.

Let us consider a linear operator A mapping some subset $\mathcal{D}_A$ of a Hilbert space H_1 into H_1.
We assume that A is positive definite, hence that $A = T^*T$, that the domain of T is dense in H_1, and therefore T^* is uniquely defined. T and T^* are the linear maps :

$$T : \mathcal{D}_T \quad H_1 \to H_2 \quad \text{and} \quad T^* : H_2 \to H_1 \quad .$$

Hence, the equation $Ax = f$ (in the space H_1) can be rewritten as a pair of equations :

$$(2.14^a) \qquad Tx = p \qquad (\text{in } H_2)$$

$$(2.14^b) \qquad T^*p = f \qquad (\text{in } H_1) \; .$$

We introduce a new Hilbert space $\mathcal{H} = H_1 \oplus H_2$, such that every element h of $\mathcal{H}$ is an ordered pair $h = (x,p)$ with x in H_1 and p in H_2. The inner product in $\mathcal{H}$ is defined by the formula :

$$(2.15) \qquad \{h_1,h_2\} = (x_1,x_2)_{H_1} + \langle p_1,p_2\rangle_{H_2} \; ,$$

where $(\, , \,)_{H_1}$ is the inner product in H_1, while $\langle \, , \, \rangle_{H_2}$ is the inner product in H_2.

Let $\tilde{p} \in H_2$, $\tilde{x} \in H_2$, $x \in H_1$, be arbitrary vectors $\tilde{w} = (\tilde{x},\tilde{p})$ a vector in $\mathcal{H}$ and the corresponding value of the functional $L: H \to \mathbb{R}$ be given by

$$(2.16) \quad L(x,Tx) = \langle T\tilde{x},\tilde{p}\rangle - \frac{1}{2} \langle \tilde{p},\tilde{p}\rangle - (f,\tilde{x}) \; .$$

We shall state a fairly trivial result. If the gradient L_W is uniquely defined at the point

$w_0 = \{x_0, p_0\} \in \mathcal{H}$, then the equations (2.14^a)

and (2.14^b) are necessary and sufficent conditions for vanishing of L_W. Hence, the functional L has a critical point at w_0 if and only if the equations (2.14^a) and (2.14^b) are satisfied.

Again the condition $x \in \mathcal{D}_A$ has been replaced by condition $x \in \mathcal{D}_T$, which is in general easier to satisfy, since $A = T^*T$. (See the next section for a more rigorous statement.) A more detailed look at equation (2.14^a) and (2.14^b) show that if we fix $\tilde{p} \in H_2$ and vary only $x \in H_1$, $L_x = \emptyset$ (in H_1) if and only if (2.14^b) is satisfied and vice versa if we fix $\tilde{x} \in H_1$ and vary $p \in H_2$, $L_p = \emptyset$ (in H_2) if and only if (2.14^a) is satisfied.

These observations have been originally implied by Kato [25],[26].We are using the same notation L_x denoting the gradient of the functional $L(x)$ (with p a fixed vector of H_2), always assuming that this gradient exists, (in a fixed topology of the space!) and if $L_x = \emptyset$ at $x = x_0 \in H_1$ then L_x is defined in some neighborhood of the point x_0 in the appropriate topology. Unless otherwise specified, we shall use the topology induced by the norm. Again, we observe that the vector $w_o = \{x_o, p_o\}$, which is the critical point for the functional L, does not have to be a "genuine solution" of the original equation : $Ax = f$, because x_o may lie in the complement of the domain of A. The system (2.14^a)-(2.14^b) has been designated by B.Noble (see [38], [39] and L.B.Rall [42] as a Hamiltonian system. This name is easily explained if we call the functional

$\mathcal{H} = \frac{1}{2}\langle p,p\rangle + (f,x) = \mathcal{H}(x,p)$ "the Hamiltonian" and observe that the equations (2.14^a),(2.14^b) are equivalent to the system of equations :

$$(2.17^a) \quad Tx = \mathcal{H}_p \quad ,$$

$$(2.17^b) \quad T^*p = \mathcal{H}_x \quad .$$

This is Hamilton's system of canonical equations in the special case when

$$T = \frac{d}{dt} \quad , \text{ and } \quad T^* = -\frac{d}{dt} \; .$$

Here x denotes the vector of generalized displacements, and p of generalized momenta. In analogy with the terminology of classical mechanics we shall call the functional $\mathcal{L}$,

$\mathcal{L} = \langle Tx,p\rangle - \mathcal{H}$ the action functional, or the Lagrangian <u>action integral</u>. We observe the following consequence of our discussion. The equation $Ax=f$ is an equation stating an equality of two vectors in the Hilbert space H_1. The introduction of the splitting space H_2 became necessary when we decomposed the oparator $A: H_1 \to H_1$, into T^*T ; where

$$T : H_1 \to H_2 \ , \ T^* : H_2 \to H_1 \ .$$

We could ask the question: "is H_2 unique?" If the answer is "no" ,then could H_2 be chosen to facilitate the analysis of the variational problem?
The work of Browder and Gupta indicates that while H_1 is a Banach space,under certain conditions H_2 may be a Hilbert space,which is generally very advantageous. In this monograph H_1 is assumed to be a Hilbert space and this discussion appears to be of no consequence.
To make more rigorous some loose ideas we offer the following definitions.
Definition 2.1 Let T be a linear operator mapping some subset of a Banach space B_1 into a Banach space B_2. The graph of T is a subset of $B_1 \times B_2$ consisting of all ordered pairs of the form : $\{u, Tu\}$, with u in the domain of A.

The following statement is known as *the closed graph theorem* : Let $D_T \subset B_1$. Then T is bounded if and only if the graph G(T) of T is closed.

Definition (Kato [26]) Any linear manifold C contained in $D_T \subset B_1$ is called the core of T if the set $\{u, Tu\}$, $u \in C$, is a dense subset of G(T). An important theorem due to von Neumann asserts the following. Let T be a closed operator $T: H_1 \to H_2$; D_T dense in H_1. Then $A = T^*T: H_1 \to H_1$ is a selfadjoint operator and D_A is a core of T. Hence $D_T \subseteq D_A$. If the containment is proper i.e. $D_T \subset D_A$ then the corresponding variational solutions describing the original problem in terms of critical points of some functionals in H_2 will exhibit solutions which are not "genuine" solutions, that is they are not in the domain of A.

2.5 Duality and the Legendre transformation.
The oldest form of duality observed even by the ancient Greeks is the duality between the points and lines in the Euclidean plane. The axioms of

Euclides may be restated in modern terminology by introducing the undefined concepts of points and lines in a plane. (See a popular translation of Euclid's Elements, Dover Publications, New York, 1964). Thus, we can start an axiomatic postulation of a geometry by introducing the (undefined) concepts of points and lines. Two points define a unique line, two lines define a unique point in a plane or "a point at infinity". Almost all statements of Euclidean geometry remain correct if in an arbitrary sentence about points and lines in a plane we substitute "line" for a "point" and "point" for a "line".

Let us consider a family of curves described by either of the two functional relations:

$$(2.17^a) \quad y = \tilde{y}(x),$$

$$(2.17^b) \quad x = \tilde{x}(y) .$$

Each of these curves may be regarded as either a locus of points (x_i, y_i) in the Cartesian plane or as an envelope of straight lines whose slope at a coordinate x_i is given by

$$\left.\frac{d\tilde{y}}{dx}\right|_{x=x_i} .$$

Let X denote $X(x_i) = \left.\frac{d\tilde{y}}{dx}\right|_{x=x_i}$.

We define the <u>Legendre transformation</u>

$(x_i, y_i) \rightarrow (X, Y)$ by the formulas

$$(2.18) \begin{cases} X(x_i y_i) = \left.\dfrac{d\tilde{y}}{dx}\right|_{x=x_i} \\ Y(x_i, y_i) = X(x_i)\, x_i - \tilde{y}(x_i). \end{cases}$$

If the second derivative $\frac{d^2\tilde{y}}{dx^2}$ exists and it does not vanish then this transformation is invertible.

In space of higher dimensions, that is, if x and y are vectors in an n-dimensional space, the corresponding Legendre transformation is invertible if the Hessian determinant :

$\left| \frac{D^2 y}{Dx^2} \right|$ is not equal to zero .

We may regard the invertible Legendre transformation as a change of coordinates from (x,y) to (X,Y).

The relations defining this transformation are:

(2.20) $X = \frac{dy}{dx}$,

(2.21) $Y = Xx - y$,

(2.22) $x = \frac{dY}{dX}$,

(2.23) $y = -Y(X) + Xx$,

provided $\frac{d^2y}{dx^2} \neq 0$.

In two dimensions we can draw a simple picture that illustrates the geometric interpretation of this transformation

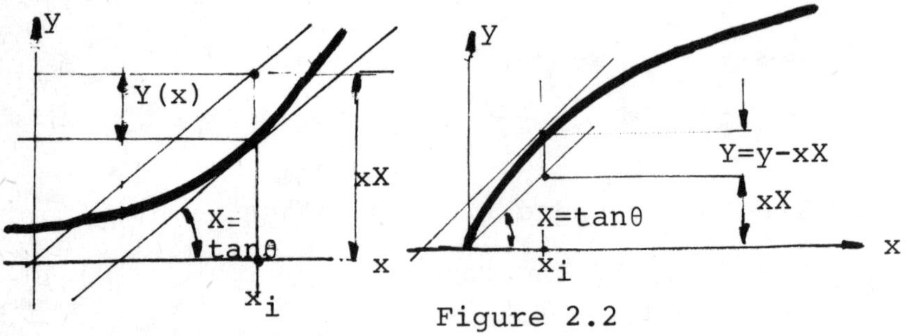

Figure 2.2

We observe a complete symmetry. Interchange of {X,Y} {x,y} preserves all formulas. For example, let us plot the parabola

$$y = x^2, \ \frac{dy}{dx} = 2x .$$

Hence, $X = 2x$
$Y = 2x^2 - y$,

or $Y = \frac{1}{4} X^2$,

$$\frac{dY}{dX} = X/2 = x ,$$

and $y = Xx - Y = 2x^2 - x^2 = x^2$,
completing the inverse transformation.

Since $\frac{d^2y}{dx^2} \neq 0$ and $\frac{d^2Y}{dX^2} \neq 0$ there are no exceptional points.

In purely geometric terms this is a duality between the description of a curve in terms of the Cartesian coordinates (x,y) of the points on the curve ,or a description of a family of straight lines which are the envelopes of the curve.Each straight line is given uniquely by its slope and the point at which it intersects the y-coordinate. An elementary thermodynamic interpretation can be found in almost any textbook on engineering thermodynamics.We identify a constitutive dependence of pressure on the specific volume in an isothermal process:

$$p = p(v) \ , \quad \frac{dp}{dv} \neq 0 \ .$$

We equate the heat production with the work done: $dQ = p \cdot dv$,assuming that only the compression work is performed.Also we introduce a new variable H :

(2.25) $H = pv - Q$

so that

(2,26) $dH = p\,dv + v\,dp - dQ = v\,dp$.

We shall return to this topic later.

2.6 The Legendre transformation for the elastic membrane problem.

Many physical phenomena are modelled by the Laplace equation.They include the deflection of a uniform membrane (that is a membrane having a constant thickness,density,and elastic properties) the flow of an incompressible fluid,the pure torsion of an elastic shaft made of homogeneous material obeying Hooke's law,or the flow of an electric current in a thin metal sheet.Related boundary value problems are reduced to mathematical statements of the Dirichlet problem,or the Neumann problem.

Let us consider the system :

$$\text{(2.27)} \qquad \Delta u = f(\underset{\sim}{x}) \quad \text{in } \Omega \subset \mathbb{R}^n ,$$

$$\Delta = \sum \frac{\partial^2}{\partial x_i^2} ,$$

and either $u = 0$ on $\partial\Omega$,
or grad $u = 0$ on $\partial\Omega$.

The physically important cases are n=2,and n=3. In mechanics of fluids it is common practice to introduce the *hodograph transformation* by defining new dual variables

$$\text{(2.28)} \qquad p_i = u_{x_i} \qquad \text{(subscript } x_i \text{ denotes } \partial/\partial x_i).$$

In two dimensions it is common to define :

$$\text{(2.29)} \quad v = -xy + u$$

$$\text{(2.30)} \quad y = u_x .$$

In a higher number of dimensions we can select some variable say x_0 and introduce the transformation

$$\text{(2.31)} \quad y_\alpha = x_\alpha \quad \text{if } \alpha > 0$$

and

$$\text{(2.32)} \quad y_0 = u_{x_0}$$

$$\text{(2.33)} \quad v = -x_0 y_0 + u(\underset{\sim}{x})$$

$$\text{(2.34)} \quad dv(y_0) = -x_0\, dy_0 - y_0\, dx_0 + du .$$

For the Dirichlet problem in two dimensions x_1, x_2 we set

$$y_2 = x_2 \quad , \; y_1 = -\partial u/\partial x_1$$

$$v(y) = x_1 y_1 + u(x)$$

$$dv(y) = x_1 dy_1 + y_1\, dx_1 + du =$$

$$x_1 dy_1 - (\partial u/\partial x_1)\, dx_1 + \sum_{i=1}^{2} \frac{\partial u}{\partial x_i}\, dx_i$$

$$= x_1 dy_1 + \partial u/\partial x_2 \cdot dx_2$$

$$\frac{\partial v}{\partial y_1} = x_1 , \quad \frac{\partial v}{\partial y_2} = \frac{\partial u}{\partial x_2} .$$

We can apply this transformation to free boundary problems by identifying an open set $S \subset \mathbb{R}^2$ such that $y_1 \equiv 0$ on S.

Then $x_1 = \frac{\partial v}{\partial y_1} \in (0, x_2)$ is locally parametrizing the free boundary (See Avner Friedman [83] for details).

The similarity between this hodograph transformation and the Legendre transformation is apparent. In his monograph [83] Friedman refers to it as the hodograph-Legendre transformation. It is commonly used in gas dynamics and in compressible fluid flow theory.

(2.7) Higher order Legendre transformations.

Such transformations were introduced by physicists in the construction of Green's functions

$W_k = W_k(A_1, A_2, \ldots A_n)$,where $A_i, i=1,2\ldots,n$ are some potentials. For example, in the so called $\lambda\phi^4$ theory such potentials may be of the type

$A_4 = \lambda\delta(x_1-x_2)\ \delta(x_2-x_3)\ \delta(x_3-x_4)$, δ denoting the Dirac delta.

The new variables (generalized momenta) introduced by this transformation are given by $\alpha_1, \alpha_2 \ldots \alpha_m$, and more generally by $\alpha_{k\ell}$, where $i\alpha_j = \frac{\partial W_0}{\partial A_j}$, $j = 1,2\ldots m$, and $i\alpha_{k\ell} = \frac{\partial W_k}{\partial A_\ell}$, with $i^2 = -1$.

For the details see for example A.N.Vasil'ev and A.K.Kazanskii [57].

To the best of my knowledge such generalizations had so far no applications in either the classical or applied mechanics.

(2.8) A glance at some consequences of the Legendre transformation.

Let the Lagrangian for a mechanical system be of the form : $L = L(q,\dot{q},t)$. The stationary property of the action integral is symbolically given by:

$$\delta W = \delta \int_{\tau_0}^{\tau_1} L(\underset{\sim}{q},\underset{\sim}{\dot{q}},t)\, dt = 0 \quad .$$

Varying simultaneously $\underset{\sim}{q}$, $\underset{\sim}{\dot{q}}$ and the position of the end points τ_0, τ_1 we derive a formal expression known as Caratheodory's variational form.

$$(2.35) \quad \delta W = L(\tau_1)\delta\tau_1 - L(\tau_0)\delta\tau_0 + \left[\sum_i \frac{\partial L}{\partial \dot{q}_i}\delta q_i\right]_{\tau_0}^{\tau_1} + \int_{\tau_0}^{\tau_1} \{\sum_i \left[\frac{\partial L}{\partial q_i} - \frac{d}{dt}\frac{\partial L}{\partial \dot{q}_i}\right]\delta q_i\}\, dt \, .$$

Formally we can write:

$$\delta\dot{q}_i \cdot dt \simeq d(\delta q_i)) .$$

If τ_0, τ_1 are fixed end points we can apply the Legendre transformation to derive

$$\delta \int_{\tau_0}^{\tau_1} (\sum_i \dot{q}_i \frac{\partial L}{\partial \dot{q}_i} - H)\, dt = 0 \quad .$$

If $H \equiv$ constant we rederive the following form of the Maupertuis principle

$$(2.36) \quad \delta \int_{\tau_0}^{\tau_1} (\sum_i \dot{q}_i \frac{\partial L}{\partial \dot{q}_i})\, dt = 0 \quad .$$

Several integral invariants can be directly derived from (2.36). For example, (2.37) :

$\{\int_{\tau_0}^{\tau_1} (\sum \frac{\partial L}{\partial \dot{q}_i}\delta q_i)\, dt\}$ is the integral invariant of Poincaré([24], [72], [67],[88], [7].

(2.37) $\omega = \sum \frac{\partial L}{\partial \dot{q}_i} \cdot \delta q_i - H \cdot \delta t$ is the momentum energy tensor of Cartan [109].

(2.35) can be shown to be equivalent to

$$\delta W = \int_{\tau_0}^{\tau_1} \left[\sum_i \left(\frac{\partial L}{\partial q} - \frac{d}{dt}\frac{\partial L}{\partial \dot{q}_i}\right) dq_i\right] dt +$$

$(\omega_{p_1} - \omega_{p_0})$, along any admissible path γ originating at p_0 (at τ_0) and terminating at p_1 (at the time τ_1). The first term vanishes if the path is the trajectory of motion.

A curve γ originating at some point p in the (n+1) dimensional space spanned by $\{q_1, q_2 \ldots q_n, t\}$ and following the direction at each τ such that

$$(2.39) \quad \sum_i \frac{\partial L}{\partial \dot{q}_i} \delta q_i - H\delta t \equiv 0$$

at every point of γ, is called transversal to the trajectories of the system.

The collection (i.e. the one-parameter bundle) of all transversal curves forms the transversal manifold M_ν.

Thus, the property

$$(2.40) \quad W(\underset{\sim}{q}, t) = \int_{p_0}^{p_1} (\Sigma \frac{\partial L}{\partial \dot{q}} \delta q_i - H\delta t) = \int_{p_0}^{p_1} [\omega] = C =$$

constant (independent of the path) - defines the transversal manifold.

Substitution of the Legendre transformation allows us to manipulate this property and to claim that

$$(2.41) \quad W = \int_{p_0}^{p_1} \{ L + \frac{\partial L}{\partial \dot{q}_i} (\frac{\delta q_i}{\delta t} - \dot{q}_i) \} dt$$

is independent of the path.

This is the Hilbert "independency" integral (see, for example, [1], [7], or [9]).

Note: the ratio of the two independent variations $\frac{\delta q_i}{\delta t}$ is identically equal to $\dot{q}_i$ only along the trajectory of the system.

Therefore, along any trajectory of motion $L \equiv \omega$, any state p can be reached by first following one of the trajectories of the system and then following a curve on the transversal manifold.

This is a restatement of the Caratheodory accessibility theorem (see [72] .

Let us manipulate these relations:

$$\int_{p_0}^{p_1} L\, dt = W(q,t).$$

We substitute $\omega = \frac{\partial L}{\partial \dot{q}_i} \delta q_i - H \cdot \delta t$,

$$\delta W = \omega_p - \omega_{p_0} + \int_{\tau_0}^{\tau_1} \Big(\sum_i \Big(\frac{\partial L}{\partial q_i} - \frac{d}{dt}\frac{\partial L}{\partial \dot{q}}\Big)\delta q_i\Big) \cdot dt \quad .$$

Therefore, along a trajectory of motion

$\delta W = \omega_p - \omega_{p_0}$. If we set $\omega_{p_0} = 0$ on M_ν, we could derive the Hamilton-Jacobi relations.

(2.42) $$\frac{\partial W}{\partial q_i} = \frac{\partial L}{\partial \dot{q}} = p_i, \; i = 1,2,\ldots n,$$

(2.43) $$\frac{\partial W}{\partial t} = -H \, .$$

See Caratheodory [72] , or Kneser [152].

These relations could be combined in the form of the Hamilton-Jacobi partial differential equation

(2.44) $$\frac{\partial W}{\partial t} + H\left(q_i, \frac{\partial W}{\partial q_i}, t\right) = 0.$$

$$\left(\frac{\partial W}{\partial q_i} = p_i\right)$$

This form has been traditionally used for determining the constants of motion and the symmetries, and to generate the contact transformations.

For a detailed insight see [67], [68], [72], [1], [7] , [153].

For a conservative system the kinetic energy of a particle expressed in the spherical coordinate $\{r,\theta,\Phi\}$ is given by

$$T = \frac{1}{2m}\left[p_r^2 + \frac{p_\theta}{r^2} + \frac{p_\Phi^2}{r^2 \sin^2\theta}\right] .$$

The Hamiltonian is given by

$$H = T + U(r,\theta,\Phi) .$$

The Hamilton-Jacobi (H-J) equation is

[2.45] $$\frac{\partial \mathcal{W}}{\partial t} + H = 0 .$$

For a conservative system ($H \equiv h$ = constant) the H-J. equation assumes the form

$$\tilde{W} = ht + \mathcal{W}_o(q_i) . \quad [2.46]$$

2.9 <u>Jacobi's theorem and its corrolaries.</u>

A function $W(q_i, p_i, t)$ satisfying the Hamilton-Jacobi (H-J) equation (in n-dimensions)

$$\frac{\partial W}{\partial t} + H\left(\underset{\sim}{q}, \frac{\partial W}{\partial q_i}, t\right) = 0 ,$$

which contains n-arbitrary independent constants $\{\alpha_1, \alpha_2 \ldots \alpha_n\}$ is called a <u>complete</u> solution, or a <u>complete integral</u> of motion.

Jacobi's theorem asserts that a complete solution of the H-J equation may be derived by solving the system of equations.

$$p_i = \frac{\partial W}{\partial q_i} , \quad \beta_i = \frac{\partial W}{\partial \alpha_i} .$$

As a corrolary we conclude that whenever H is constant along any trajectory of motion, that is, $H(t) \equiv h$ = constant, then the complete integral is

$$\tilde{\mathcal{W}} = - ht + V(q_1, q_2 \ldots q_n, \alpha_1, \alpha_2 \ldots \alpha_{n-1}, h) ,$$

with h chosen as α_n.

[2.10] <u>Separation of variables in spherical coordinates.</u>

We follow the arguments of Landau and Lifshitz in [67]. Let the Hamiltonian be of the form

$$H = T + V(r,\theta,\Phi) = \frac{1}{2m}\left(p_r^2 + \frac{p_\theta^2}{r^2} + \frac{p_\Phi^2}{r^2\sin^2\theta}\right)$$

$$+ V(r,\theta,\Phi).$$

The variables can be separated if

$$V(r,\theta,\Phi) = R(r) + \Theta(\theta)/r^2 + \Xi(\Phi)/(r^2\cdot\sin^2\theta),$$

where $R(r)$ is a function of r only,

$\Theta(\theta)$ of θ only

and $\Xi(\Phi)$ of Φ only .

The Hamilton-Jacobi equation assumes the form

$$\frac{1}{2m}\left(\frac{\partial W_o}{\partial r}\right)^2 + R(r) + \frac{1}{2mr^2}\left[\left(\frac{\partial W_o}{\partial \theta}\right)^2 + 2m\,\Theta(\theta)\right]$$

$$+ \frac{1}{2mr^2\sin^2\theta}\cdot\left(\frac{\partial W_o}{\partial \Phi}\right)^2 = h.$$

We can write:

$$W_o = p_\phi\,\phi + W_1(r) + W_2(r) + W_2(\theta),$$

$$\left(\frac{\partial W_2}{\partial \theta}\right)^2 + 2m\,\Theta(\theta) + \frac{p_\Phi^2}{\sin^2\theta} = \beta\,,$$

$$\frac{1}{2m}\left(\frac{\partial W_1}{\partial r}\right)^2 + R(r) + \frac{\beta}{2mr^2} = h,$$

thus recovering

$$W(r,\theta,\Phi,t) = ht + p_\Phi\cdot\Phi + \int[\beta - 2m\,\Theta(\theta) - p_\Phi^2/\sin^2\theta]^{\frac{1}{2}}\,d\theta + \int[2m(h-R(r)) - \beta/r^2]^{\frac{1}{2}}\,dr.$$

2.11 <u>An Application of the Hamilton-Jacobi equation.</u>

Find an invariant of motion for the trajectories of particles scattered in the force field with the central field potential

$V(r) = \frac{K \xi}{r}$,where $r = \|\mathbf{r}\|$.

Before encountering the force field (that is at large distance from the origin) the flow field consists of parallel trajectories.
The complete integral of motion is

$$W = -ht + \int(\beta - 2mk\cos\theta)^{\frac{1}{2}}\, d\theta + \int(2mh - \beta/r^2)^{\frac{1}{2}}\, dr \quad ,$$

with $\frac{\partial W}{\partial r} = (2mh - \beta/r^2)^{\frac{1}{2}} = p_r = m\dot{r}$,

$\frac{\partial W}{\partial \theta} = (\beta - 2mk\cos\theta)^{\frac{1}{2}} = p_\theta = mr^2\dot{\theta}$.

$\frac{\partial W}{\partial \beta}$ = constant = C is an equation that is satisfied along a trajectory of the system.

Therefore

$$\int_0^{\theta=\tilde{\theta}} \frac{d\theta}{(\beta - 2mk\cos\theta)^{\frac{1}{2}}} + \int_{r=\infty}^{r=r} \frac{dr}{r^2(2mh-\beta/r^2)^{\frac{1}{2}}}$$

= constant $=C$ is an invariant of motion .

2.12 An excursion into relativistic mechanics.

The H-J equation

$$\frac{\partial S}{\partial t} + H = 0$$

can be modified by inclusion of the rest energy term mc^2 for the particle :

$$\frac{\partial \tilde{S}}{\partial t} = \frac{\partial S}{\partial t} + mc^2 \tag{2.47}$$

Then

$$\frac{1}{c^2}\left(\frac{\partial S}{\partial t}\right)^2 = \text{grad } S^2 + mc^2$$

and

$$\frac{1}{2mc^2}\left(\frac{\partial \tilde{S}}{\partial t}\right)^2 - \left(\frac{\partial \tilde{\tilde{S}}}{\partial t}\right) - \frac{1}{2m} \text{ grad } S^2 \tag{2.48}$$

Equation (2.48) is invariant under the Lorentz transformation in one space dimension :

$$t = \gamma\cdot(t' + (V/c^2)x'),$$
$$x = \gamma\cdot(x' + Vt') \quad ,$$
$$\gamma = [\,1 - (V/c)^2]^{-\frac{1}{2}}$$

Equation (2.48) is invariant under a general 4-dimensional Lorentz transformation, but we considered deliberately the simplest one-dimensional problem.

The difficulty arises from an unexpected source. The Hilbert integral is no longer invariant and $\tilde{S}(\underset{\sim}{x}) \neq \tilde{S}(x')$ when coordinates $\mathbf{x}$ are replaced by x'. This difficulty was partially overcome in papers of R.P. Gaída and others, who used the Dirac-Fock-Podolski multiple time scales formalism. Lorentz invariance is exact after replacing

$$\frac{\partial S}{\partial t} \text{ by } \sum_{\alpha} \left(\frac{\partial S(\underset{\sim}{x}_i, t_\alpha)}{\partial t_\alpha} \right)$$

and $$\frac{\partial S}{\partial x_\alpha} \text{ by } \frac{\partial S'}{\partial x'_\alpha} + m_\alpha V, \text{ with } \frac{\partial S}{\partial x_\alpha} \approx \frac{\partial \tilde{S}}{\partial x_\alpha} + m_\alpha V.$$

These difficulties disappear in the non-relativistic limit when Lorentz transformation is replaced by Galileo's transformation $t = t'$, $x = x' - vt$. See [91].

2.13 The Young-Fenchel transformation.

We associate with a $C^1(\Omega)$ function $f(\underset{\sim}{x})$, $x \in \Omega \subset \mathbb{R}^n$, $f: \mathbb{R}^n \to \mathbb{R}$, a "dual" function

$$f^*(\underset{\sim}{y}) = \sup_{\underset{\sim}{x} \in \Omega} \{ \Sigma\, y_i x^i - f(\underset{\sim}{x}) \} \quad ,\text{where}$$

$$y_i = \frac{\partial f}{\partial x^i} \; .$$

The function $f^*(y)$ is called the Young-Fenchel transformation of the function $f(x)$.

Obviously $$-f^*(y) = \inf_{\underset{\sim}{x} \in \Omega} \{f(\underset{\sim}{x}) - \Sigma y_i x^i\} \; .$$

<u>A simple example</u> :

Let $f(x) = |x|$, $x \in \mathbb{R}^1$.

Then $-f^{*}(y) = \inf_{x \in R} \{|x| - yx\}$.

Therefore $f^{*}(y) = 0$ if $-1 \leq y \leq +1$
and $f^{*}(y) = +\infty$ if $|y| > 1$.

2.14 The Young-Fenchel duality of variational problems related to the Laplace equation.

The Dirichlet problem may be reduced to finding a minimum in the Sobolëv class of functions $u \in H^1_0(\Omega)$ for the functional

$$J(u)= \tfrac{1}{2}\int_{\Omega}[(u_x)^2+ (u_y)^2]\, dx\, dy,\ \{x,y\} \in \Omega \text{ , with}$$

$u(x,y) = g(x,y)$ or $\partial\Omega$.

We introduce $(p_1, p_2) = \underset{\sim}{p}$ as the dual variable and consider

$$\inf_{\substack{\{u \in H^1(\Omega)\} \\ u=g \text{ on } \partial\Omega}} \{\sup_{\underset{\sim}{p}} \int_{\Omega}[p_1 u_x + p_2 u_y - \tfrac{1}{2}(p_1^2 + p_2^2)\ ] .$$

If we denote $p_1 = \frac{\partial\psi}{\partial x}$, $p_2 = -\frac{\partial\psi}{\partial y}$

$$L(\psi) = - \int_{\partial\Omega} \frac{\partial\psi}{\partial s} g(x,y)\cdot ds\ ,$$

and $$F^{*}(\underset{\sim}{p}) = \tfrac{1}{2} \int_{\Omega} (p_1^2 + p_2^2)\, dx\, dy\ ,$$

then $\inf_{\psi \in H^1(\Omega)} \{ L(\psi) - F^{*}(\psi)\}$ is attained by a function $\psi(x,y)$ such that:

$$\Delta\psi = 0 \qquad \text{in } \Omega$$

and

$$\frac{\partial\psi}{\partial\nu} = \frac{\partial g}{\partial s} \qquad \text{on } \partial\Omega \ ,$$

where ν denotes the direction normal to the boundary.

This confirms a known result.The Young-Fenchel dual relation transforms the Dirichlet problem into the Neumann problem for the Laplace equation.

2.15 Some properties of the Young-Fenchel transformation.

a) Let f(x) be an arbitrary function defined at

all points of a convex open set $\Omega \subset R^n$. Then $f^*(\underset{\sim}{y})$ is a convex function. Proof: $f^*(\lambda \underset{\sim}{y}_1 + (1-\lambda)\underset{\sim}{y}_2)$

$$\doteq \sup_{\underset{\sim}{x} \varepsilon \Omega} \{ f^*(\Sigma \lambda y_{1i} x^i) + (1-\lambda)[\Sigma y_{2i} x^i - f(x)] - \lambda f(\underset{\sim}{x}))\} \leq \lambda \sup_{\underset{\sim}{x}} \{\Sigma\, y_{1i} x^i - f(\underset{\sim}{x})\} + (1-\lambda) \sup_{\underset{\sim}{x}} \{(\Sigma y_{2i} x^i - f(x))\} =$$

$= \lambda f^*(\underset{\sim}{y}_1) + (1 - \lambda(f^*(y_1)$, which agrees with the definition of convexity of $f^*(y)$.

Corrolary: $f^{**}(x)$ is also convex.

b) $f^{**}(\underset{\sim}{x}) \leq f(\underset{\sim}{x})$

c) If $f(\underset{\sim}{x})\underset{\sim}{x} \in \Omega$ is lower semicontinuous and convex then $f^{**}(\underset{\sim}{x}) = f(\underset{\sim}{x})$ in Ω .

If $f(x)$ is strictly convex

(i.e. $f(\lambda x_1 + (1-\lambda)x_2) < \lambda f(x_1) + (1-\lambda)\, f(x_2))$,

and $f(x)$ is continuously differentiable, than the hodograph transformation is identical with the Young-Fenchel transformation. However, the Young-Fenchel transformation is defined for functions for which neither the definition of the hodograph nor the classical form of Legendre transformation makes any sense.

Relation with Pontryagin's maximality principle and the problem of Bolza.

The problem of Bolza consists of finding an admissible function $u(t)$, where, generally $u(t)$ is regarded as an ẽlement of a bounded convex s̃ubset of some Banach space (could be L_∞, L_1 or L_p - $1 \leq p < \infty$ depending on the specific nature of the problem) that minimizes an integral

(2.49) $J(u) = \int_{t_0}^{t_1} [fo(t, \underset{\sim}{x}(t), \underset{\sim}{u}(t)]dt$, subject to constraints

(2.50) $\dot{x}_i = f_i(t, \underset{\sim}{x}, \underset{\sim}{u})$, $i = 1,2....n$,

(2.51) $\int_{t_0}^{t_1} \ell_i(t, \underset{\sim}{x}, \underset{\sim}{u})\, dt\ 0$, $i = 1,2,...r$,

(2.52) $g_i(t,\underset{\sim}{x},\underset{\sim}{u}) = 0$, $i = r+1, r+2,...S.$

2.16 Pontryagin's principle.

Pontryagin et al [71] investigated in 1948 the control problem that appears to be similar to the problem of Bolza. The terminology they offered is different and the formulation and terminology cleared up unnecessary complications that bothered a generation of mathematicians.

The variational problem discussed in part 1 of [71] consists of minimizing a functional

$$J(\underset{\sim}{x},\underset{\sim}{u}) = \int_{t_o}^{T} f^o(t,\underset{\sim}{x}(t),\ \underset{\sim}{u}(t)\ )\cdot dt\ ,$$

with constraints

$$\dot{x}^i = f^i(t,\underset{\sim}{x}(t),\ \underset{\sim}{u}(t)),\ \underset{\sim}{f} \in H\ ,\quad i=1,2,...n\ ,$$

$$\underset{\sim}{x}\ (t_o) = \underset{\sim}{x}_o,$$

$\underset{\sim}{u} \in U$, where U is an admissible set,

$$\underset{\sim}{x} \in \mathbb{R}^n,\ \underset{\sim}{f} \in \mathbb{R}^{n+1},\ \underset{\sim}{u} \in \mathbb{R}^m\ .$$

$u^i(t)$ are measurable functions of t on $[t_o,T]$ which are vectors in a bounded, closed, convex subset of a Banach space $\mathbb{B}$.

Following Pontryagin we introduce a "state"

variable $x^o(t) = \int_{t_o}^{t} f^o(\xi,\underset{\sim}{x}(\xi),\ \underset{\sim}{u}(\xi))\ d\xi,$

so that $\dot{x}^o = f^o(t)$

$$\left.\begin{array}{l}\underset{\sim}{u}(t) \in \mathbb{R}^m\\ \underset{\sim}{x}(t) \in \mathbb{R}^n\end{array}\right\}\qquad t \in [t_o,t_1]\ .$$

The functions f^i are piecewise continuous in t, and differentiable functions of x_i,u_i except at countably many points t_k at which $f(t_k,\ \underset{\sim}{x}(t_k),\underset{\sim}{u}(t_i))$ may have a jump discontinuity.

The variable $\underset{\sim}{x}$ is called the state variable, x is a function of t, and also of the initial conditions: $\underset{\sim}{x} = \underset{\sim}{x}(t; t_o, \underset{\sim}{x}_o)$.

In the original formulation of the problem of Bolza $\underset{\sim}{u}$ was directly identified with $\dot{\underset{\sim}{x}}$.

Here we use a modern terminology of control theory. The vector $\underset{\sim}{u}(t)$ is regarded as "control" and is used independently of the state variable $\underset{\sim}{x}(t)$. The space in which the control vector "lives" depends on our choice. If the system represents a "real world" problem, then the physics of the system dictates the admissibility criteria for u(t) and consequently for admissibility of trajectories for the system $\underset{\sim}{x}(t; t_o, \underset{\sim}{x}_o)$. The function $\underset{\sim}{f}(t,x,u)$ could be discontinuous reflecting a sudden reversal of thrust, of voltage, or of the direction or of the magnitude of the applied generalized force vector.

If $\underset{\sim}{x}(t, t_o, x_o)$ represents a trajectory of a particle then the physical interpretation demands continuity. If it represents deflection of a beam then differentiability is necessary. Existence of barriers or obstacles imposes constraints that may cause lack of smoothness in both controls and trajectories. While it is physically impossible to produce infinitely many discontinuities in the control on a finite time interval, it is important to discuss such controls, known as chattering controls, as possible limits of controls that can be physically applied. For this reason the control vector $\underset{\sim}{u}(x)$ is generally assumed to be only measurable. The engineers realized in the middle 1930s that optimal controls are generally discontinuous. The term "bang-bang" control was used long before Pontryagin's work was published by Ms. Flügge-Lotz and has been attributed to Goddard. The mathematical community took longer to appreciate this fact. Some of the difficulties encountered in mathematical papers dating to the Chicago days of the 1930-s may be related to unwillingness to abandon mathematical convenience and traditional outlook.

The vector f belongs to a Hilbert space F, or only to a subclass of functions in a Hilbert space. The minimization problem assumes the following form:

minimize $x^o(T)$,

subject to conditions

(*) $\dot{x}^i = f^i(t,\underset{\sim}{x},\underset{\sim}{u})$,

(**) $\begin{cases} x^o(0) = 0, \\ x^i(0) = x^i_o, \end{cases}$

(***) $\underset{\sim}{u}(t) \in U$.

Pontryagin introduced an element $\psi(x)$ of the space F*, which is a Lagrangian multiplier for the set of constraints(*). One could consider the problem of Bolza, or Pontryagin's control problem as an analog of the classical problem of Lagrange with holonomic constraints. We recall that constraints are holonomic if they can be reduced to the form $g(\underset{\sim}{x}) \leq 0$, or

$\int_\Omega g(\underset{\sim}{x})\, d\underset{\sim}{x} \leq 0$. They are nonholonomic otherwise.

For example, the constraints $\dot{\underset{\sim}{x}} = f(\underset{\sim}{x})$ are not holonomic, if they can not be reduced to the form $g(\underset{\sim}{x})=0$.

Instead of the constrained problem Pontryagin substitutes the unconstrained minimization problem for :

$$J(\underset{\sim}{x},\underset{\sim}{\psi},\underset{\sim}{u}) = \int_{t_0}^{T} f^o(t,\underset{\sim}{x},\underset{\sim}{u})\, dt + \int_{t_0}^{T} \sum_i \{\psi_i(t)\cdot[\dot{x}_i - f^i(t,\underset{\sim}{x},\underset{\sim}{u})]\}dt .$$

This problem may be restated as a min-max problem :

Find $$\inf_{\underset{\sim}{x}\in X}\left\{ \sup_{\underset{\sim}{\psi}} \{ \int_{t_0}^{T} f^o(t.\underset{\sim}{x},\underset{\sim}{u})\, dt + \int_{t_0}^{T} \sum_i \psi_i(t)\cdot [\dot{x}_i - f^i(t,\underset{\sim}{x},\underset{\sim}{u})]\, dt \right\}$$

$$= \sup_{\underset{\sim}{\psi}}\left\{ \inf_{\substack{\underset{\sim}{x}\in X \\ \underset{\sim}{u}\in U}} \int_{t_0}^{T} \{\sum_{i=1} (\psi_i \dot{x}^i) - H(t,\underset{\sim}{x},\underset{\sim}{\psi},\underset{\sim}{u})\}\, dt \right\},$$

where $H(t,\psi,x,u) = \sum_{i=1}^{n} \psi_i f^i - \psi_0 f^0$, ψ_0 = constant.

Without any loss of generality we can take $\psi_0=1$. The functional $H(t,\psi,x,u)$ is called the Pontryagin Hamiltonian.

2.17 The optimal control of a vibrating Euler-Bernoulli beam.

We discuss the simplest control problem for a transversely vibrating beam. The optimal control is defined in terms of Pontryagin's maximality principle.

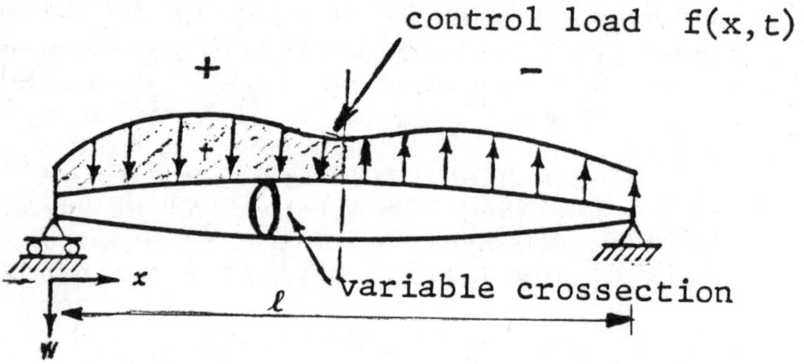

Figure 2.3 : Simply Supported Beam

2.17.1 The basic equation of motion and related hypotheses for a transversely vibrating beam.

Given the usual simplifying assumptions, the Euler-Bernoulli equation describing the transverse vibration of a beam is :

$$(2.53)\quad L(w) = \rho(x)\cdot A(x)\cdot \frac{\partial^2 w(x,t)}{\partial t^2} + \frac{\partial^2}{\partial x^2}\left(E(x)\cdot I(x)\cdot \frac{\partial^2 w(x,t)}{\partial x^2}\right) = f(x,t),$$

$$-\ell/2 \leq x \leq +\ell/2,\; t \geq 0.$$

The corresponding homogeneous equation is

$$(2.53^a) \quad Lw_H = 0.$$

We make the usual assumptions.

$$\left.\begin{array}{ll} \text{a)} & \sqrt{A\rho}\ \dfrac{\partial w}{\partial t} \in H_0^1(\Omega) \\ & \sqrt{EI}\ \dfrac{\partial^2 w}{\partial x^2} \in H_0^2(\Omega) \end{array}\right\}$$

We restrict the domain: $\Omega = [-\frac{\ell}{2}, \frac{\ell}{2}] \times [0,T]$.

The physical meaning of the symbols is

- $\rho(x)$: the material density;
- $A(x)$: the cross-sectional area;
- $w(x,t)$: the transverse displacement;
- $E(x)$: Young's modulus; (here it is assumed to be constant)
- $I(x)$: the moment of inertia of the cross-sectional area about the neutral axis;
- $f(x,t)$: the applied load.

Because of the physical meaning we must assume that the displacement $w(x,t)$ is a continuously differentiable function of x and t. On the interval $[-\ell/2,+\ell/2]$, ρ, A, E, I are uniformly bounded, piecewise smooth, positive functions of x. An additional assumption of material homogeneity would imply that ρ and E are constant. ($A(x)$, $I(x)$ may still vary along the length of the beam.) We assume the correctness of Hooke's law and equate the strain energy with the complementary energy of the beam.

In the remainder of this section we shall consider only the class of weak solutions of (1) obeying the following conditions:

(a) $w(x,t)$ is a continuously differentiable function of x and of t on $\Omega = [-\ell/2,+\ell/2] \times [0,T]$ (i.e., $\partial w(x,t)/\partial t$ and $\partial w(x,t)/\partial t$ and $\partial w(x,t)/\partial x$ are continuous functions of x and t on Ω).

$$(2.53^b) \quad \frac{\partial w(\pm \ell/2,t)}{\partial x} = 0 = w(\pm \ell/2,t)$$

(a built-in end) ,
or

$$(2.53^c)\quad \left.\begin{array}{l} w(\pm\ \ell/2,t) = 0 \\ EI\dfrac{\partial^2 w(\pm\ \ell/2,t)}{\partial x^2} = 0 \end{array}\right\}$$

(a freely supported end);
or

$$(2.53^d)\quad \left.\begin{array}{l} EI\ \dfrac{\partial^2 w(\pm\ \ell/2)}{\partial x^2} = 0 \\ \dfrac{\partial}{\partial x}\left[EI\ \dfrac{\partial^2 w(\pm\ \ell/2,t)}{\partial x^2}\right] = 0 \end{array}\right\}$$

(a free end);

(2.53^e) $w(x,t)$ obeys given initial conditions of the form

$$\left.\begin{array}{l} w(x,0) = \psi(x), \\ \dfrac{\partial w(x,0)}{\partial t} = \eta(x), \end{array}\right\}\ \psi(x),\eta(x)\in C^1[-\ell/2,+\ell/2].$$

This system for $w(x,t)$ will be denoted as MBVP (mixed boundary value-initial value problem).

We also assume that the inhomogeneous term $f(x,t)$ satisfies either of the conditions:
either $f(x,t)$ is a square integrable (hence it is absolutely integrable) function of the variable x in the region Ω or else
$f(x,t) = \phi_1(t)\cdot\phi_2(x)$, where $\phi_1(t)$ is an $L_1[0,T]$ function and $\phi_2(x)$ is an $H^{-2}[-\ell/2,+\ell/2]$ function, that is, possibly a Dirac delta, or its first derivative multiplied by a bounded $L_2[-\ell/2,+\ell/2]$ function.
Note : The integral symbol $\int(\phi(x)f(x))dx$ is used even if $\phi(x)$ is a generalized function that is not locally integrable or square integrable. We follow the usual practice in physics and engineering and avoid unnecessary complicated notation and a discussion of separate cases ([36]).

The solution of the inhomogeneous equation (2.53) is known to obey the Duhamel principle :

$$(2.54)\quad w(x,t) = w_H(x,t) + \int G(x,\xi,t,\tau)\cdot\phi(\tau,\xi)d\xi d\tau,$$

where $w_H(x,t)$ is the solution of the homogeneous equation, while $G(x,\xi,t,\tau)$ depends only on the coefficients ρ, A, E and I and on the boundary conditions (2.53b - 2.53d) but does not depend on either $\phi(x,t)$ or on the initial value functions $\psi(x)$ and $\eta(x)$.

Again, an elementary argument concerning delta-convergent sequences shows that this statement may be extended to cover a more general situation. See Mikusinski's approach in [115].

We observe that the admissible controls form a convex set, i.e., if $f_1(x,t)$ and $f_2(x,t)$ are admissible controls, then $\Lambda f_1+(1-\Lambda)f_2$ is also an an admissible control for any Λ, $0 \le \Lambda \le 1$.

2.17.2 <u>The energy terms</u>. The kinetic energy of the beam is given by the formula :

$$(2.55)\qquad T = \tfrac{1}{2}\int \rho(x)A(x)\left(\frac{\partial(x,t)}{\partial t}\right)^2 dx,$$

and the strain energy is given by :

$$(2.56)\qquad V = \tfrac{1}{2}\int_{-\ell/2}^{\ell/2} E\, I(x)\left(\frac{\partial^2 w(x,t)}{\partial x^2}\right)^2 dx .$$

The total energy $E(t)$ is the sum of the kinetic and the strain energy :

$$(2.57)\qquad E = T + V = \tfrac{1}{2}\int_{-\ell/2}^{+\ell/2} \rho(x)A(x)\left[\frac{\partial w(x,t)}{\partial t}\right]^2 + E(x)I(x)\left[\frac{\partial^2 w}{\partial x^2}\right]^2 dx.$$

The physical interpretation of (2.55) and (2.56) implies that $\partial w/\partial t$ and $\partial^2 w/\partial x^2$ have to be square integrable on the interval $[-\ell/2,+\ell/2]$. We also assume that they are square integrable on $[0,T]$. We introduce the following energy product of two functions $u(x,t)$, $v(x,t)$, whose derivatives $\partial u/\partial t, \partial^2 u/\partial x^2$, $\partial v/\partial t$, $\partial^2 v/\partial x^2$ are square integrable functions in the interval $-\ell/2 \le x \le + \ell/2$:

$$(2.58)\quad \langle u,v\rangle = \tfrac{1}{2}\int_{-\ell/2}^{+\ell/2} [\rho(x)\cdot A(x)\ \frac{\partial \bar{u}}{\partial t}\frac{\partial \bar{v}}{\partial t} \ +$$

$$+ E(x)I(x) \frac{\partial^2 u}{\partial x^2} \quad \frac{\partial^2 v}{\partial x^2} \,] \, dx.$$

$\langle u,v \rangle$ is clearly a function of t only. If $u = v$, then $\langle u,v \rangle$ is the total energy, as defined by the formulas (2.55), (2.56) and (2.57).

2.18 Statement of the Control Problems.

Given the initial conditions and one of the admissible boundary conditions at each boundary point $x = -\ell/2$, $x = +\ell/2$ and given $T > 0$, find an admissible control $\tilde{\phi}(x,t)$ such that the total energy of the beam obeys the inequality

$$\mathcal{E}(\tilde{\phi}(x,t), t=T) \leq \mathcal{E}(\phi(x,t), t=T,$$

where $\phi(x,t)$ is any other admissible control. The control $\tilde{\phi}(x,t)$ will be called an optimal control for the interval $[0,T]$.

The control problem stated above is called the fixed time interval control problem. Closely related to it is the minimal time control problem. Given the same initial and boundary conditions and given a nonnegative number E, such that $E < \mathcal{E}(t=0)$, find the control $\hat{\phi}(x,t)$ which reduces the total energy of the beam to the value E in the shortest possible time.

(2.19) Pontryagin's principle. The principle stated in the Theorem A below is in complete agreement with the maximum principle of Pontryagin (see [71], [90], [92], [93], [94], [146]).

Theorem A Let $\phi(x,t)$ be an optimal control for the fixed interval control of the MBVP, as stated in the preceding section. We assume that $\mathcal{E}(\tilde{\phi}(x,t),t) > 0$ if $t \in [0,T]$. Let $\tilde{w}(x,t)$ be the solution of the MBVP corresponding to $\tilde{\phi}(x,t)$. Let $v(x,t)$ be a solution of the homogeneous equation satisfying the same boundary conditions, and such that $v(x,T) = \tilde{w}(x,T)$, i.e.

$$v(x,T) = \tilde{w}(x,T),$$

$$\frac{\partial v(x,T)}{\partial t} = \frac{\partial \tilde{w}(x,T)}{\partial t}.$$

Then

$$\text{(2.59)} \quad \int_{-\ell/2}^{+\ell/2} [-\tilde{\phi}(x,t)\frac{\partial v(x,t)}{\partial t}]\, dx =$$

$$\max \int_{-\ell/2}^{+\ell/2} [-\phi(x,t)\cdot \frac{\partial v(x,t)}{\partial t}]\, dx$$

for all admissible controls $\phi(x,t)$.

For the proof see [76], [34].

$\phi\cdot\frac{\partial v}{\partial t}$ is the Pontryagin's Hamiltonian for this problem. For similar result in thin plate theory see [77].

2.19. Pontryagin's principle for coupled transverse and torsional vibrations of a beam

Theorem

Let $u = \begin{bmatrix} u_1(t) \\ u_2(t) \end{bmatrix}$ be the optimal control vector for the fixed interval problem $t \in [0,T]$. Let $\min_{u\in U} E(T) > 0$. Let $\tilde{w}_H$ ($\tilde{w}_H = [\begin{matrix} y_H \\ \theta_H \end{matrix}]$) denote the solution of the homogeneous equation satisfying the same final conditions as the optimal solution $[\begin{matrix} y(\tilde{u}) \\ \theta(\tilde{u}) \end{matrix}]$. This is :

$$y(\tilde{u}(t),x,T) = y_H(x,T)$$
$$\dot{y}(\tilde{u}(t),x,T) = \dot{y}_H(x,T)$$
$$\theta(\tilde{u}(t),x,T) = \theta_H(x,T)$$
$$\dot{\theta}(\tilde{u}(t),x,T) = \dot{\theta}_H(x,T).$$

Then, for all $t \in [0,T]$, we have the inequality:

$$-\int_0^{\ell} [\dot{y}_H\phi_1(x)\tilde{u}_1(t) + \dot{\theta}_H\phi_2(x)\tilde{u}_2(t)]\, dx$$

$$= \max_{u\in U} \{-\int_0 [\dot{y}_H\phi_1(x)u_1(t) + \dot{\theta}_H\phi_2(x)u_2(t)]\, dx\}$$

The proof is lengthy but follows exactly the line of argument used by the author in [77].

Note: The uniqueness of optimal control depends entirely on the properties of the norm we assign on U, and has little to do with the properties of the linear differential operator L. To state this result we define the concept of strict convexity of the norm $|| \; ||$. The norm $|| \; ||$ is strictly convex (or we say that the unit ball of U is strictly convex), if for any two points u_1, u_2 in U (the closure of U) such that $||u_1|| = ||u_2|| = 1$ the open line segment connecting u_1, u_2, that is the set $\tilde{\ell}$ is defined by $\ell = \{z | z = \lambda u_1 + (1-\lambda)u_2, \; 0 < \lambda < 1)\}$ contains only points of norm strictly less than one; i.e. $z \in \tilde{\ell}$ implies $||z|| < 1$. In other words, any point lying in the interior of the line segment connecting the boundary points u_1, u_2 of the unit ball lies entirely in the interior of the unit ball. Hence, the unit ball is convex, and contains no straight line segments on its boundary.

2.20 Equivalence of Pontryagin's principle and the Lagrange-Hamilton principle.

For a dynamic system with unconstrained Lagrangian $\mathbf{L}(q, \dot{q}, t)$ and non-holonomic constraints $G_i(t, \underset{\sim}{q}, \underset{\sim}{\dot{q}}) = 0$, $i=1,2,\ldots m$, that satisfy the Četaev relations $\sum \frac{\partial \Phi}{\partial \dot{q}_i} \cdot \delta q_i = 0$, the Lagrange equation becomes

$$\frac{d}{dt}\left(\frac{\partial \mathbf{L}}{\partial \dot{q}_i}\right) - \frac{\partial \mathbf{L}}{\partial q_i} + \sum_k \lambda_k \frac{\partial G_k}{\partial \dot{q}_i} = 0, \tag{2.60}$$

where $\lambda_k(t)$ are the Lagrange multiplier functions.

Regarding $\dot{q}_i$ as components of the control vector $\underset{\sim}{u} \in \mathbb{R}^m$ we restate the problem of minimizing of the Lagrangian as a control problem:

Minimize

$$\int_0^T \mathbf{L}(t, \underset{\sim}{q}, \underset{\sim}{u}) \, dt$$

with constraints $G_i(t, \underset{\sim}{q}, \underset{\sim}{\dot{q}}) = 0$, $i=1,2\ldots m$.

Under fairly mild conditions the constraint equations can be reduced to a more manageable form

$$\dot{q}_i = f_i(t, \underset{\sim}{q}, \underset{\sim}{\dot{q}})$$

Note that we do not have to solve for $\dot{q}$ in terms of q and t, therefore we do not need the explicit function theorem (See [127]). For example, the constraint condition:

$$-(x+y)t^{\frac{1}{2}} + \dot{x}\,(\sqrt{\dot{x}^2+\dot{y}^2+1})\cosh^2 x + \dot{y}(\sqrt{\dot{x}^2+\dot{y}^2+1})\cdot$$

$\cosh^2 y = 0$ is not holonomic and we cannot "solve" for $\dot{x}$ and for $\dot{y}$ in terms of (t,x,y). But

$$\dot{x} = [-\dot{y}\sqrt{\dot{x}^2+\dot{y}^2+1}\sin x + (x+y)t^{\frac{1}{2}}]\cdot (\dot{x}^2+\dot{y}^2+1)^{-\frac{1}{2}}\cosh^{-2} x = f(t,x,u)$$

is a convenient and a reasonable relation.

Thus, we can obtain a classical formulation of the Pontryagin variational problem :

$$\text{Minimize} : \int_0^T L(\underset{\sim}{q},\dot{\underset{\sim}{q}},t)\,dt = \int_0^T f_o(\underset{\sim}{q},f(\underset{\sim}{q},\underset{\sim}{u},\underset{\sim}{y}),t)\,dt$$

with constraints :

$$\dot{q}_i = f_i(t,\underset{\sim}{q},\underset{\sim}{u}) \ .$$

We construct Pontryagin's Hamiltonian functional:

$$H = \psi_0 f_0 + \sum_{i=1}^{n} \psi_i f_i \ .$$

The Hamilton's canonical equations are:

$$\frac{\partial H}{\partial \psi_i} = \dot{q}_i \ , \ i = 1,2,\ldots,n \ ,$$

with $q_o(t) = \int_0^t f_o(\underset{\sim}{q},\underset{\sim}{u},\tau)\,d\tau$

$$\frac{\partial H}{\partial q_i} = -\dot{\psi}_i + \sum_{k=1}^{r} \lambda_k \frac{\partial G_k}{\partial q_i} \ , \ i = 1,2\ldots n.$$

The optimality of $\underset{\sim}{u}$ is asserted in the relations $\frac{\partial H}{\partial u_i} = 0$. This is Pontryagin's formulation of the control problem, if we do not forget the original interpretation of the vector $u = \dot{q}$, i.e. the constraints $u_i - \dot{q}_i = 0$. Thus, Pontryagin's formulation is a direct consequence of the classical variational formulation even if we assign nonholonomic constraints. Vice-versa, reversing the whole chain of arguments we derive Hamilton's canonical equation and Hamilton's principle in the form (2.60) from Pontryagin's principle.

This development can be found in R. van Doren's paper [79]. A crucial relation in van Doren's paper is

$$\delta H = \sum \dot{q}_i \cdot \delta \psi_i + \{ \sum_j (\psi_j - \frac{\partial f_o}{\partial \dot{q}_j}) \cdot \frac{\partial f_j}{\partial q_i} - \frac{\partial f_o}{\partial q_i}\} \cdot \delta q_i$$

+ (an expression that vanishes identically along an optimal path) · δ u.

2.21. The general form of Legendre transformation.

The so-called canonical decomposition of a positive definite operator gave rise to an extensive theoretical development in duality theory based on Legendre transformation. The idea originated in 1950s, with credit given to von Neumann and Kato [25], [26] and [37]. Much earlier, Korn's duality was introduced in classical elasticity (see [73]).

Numerous papers such as [5] [8], [27] [28], [29], [42], [43] contained details of several variants of canonical decomposition and its applications to elasticity, electromagnetic field theory, theory of networks, and to partial differential equations in general. It is a trivial statement that an operator of the form AA^* where $A : H_1 \to H_2$, A^*. (the adjoint of A) : $H_2 \to H_1$ are mapping between Hilbert spaces H_1, H_2, is positive.

It is more difficult to show that any positive operator T mapping a Hilbert space H into itself can be written in the form A^*A. If T has discrete spectrum, the idea of this development is illustrated in Gel'fand and Vilenkin [20] (Volume IV of the theory of generalized functions; also see [19]). This is true in general, but the proof of this statement lies outside the scope of this monograph.

Papers such as [30], [25], [27] deal with the canonical decomposition of equations

(2.61) $A^*Aq(t) = f(q,t)$,

where q is an element of a Hilbert space H_1, A - is a densely defined linear operator mapping H_1 into a Hilbert space H_2, and A*-is an adjoint of A, $A^* : H_2 \to H_1$.

This is of course, the simplest scenario and many generalizations have been offered. See [61], [62] or [84] as examples of theoretical developments where the Hilbert space structure has been discarded [5], [8] [29] as examples of the corresponding theories where A* is only a formal adjoint of A, or [62] where additional spaces have to be introduced before the results of this theory can be applied. All these refinements can be ignored for the time being, permitting us to introduce the simplest possible version of our theory.

A standard "trick" introduced in this fairly abstract form by T. Kato ([25]) is to rewrite the equation (2.61) as a system of two equations.

(2.61) $A^*Aq = f$,

(2.62^a) $Aq = p(t)$,

(2.62^b) $A^*p = f(q(t),t)$.

The Hamiltonian for the system (2.61) is given by

$$(2.63)\quad H = \frac{1}{2}\langle p,p\rangle + V(q,t),$$

where $\frac{\partial V}{\partial q} = f(q,t)$.

The system (2.62) is easily recognized in the case when A is the differential operator $(\frac{d}{dt})$ and A* is the formal adjoint of A, that is $(-\frac{d}{dt})$.

The Legendre transformation is defined by mapping:

$$(2.64)\quad \{q,Aq\} \to \{q,p\}$$

with $p = \frac{\partial L}{\partial (Aq)}$,

and $H = L - \langle Aq,p\rangle$,

thus establishing a duality between a Lagrangian and Hamiltonian formulation. Equations (2.62^a) (2.62^b) go back to Korn who observed dual variational principles in theory of elasticity of materials that do not necessarily obey Hooke's law. Friedrichs has interpreted Korn's result in terms of modern operator theory [15].

An intuitive approach to the postulation of duality and formulation of dual variational principles originated with Noble [38], [39]. It has been successfully applied by a number of authors to problems of mathematical physics (see, for example, the monograph of A.M. Arthurs [4], and articles [42] - [47], [64], [5], [63], [108].

Comments on Generalized Solutions, Energy Norms and Extremal Points of Functionals (expository discussion).

There is a well-known close relation between a generalized solution of the equation:

$$Au - f = \emptyset \tag{2.65}$$

in Hilbert space and extreme value of a functional, when A is a positive definite operator, bounded away from zero, and the domain of A is dense in H. In this case we can introduce a new inner product $[u,v] = \langle Au,v\rangle$, and a new norm $|||u|||^2_{(A)} = \langle Au,u\rangle$. Let H_A denote the closure of the new inner product space in this norm topology. We examine the corresponding functional:

$$F(u) = \langle Au,u\rangle - 2\langle u,f\rangle = |||u|||^2_{(A)} - 2\langle u,f\rangle$$

(Since A is bounded away from zero in the $||\ ||$ norm topology meaning $\langle Au,u\rangle \geq C^2\, ||u||$ for some $C > 0$, or $|||u||| > C\,||u||$) it follows that

$$|\langle u,f\rangle| \leq \frac{||f||}{C}\, ||u||.$$ Hence, by the Riesz representation theorem there exists $u_0 \in H_A$ such that $\langle u,f\rangle = [u,u_0]$, and $F(u) = |||u - u_0|||^2 - |||u_0|||^2$. Therefore, F(u) attains a minimum at u_0, and

$$\min_{u \in H_A} F(u) = - \|u_0\|^2 .$$

As before, H_A denotes the Hilbert space obtained by closure of the inner product space with the product $[\cdot,\cdot]$. The point $u_0 \in H_A$ is called the generalized solution of the equation (2.4). If u_0 was in the domain of A, then it is a "genuine" (or classical) solution. Otherwise it is called a generalized solution of (2.65). In fact, since A is a positive definite, there exists an operator B, such that $A = B^*B$, and for all $u \in D_A \subset D_B$ and $\langle Au,u \rangle = \langle Bu,Bu \rangle$. We have the containment $D_A \subset D_B$. However, $u_0 \in D_B$ does not mean that u_0 is necessarily in the domain of A. As an example consider

$$A = -\frac{d^2}{dx^2},\ D_A = C^2[0,1] \subset L_2[0,1]\ ,\ u(0)=u(1) = 0.$$

The generalized solution of $-\frac{d^2}{dx^2} u = f$ corresponds to the minimization problem for the L_2 functional

$$\langle -\frac{d^2}{dx^2} u,\ u \rangle - 2 \langle f,u \rangle$$

$$= - \int_0^1 (u'' \cdot u)\, dx - 2 \int_0^1 (f \cdot u)\, dx$$

$$= \int_0^1 (u')^2 dx - 2 \int_0^1 (f \cdot u)\, dx.$$

We note that the existence of the functional

$$F(u) = \int_0^1 (u')^2\, dx - 2 \int_0^1 (f \cdot u)\, dx$$

requires only $u' \in L_2[0,1]$ and does not require twice differentiability.

$B = i \frac{d}{dx}$ is one of the possible corresponding square roots of the operator $-\frac{d^2}{dx^2}$. Hence, the minimization problem can be accomplished in the Sobolev space $H^1[0,1]$ instead of assigning

$u \in C^2[0,1] \subset L_2[0,1]$. The energy space H_A turns out in this case to be $H^1[0,1]$. Of course, in the above case we knew several properties of the operator A, (positive definite, bounded away from zero) which made life very easy, and gave us so easily an equivalent extremal problem for the functional F(u).

Convexity, and concavity of certain functionals are closely tied with existence theory of maxima and minima. But first, let us concentrate on sufficiency conditions which generalize the Euler-Lagrange equations of classical calculus of variations in a specific setting of applied mechanics.

(2.22) Fréchet differentiation of functionals.

We consider the Fréchet derivative of the map $L(x): \Omega \subset H_1 \to \mathbb{R}$. Let the Fréchet derivative L_x exist at $x_o \in \Omega$. Then L_{x_0} is a continuous linear map from $\Omega \subset H_1$ into $\mathbb{R}$. Therefore for some $h \in H_1$ we have $L_{x_0}(ch) = c(L_{x_0}h)$, $\forall c \in \mathbb{R}$, and by the Hahn-Banach theorem L_{x_0} can be extended to all of H_1 without changing its norm.

By the Riesz representation theorem there exists a unique $Z \in H_1$ such that $L_{x_0}(h) = \langle Z,h\rangle$ for all $h \in H_1$. Hence, for all $h \in H_1$ the Fréchet derivative of a functional f evaluated at x_0 is a continuous linear functional L_{x_0}, and $L_{x_0}(h) = \langle Z_{x_0},h\rangle$ for all $h \in H_1$. See [14], [16], [27].

The unique vector Z_{x_0} is called the gradient of f evaluated at x_0. Clearly Z depends on the definition of the inner product in H. However, the existence of the gradient depends only on the existence of the Fréchet derivatives which in turn depends on convergence of our limiting process. Convergence is a topological property, and remains invariant in equivalent topologies. Hence, the introduction of an equivalent norm in H will not affect the existence, or non-existence of a gradient (or of the Fréchet derivative.) In fact, let us introduce a new inner product (the so called energy product, and corresponding energy norm ([15]), $[x,y] = \langle Tx,y\rangle$, where T is positive definite, symmetric operator. Denoting by $|||x|||^2 = [x,x] = \langle Tx,x\rangle$, and by $||y||^2 = \langle y,y\rangle$, we have $\langle \text{grad } f,h\rangle = [\text{grad}^\# f, h] = \langle T\text{grad}^\# f, h\rangle$, where $\text{grad}^\#$ is the gradient of f in the topology induced by the norm $|||\quad|||$. Since this relation is true for all $h \in H_1$, we have the equality $T \cdot \text{grad}^\# = \text{grad}$. If T^{-1} exists, then $\text{grad}^\# = T^{-1} \cdot \text{grad}$. If the operator T is bounded away from zero (i.e. the spectrum of T does not have zero as a limit point) then there exists a constant $\gamma > 0$, such that for any $x \in H_1 \supset \mathcal{D}(T)$ $\gamma||x|| \leq |||x||| \leq \frac{1}{\gamma}||x||$. ($\mathcal{D}(T)$ denotes the domain of T). Hence, convergence (or continuity, or existence of limit points) in the new norm $|||\quad|||$ implies convergence (or continuity...) in the old norm $||\quad||$. See [15][46].

In the final dimensional case all these statements are trivial, and all norms $\langle,\rangle$, $[\,,\,]$, (where $\langle Tx,y\rangle = [x,y]$, and T is positive definite, and symmetric) are equivalent and generate the same topology.

At this point it is appropriate to make a comment about the non-existence of the Fréchet derivative in some Hilbert space H_1. Suppose the Gateaux derivative of Φ (in arbitrary direction h) exists, but fails to be a continuous functional of h in the usual (norm) topology of H_1, i.e. we can find a sequence of vectors $\{h_i\} \in H_1$ such that $\lim_{i\to\infty} ||h - h_i|| = 0$, but $\lim_i \Phi(h_i) \neq \Phi(h)$. It may be possible to introduce a different, and non-equivalent inner product $\{\cdot,\cdot\}$, with a corresponding norm $||\quad||_{(2)}$ such that $\Phi(h)$ is a continuous functional of h in the new (norm) topology.

If the inner product is fixed, and no attempt will be made to change the topology, then we can use the sloppy notation $\Phi_x(x_0)$, or equivalently $\left.\frac{\delta\Phi(x)}{\delta x}\right|_{x_0}$, to denote the gradient of Φ evaluated at x, where Φ is a Fréchet differentiable functional whose values depend on $x \in H_1$ i.e. $\Phi: H_1 \to \mathbb{R}$, and $L_{x_0}(x) = \langle Z_{x_0},x\rangle = \langle \left.\frac{\delta\Phi}{\delta x}\right|_{x_0}, x\rangle$ is the corresponding Fréchet functional. This notation is both confusing and sloppy, but it has been consistently in use, and the author of this work is also guilty of having abused it.

It is partially vindicated by a physical explanation. The Hilbert space setting has obscured our vision to a certain extent, since we identify H with its dual, and we are allowed to neglect the bookkeeping which vector is in H, and which vector is in H* (the dual of H, which is identified with H). This distinction becomes important when we try to interpret certain equations of mathematical physics. In the remainder of this work we shall use the notation gradΦ if Φ is a functional Φ: $H_1 \to \mathbb{R}$. We shall also find examples where Φ is a functional $\Phi : H_1 \oplus H_2 \oplus \ldots \oplus H_k \to \mathbb{R}$ (or $\mathbb{C}$), that is Φ maps the direct product of Hilbert spaces $H_1, H_2 \ldots H_k$ into the real (or complex) numbers. In this case the notation grad $\Phi_{(H_i)}$ is too cumbersome and we shall write Φ_{x_i} to denote

$$<\Phi_{x_i}, h> = <\lim_{t \to 0} \left(\frac{\Phi(x_0+th) - \Phi(x_0)}{t}\right), h>$$

with $h \in H_i$, $x_0 \in H_1 \oplus H_2 \oplus \ldots \oplus H_k$. That is, we generalize simultaneously the idea of a gradient and of a partial derivative, by allowing the variation of Φ to take place exclusively in H_i, and regarding the components of x in H_j, $j \neq i$, as fixed. We shall refer to this derivative as gradient of Φ restricted to the space H_i. This idea plays a crucial part in formulation of multiple variational principles. For example, the following equations of mildly nonlinear elasticity correspond to critical points of a simple functional restricted to respective subspaces of the space $H = \sum H_i$. The Maxwell and Morrera's stress function relation, the equations of equilibrium, the equations of compatibility, Ricci's equations. All becomer easily interpreted in this context (See [47],[30],[28] ,[64]).

2.23 Some rules of manipulation for gradients of functionals

We shall restrict ourselves first to the simplest case: $\Phi: H \to R$, $D(\Phi) \subset H$, Φ is Fréchet differentiable in $D \subset H$.

We shall denote by $\Phi_{x(\Omega)}$ the gradient of $\Phi = Z \epsilon D$, $\Phi = \langle Z, \eta \rangle$. $Z \epsilon D^*$ is the Fréchet derivative of Φ coinciding with the Gateaux derivative in the direction η. We compute the value of the gradient at a point $x_0 \epsilon D$. $\Phi_x(x_0) = Z(x_0)$. If $Z(x_0) = \emptyset$ (the zero vector) we say that x_0 is a critical point of the functional Φ. A basic theorem of optimization theory (Vainberg [53] [55]) states a necessary condition for stationary behavior of a functional Φ (particularly for a maximum or minimum) at the point $x_0 \in H$. If Φ is Fréchet differentiable at x_0, then the point x_0 must be a critical point of Φ. In particular this is a necessary condition for a local maximum or minimum of a Fréchet differentiable functional.

Example 1. $\Phi = \langle y_1, y_2 \rangle$
$y_1, y_2 \in H$,
$y_1 = C_1 X$,
$y_2 = C_2 X$. Then $Z = \Phi_X = (C_1 + C_2) X$.

Proof of this statement follows from the definition. We compute the Gateaux derivative of Φ in the direction of a vector $\eta \in H$.

$$\langle \eta, Z \rangle = \lim_{\varepsilon \to 0} \frac{\langle C_1 X + \varepsilon\eta, C_2 X + \varepsilon\eta \rangle}{\varepsilon} - C_1 C_2 \langle X, X \rangle$$

$$= (C_1 + C_2) \langle \eta, X \rangle = \langle \eta, (C_1 + C_2) X \rangle.$$

Hence, $\Phi_X = (C_1 + C_2) X$.

Example 2. $\Phi(X) = \Phi_1(X) = \Phi_1(X) \cdot \Phi_2(X)$ Then the gradient of Φ is:

$$Z = \Phi_X = (\Phi_{1X} \cdot \Phi_2) + (\Phi_1 \cdot \Phi_{2X}) .$$

Similarly,

$$\left.\frac{\Phi_1(x)}{\Phi_2(x)}\right|_X = - \frac{\Phi_{1X} \Phi_2 - \Phi_{2X} \cdot \Phi_1}{\Phi_2^{\,2}}$$

Example 3. $\Phi = -\frac{<AX,X>}{<X,X>}$, A: $H \to H$, $A^* = A$, $X \neq \emptyset$.

Then,

$$\Phi_X = \frac{2}{<X,X>} \cdot (\Phi(X)X - AX).$$

2.24 Critical points of functionals and the equation Ax = f.

We shall consider a fairly arbitrary linear operator A mapping $\mathcal{D}_A$ H_1 into H_1, where H_1 is a Hilbert space. We wish to solve the equation $Ax = f$, $f \varepsilon H_1$, for the state variable $x \varepsilon H_1$. We shall assume that the operator A has a representation $A = T^*T$, (A is positive definite) and that the domain of A is dense in H_1. Therefore the operator T^*, that is the adjoint of T, is uniquely defined. T and T^* are the respective linear maps:

$$T : \mathcal{D}_T \ H_1 \to H_2; \quad T^* : H_2 \to H_1 \quad .$$

Hence, the equation $Ax = f$ (in the space H_1) can be rewritten as a pair of equations :

(2.66) $\quad Tx = p \quad$ (in the space H_2) ,

(2.67) $\quad T^*p = f \quad$ (in the space H_1) .

We can also consider the Hilbert space $H = H_1 \oplus H_2$ whose elements h are ordered pairs $h = (\underset{\sim}{x}, \underset{\sim}{p})$ ($\underset{\sim}{x} \in H_1$, $\underset{\sim}{p} \in H_2$), with the product

(2.68) $\{h_1, h_2\} = (x_1, x_2) + <p_1, p_2>$

where (,) is the inner product in H_1, and <,> is the inner product in H_2. Consider arbitrary vectors $\tilde{p} \varepsilon H_2$, $\tilde{x} \varepsilon H_1$ i.e. an arbitrary $\tilde{w} = \{\tilde{x}, \tilde{p}\} \varepsilon H$ and the corresponding value of the functional L: $H \to R$

(2.69) $L(\tilde{x},\tilde{p}) = <T\tilde{x},\tilde{p}> - \frac{1}{2} <\tilde{p},\tilde{p}> - (f,\tilde{x})$.

It is a fairly trivial result that if the gradient L_W is uniquely defined at the point $W_0 = \{x_0, p_0\} \varepsilon H$, then the equations (2.66) and (2.67) are necessary and sufficient conditions for vanishing of L_W. Hence, the functional L has a critical point at W_0 if and only if the equations (2.66) and (2.67) are satisfied at W_0.

Again, the condition $x\epsilon\mathcal{D}_A$ has been replaced by condition $x\epsilon\mathcal{D}_T$, which is in general easier to satisfy, since $A=T^*T$. (See the next section for a more regorous statement.) A more detailed look at equations (2.66) and (2.67) shows that if we fix $p\epsilon H_2$ and vary only $\tilde{x} \epsilon H_1$, $L_x=\emptyset$ (in H_1) if and only if (2.67) is satisfied, and vice versa if we fix $\tilde{x}\epsilon H_1$ and vary $p\epsilon H_2$, $L_p = \emptyset$ (in H_2) if and only if (2.66) is satisfied.

These observations have been originally implied by Kato [26]. (We are using the same notation L_x denoting the gradient of the functional $L(x)$ (with p a fixed vector of H_2), always assuming that this gradient exists, (in a fixed topology of the space!) and $L_x = \emptyset$ at $x = x_0 \epsilon H_1$ then L_x is defined in some neighborhood of the point x_0 in the appropriate topology. Unless otherwise specified we shall use the topology induced by the norm. Again, we observe that $\tilde{W}_0 = \{x_0, \tilde{p}_0\}$ which is a critical point of the functional L does not have to be a "genuine" solution of the original equation $Ax = f$, since $\tilde{x}_0$ may be in the complement of the domain of A.

The system (2.66), (2.67) has been designated by Noble ([38]) and ([39]) and Rall ([43]) as a Hamiltonian system. The name is easily explained if we call the functional $\frac{1}{2}\langle p,p\rangle + \langle f,x\rangle = W(p,x)$ the Hamiltonian and observe that (2.66), (2.67) can be rewritten as

$$(2.70) \quad T x = W_p \ ,$$

$$(2.71) \quad T^*p = W_x \ .$$

This is the Hamilton system of canonical equations in the special case when $T = d/dt, T^* = - d/dt$, with x identified as the vector of generalized displacements, and p as the vector of generalized momenta.

In analogy with the terminology of classical mechanics the functional L shall be called the action functional, or the Lagrangian (integral) functional.

Let us make some analytic observations. The equation $Ax = f$ is an equation in the space H_1.

The introduction of the "splitting space" H_2 was a result of decomposing the operator $A : H_1 \to H_1$ into $A = T^*T$, $T : H_1 \to H_2$, $T^* : H_2 \to H_1$.
It is known that under certain conditions H_2 may be chosen to be a Hilbert space even if H_1 is a Banach space, making both the analysis and computations much easier. In this work H_1 and H_2 are Hilbert spaces (generally Sobolëv spaces), unless stated otherwise. For definitions of Fréchet and Gateaux derivatives see [14], [16], [3]. For a general theory of Hilbert spaces see [2].

2.25 <u>Connection with the Legendre transformation.</u>
We generalize the classical Legendre transforms

$$\underset{\sim}{p} = \frac{\partial \mathcal{L}(\dot{x},x,t)}{\partial \dot{x}}, \quad H(\underset{\sim}{x},\underset{\sim}{p},t) = \sum_{i=1}^{n} p_i x_i - \mathcal{L}(\underset{\sim}{x},\underset{\sim}{x},t),$$

"replacing" $\dot{x}$ by a new variable Tx, thus changing the "Lagrangian formalism" into the "Hamiltonian formalism". To this transformation corresponds the abstract problem of defining what is meant by

$$\frac{\partial \mathcal{L}}{\partial (Tx)}, \quad T: H \to H_1,$$

and generalizing this concept to an abstract Hilbert space setting. Again, perhaps the best starting point is the equation
$Ax = f \quad (\in H_1),\ A: \ H_1 \to H_1,$

for which we intend to establish (weak) solutions which correspond to a critical point of some functional $\mathcal{L}$, or to two (or more) critical points of functionals in some spaces possibly other than H_1. In the case we have discussed we already presumed that A is positive definite and that we can find a space H_2 such that

$x \in \mathcal{D}_T$, $Tx = p \in H_2$, $T^*p = f (\in H_1)$ with $\mathcal{D}_T$ dense in H_1, and $T^*T = A$.

Of course, the choice of H_1 is non-unique, as can be seen by studying even the simplest examples

$$-\frac{d^2}{dx^2} \; : \; H^2(\mathbb{R}) \to L_2(\mathbb{R}), \; -\frac{d^2}{dx^2} = T^*T,$$

$$T = i\frac{d}{dx}, \; T^* = i\frac{d}{dx}, \; T: H^2(\mathbb{R}) \to H^1(\mathbb{C}),$$

$$T^*: H^1(\mathbb{C}) \to L_2(\mathbb{R}),$$

$$\text{or } T = \frac{d}{dx}, \; T^* = -\frac{d}{dx}.$$

$$T: H^2(\mathbb{R}) \to H^1(\mathbb{R}), \; T^*: H^1(\mathbb{R}) \to L_2(\mathbb{R}),$$

showing that A can be factored through different Hilbert spaces H_1 and H_2. We could offer more involved examples of multiple factoring process with corresponding multiple variational principles, in which the choices of intermediate Hilbert spaces are not uniquely determined.

But first let us consider the effect of the $Tx \in H_1$ on the value of the Lagrangian functional L. We regard p as a vector in H_1, and consider $L(x,y)$ as functional, mapping the pair $x \in H_1$, $y \in H_2$ into the real line, where $y = Tx$.

L_y denotes the gradient of L (whenever it exists) restricted to H_2. ($x \in H_1$ is ignored). Similarly L_x denotes the gradient of L in H_1. It is a straightforward computation that $L_y = p$, (2.72ª) with p regarded as fixed and with $y = Tx$. We observe that

(2.72) $\langle Tx, p \rangle - L = \frac{1}{2}\langle p,p \rangle + (f,x) = W$ defines a new functional $W(x,p)$ satisfying

(2.73ª) $$\begin{cases} W_x = f, \\ W_p = p. \end{cases}$$

The notation is the same as before. W_x is the gradient of W restricted to the space H_1 and W_p is the gradient of W restricted to the space H_2. Again, we describe the relation (2.72), (2.72ª) as the Legendre transformation. (See [9] for a classical definition.)

Part 2: Nonlinear elasticity

2.26 <u>An example of multiple critical points, with applications to the general theory of solids.</u>

We shall consider the equations of classical

elasticity, assuming small strain theory, but not necessarily small rotations. The equations of equilibrium assume the form:

$$(2.74^a) \quad \frac{\partial}{\partial x}\left[(1 + e_{xx})\,\tau_{xx} + (e_{xy} - \omega_x)\,\tau_{xy} + (e_{xz} + \omega_y)\,\tau_{xz}\right] + \frac{\partial}{\partial y}\left[(1 + e_{xx})\,\tau_{xy} + (e_{xy} - \omega_z)\,\tau_{yy} + (e_{xz} + \omega_y)\,\tau_{yz}\right] + \frac{\partial}{\partial z}\left[(1 + e_{xx})\,\tau_{xy} + (e_{xy} - \omega_z)\,\tau_{yz} + (e_{xz} + \omega_y)\,\tau_{zz}\right] = 0.$$

Equations (2.74^b), (2.74^c) are obtained by cyclic permutation of letters x, y, z. Here e_{ij} is the linear strain matrix $\varepsilon_{xx} \approx e_{xx} = \frac{\partial u}{\partial x}$, $\varepsilon_{xy} \approx e_{xy} = \frac{1}{2}\left(\frac{\partial u}{\partial y} + \frac{\partial v}{\partial x}\right)\ldots$, ω_i are the rotation components:

$$\omega_x = \frac{1}{2}\left(\frac{\partial w}{\partial y} - \frac{\partial v}{\partial z}\right),$$

$$\omega_y = \frac{1}{2}\left(\frac{\partial u}{\partial z} \quad \frac{\partial w}{\partial x}\right),$$

$$\omega_z = \frac{1}{2}\left(\frac{\partial v}{\partial x} - \frac{\partial u}{\partial y}\right),$$

i.e. $\vec{\omega} = \frac{1}{2}\,\text{curl}\,(\vec{u})$.

We stress the fact that equations (2.74) are only approximate equations with only some first order non-linear terms retained. For a more complete system see either the classical work of V.V. Novozhilov, or more recent text of de Vreubeke [58].

If we denote by α the following stress tensor:

(2.75)

$$\alpha = \begin{bmatrix} \alpha_{xx} & \alpha_{xy} & \alpha_{xz} \\ \alpha_{yx} & \alpha_{yy} & \alpha_{yz} \\ \alpha_{zx} & \alpha_{zy} & \alpha_{zz} \end{bmatrix} = (I + e) \begin{bmatrix} \tau_{xx} & \tau_{xy} & \tau_{xz} \\ \tau_{yx} & \tau_{yy} & \tau_{yz} \\ \tau_{zx} & \tau_{zy} & \tau_{zz} \end{bmatrix}$$

where e is the Jacobian

(2.76)

$$e = \frac{\partial\ (u,v,w)}{\partial\ (x,y,z)} = \begin{bmatrix} \frac{\partial u}{\partial x} & \frac{\partial u}{\partial y} & \frac{\partial u}{\partial z} \\ \frac{\partial v}{\partial x} & \frac{\partial v}{\partial y} & \frac{\partial v}{\partial z} \\ \frac{\partial w}{\partial x} & \frac{\partial w}{\partial y} & \frac{\partial w}{\partial z} \end{bmatrix} ,$$

we can formulate the following sets of equations of equilibrium.

(2.77)

$$B\,\alpha = \begin{bmatrix} \partial x & 0 & 0 & \frac{\partial}{\partial y} & 0 & \frac{\partial}{\partial z} \\ 0 & \frac{\partial}{\partial y} & 0 & \frac{\partial}{\partial x} & \frac{\partial}{\partial z} & 0 \\ 0 & 0 & \frac{\partial}{\partial z} & 0 & \frac{\partial}{\partial y} & \frac{\partial}{\partial x} \end{bmatrix} \begin{bmatrix} \alpha_{xx} \\ \alpha_{yy} \\ \alpha_{zz} \\ \alpha_{xy} \\ \alpha_{yz} \\ \alpha_{xz} \end{bmatrix} = \emptyset ,$$

$$(\alpha_{ij} = \alpha_{ji}).$$

Because of smallness of strains, we identify the Eulerian strain ε_{ij} with the linear strain e_{ij}. We can define the stress components α_{ij} in terms of appropriate Maxwell and Morrera stress functions : $\underset{\approx}{\chi}$.

(2.78)
$$\begin{bmatrix} \chi_{xx} & \chi_{xy} & \chi_{xz} \\ \chi_{yx} & \chi_{yy} & \chi_{yz} \\ \chi_{zx} & \chi_{zy} & \chi_{zz} \end{bmatrix} = \underset{\approx}{\chi}, \quad \chi_{ij} = \chi_{ji} ,$$

(2.79)
$$\underset{\sim}{\alpha} = A\,\chi,$$

where

(2.80)
$$A = \begin{bmatrix} 0 & \frac{\partial^2}{\partial z^2} & \frac{\partial^2}{\partial y^2} & 0 & -\frac{\partial^2}{\partial y \partial z} & 0 \\ \frac{\partial^2}{\partial z^2} & 0 & \frac{\partial^2}{\partial x^2} & 0 & 0 & -\frac{\partial^2}{\partial y \partial z} \\ \frac{\partial^2}{\partial y^2} & \frac{\partial^2}{\partial x^2} & 0 & -\frac{\partial^2}{\partial x \partial y} & 0 & 0 \\ 0 & 0 & -\frac{\partial^2}{\partial x \partial y} & -\frac{\partial^2}{\partial x^2} & \frac{\partial^2}{\partial x \partial z} & \frac{\partial^2}{\partial y \partial z} \\ -\frac{\partial^2}{\partial y \partial z} & 0 & 0 & \frac{\partial^2}{\partial x \partial z} & -\frac{\partial^2}{\partial y^2} & \frac{\partial^2}{\partial x \partial y} \\ 0 & -\frac{\partial^2}{\partial x \partial z} & 0 & \frac{\partial^2}{\partial y \partial z} & \frac{\partial^2}{\partial x \partial y} & -\frac{\partial^2}{\partial z^2} \end{bmatrix}$$

The equations of equilibrium become

(2.81) $B\,A\,\chi = q$,

where $q = \{q_1, q_2, q_3\}$ is the vector of body forces. The equation (2.81) can be regarded as a vector equation in the Hilbert space $L_2\ (\Omega)$, where Ω is the region occupied by the elastic body.

We observe that A is symmetric and its conjugate transpose A* is equal to A. If we choose an arbitrary strain distribution $\varepsilon_{ij}(x)$, $x \in \Omega$, we observe that

(2.82) $A^*\ \varepsilon_{ij} = \kappa$,

where κ is the incompatibility tensor. The compatibility equations are $\kappa \equiv \emptyset$, $\kappa = (\kappa_{11}, \kappa_{12}, \ldots, \kappa_{33})$. Introducing Hilbert space H_1 with the product

$$\langle z,y\rangle_{H_1} = \int_\Omega \left(\sum_{i=1}^{6} z_i y_i \right) dx,$$

and a space H_2 with an identical definition of an inner product $\langle\, ,\rangle_{H_2}$, we set A: $H_1 \to H_2$ A*: $H_2 \longrightarrow H_1$.

We have assumed the existence of a functional W, which we shall call the potential energy, such that $W_\alpha = \varepsilon$ (Principle of complementary virtual work). Equivalent statement is that we have assumed the existence of Gibbs' thermodynamic potential.

(2.83) $2W = \langle \kappa, \chi\rangle_{(H_1)} = \langle \chi,\ A^*\varepsilon\rangle_{(H_1)} = \langle A\chi, \varepsilon\rangle_{(H_2)}$

$$= \langle \alpha, \varepsilon\rangle_{(H_2)} = \langle B^*U, \alpha\rangle_{(H_2)}$$

$$= \langle U, B\alpha\rangle_{(H_3)}$$

However, the pairs $\{U,\phi\}$, $\{\alpha,\varepsilon\}$, $\{\kappa,\chi\}$ do not contain vectors (in appropriate spaces) which can be picked independently of each other. If U denotes the displacement vector, $B^*U = \varepsilon$ is the definition of the linear strain.

(2.84) $B\ A^*B^*U = \emptyset$

is exactly the set of Ricci's equations. (see Washizu [59] for explanation of their importance in linear elasticity.)

The corresponding Legendre transformation is

given by the following set of equations, labelled (2.85) (a, b, c, d, e, f, g).
The following diagram illustrates the relations (2.75) - (2.85)

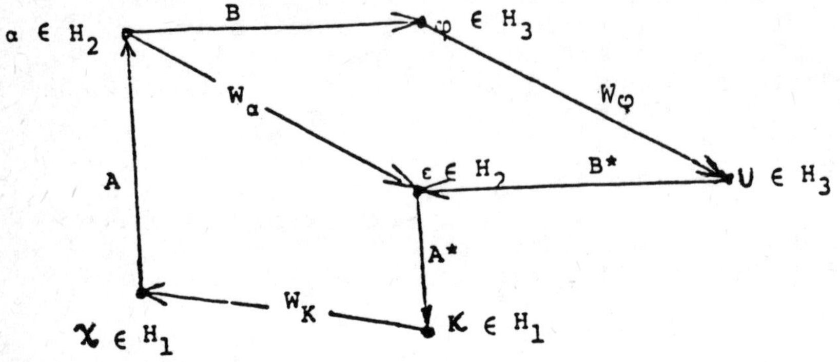

Figure (2.4)

They are related to each other through the constitutive equations of the solid. For example, if Hooke's law is assumed, then we have

(2.86) $\alpha_{ij} = C_{ijk\ell}\varepsilon_{k\ell}$, ($C_{ijk\ell}$-non-singular 9 x 9 matrix of anisotropic coefficients),
or
(2.87) $\varepsilon_{ij} = \gamma_{ijk\ell}\,\alpha_{k\ell}$,

where $\gamma_{ijk\ell}$ is a constant 9 x 9 matrix such that $\gamma_{ijk\ell}\,C_{ijk\ell} = I$ (the 9 x 9 identity matrix), that is $\gamma = C^{-1}$.

A well-known set of variational principles in classical elasticity can be derived by
i) Assuming a constitutive equation of a solid
ii) Restating all basic equations of elasticity as the existence of a critical point of the corresponding functionals: (2.85) (a), (b), (c), (d), (e), (f), (g)

2.85$^{(a)}$ $\Phi_1 = W - \frac{1}{2} \langle K,\chi\rangle$, in H_1

2.85(b) $\Phi_2 = W - \frac{1}{2} \langle \chi, A^*\varepsilon \rangle$, in H_1

2.85(c) $\Phi_3 = W - \frac{1}{2} \langle A\chi, \varepsilon \rangle$, in H_2

2.85(d) $\Phi_4 = W - \frac{1}{2} \langle \alpha, \varepsilon \rangle$, in H_2

2.85(e) $\Phi_5 = W - \langle U, B\alpha \rangle$, in H_3

2.85(f) $\Phi_6 = W - \langle B^*U, \alpha \rangle$, in H_2

2.85(g) $\Phi_7 = W - \langle U, \Phi \rangle$, in H_3

We have deliberately ignored the problem of solving the basic boundary value problems of elasticity (using Muskhelishvili's terminology) and concentrated on the formulation of fundamental principles with natural boundary conditions. (See definition and discussion in[53],[54],[55],[56].)

In fact, should another loop be added to our mapping diagram (2.4), we can immediately formulate the corresponding critical point statement, hence a variational principle for the corresponding functional.

2.27 The Second Derivative: Heuristic Comments

Suppose $x = (x_1, x_2, \ldots, x_n)$, $= (y_1, y_2, \ldots, y_m)$ are vectors in $\mathbb{R}^n$, $\mathbb{R}^m$ respectively, and that we have the functional relation $x_1 = x_1(y)$, $x_2 = x_2(y) \ldots x_n = x_n(y)$. i.e. a map $\mathbb{R}^m \to \mathbb{R}^n$. What meaning do we assign to $\frac{\partial x}{\partial y}$? In this case, we have suggested that the n x m (Jacobian matrix of partial derivatives

$$\left(\frac{\partial x_i}{\partial y_j}\right) \quad \begin{matrix} i = 1,2,\ldots,n, \\ j = 1,2,\ldots,m, \end{matrix}$$

plays the part of the Fréchet derivative. If $x \in H_1$, $y \in H_2$, where H_1, H_2 are infinite dimensional Hilbert (or Banach) spaces, then the meaning of the Jacobian matrix has to be redefined. However, we have to recognize that even in the finite dimensional case $x \in \mathbb{R}^n$, $y \in \mathbb{R}^m$, the Jacobian matrix is not an element of either $\mathbb{R}^n$ or $\mathbb{R}^m$, but a linear mapping from $\mathbb{R}^n$ to $\mathbb{R}^m$. Its conjugate transpose is $\left(\frac{\partial y_j}{\partial x_i}\right)$.

Tensor products.

For purposes of clarity we shall carefully distinguish between elements of a Banach or Hilbert space $\mathbb{H}$, and its algebraic dual. In the definition of the algebraic dual we use the assumption of linearity only. No boundedness (or continuity) properties are assumed for the functionals in $\mathbb{H}^*$.

Let E be a Banach space, and E^* its algebraic dual. We consider a Cartesian product of $E^* \times E^*$, i.e. all possible sets of ordered pairs $\{\rho,\psi\}$ $\psi \epsilon E^*$, $\rho \varepsilon E^*$ (denoted by $(\rho \otimes \psi)$), and define the map $E \to \mathbb{R}$ by the operation $\psi(x) = \langle\psi, x\rangle$ which is defined for any $x \epsilon E$. We shall use the following notation. Vectors subscripted by indices i,j,k,ℓ ... will denote elements of E^*, while superscripts will denote elements of E. Our algebraic operation then assumes the formal representation:

$$(2.88) \quad (\rho_i \otimes \psi_j) X^j = \langle\psi_j, X^j\rangle \rho_i \; ,$$

where $\langle\psi_j, X^j\rangle$ is a real number obtained by applying the linear map $\psi_j \epsilon E^*$ to the vector $X^j \epsilon E$. Hence, $(\rho_i \otimes \psi_j)$ is a linear map whose domain is E and whose range is contained in E, and is of dimension one, or zero. In the same manner we define the tensor products

$(\rho^i \otimes \psi^j)$: $E^* \to E^*$,

$(\rho^i \otimes \psi_j)$: $E \to E^*$,

$(\rho_i \otimes \psi^j)$: $E^* \to E^*$, by the algebraic rules:

$$(2.89) \quad (\rho^i \otimes \psi_j) X^j = \langle\psi_j, X^j\rangle \rho^i ,$$

$$(2.90) \quad (\rho_i \otimes \psi^j) X_j = \langle X, \psi^j\rangle \rho_i \, .$$

Again we see that the range of this map is of dimension one. Higher order tensor products are defined similarly. For example, $(\rho_i \otimes \psi_j \otimes \eta_k) \in E^* \times E^* \times E^*$ is defined as a map from ${}^jE \times E^k$ into E, or as a map from E into E × E.

(2.91) $$(\rho_i \otimes \psi_j \otimes \eta_k)\ (X^k \otimes Y^j) = <\eta_k, X^k> \cdot <\psi_j, Y^j> \rho_i .$$

We assumed associate property of this algebraic operation in the following sense:

(2.92)
$$\begin{aligned}
&((\rho_i \otimes \psi_j \otimes \eta_k) X^k)\ Y^j \\
&= <\eta_k, X^k>\ (\rho_i \otimes \psi_j)\ Y^j \\
&= <\eta_k, X^k>\ <\psi_j, Y^j>\ \rho_i \\
&= (\rho_i \otimes \psi_j \otimes \eta_k)\ (X^k \otimes Y^j).
\end{aligned}$$

We shall not assume that such compositions of mappings are commutative, and indeed such assumption can not be consistently supported in the general case. Moreover, we try to pattern our discussion to agree with the usual engineering ideas of "what a tensor is".

We find no reason why the notation α^{ij} cannot be used to denote a specific tensor product $(\rho^i \otimes \psi^j)$. At this point we are going to make further assumptions (which in general, may exclude some important considerations in classical mechanics, but seem to be justified in solid mechanics.) We are going to restrict the spaces E, E*, (E**) to be topological duals of each other. Hence, all linear maps defined above are now assumed to be continuous. This could be called "the finite energy hypothesis". Thus, if the discussion is restricted to a Hilbert space setting, we can apply the Riesz representation theorem. This simplifies the definitions of the higher order derivatives.

2.28 Definitions of the higher order derivatives in a Hilbert space setting.

Let $f: E \to E$ be a continuous function. We use the notation $Y = f(X)$, or $Y = Y(X)$, $X \in \mathcal{D}_f$. We define the Gateaux derivative of f (which may be called the directional derivative of Y with respect to X in the direction of h) as

$$(2.93)\quad \frac{\partial Y}{\partial X_h} = \lim_{t\to 0} \frac{Y(X+th) - Y(X)}{t} = \psi(X,h) \in \mathbf{E}.$$

provided this limit exists. We claim that for any $Z \in E^*$, $\langle Z,\psi\rangle$ is a linear functional of h, i.e. $\langle Z,\psi(X,ch)\rangle = c\langle Z,\psi(X,h)\rangle$ and $\langle Z,\psi(X,h_1+h_2)\rangle = \langle Z,\psi(X,h_1)\rangle + \langle Z,\psi(X,h_2)\rangle$. Each follow easily from assumption of the existence of the limit (2.93) in some open region of $\mathbf{E}$. If we assume continuous dependence of ψ on h, then by Milgram - Lax theorem (the bilinear version of Riesz representation theorem) $c\langle Z,\psi\rangle$ can be written as $\langle AZ,h\rangle$, but in this case the operator A is the operator of multiplication by a constant and

$$(2.94)\quad \langle Z, \frac{\partial Y}{\partial X_h}\rangle = c\langle Z,h\rangle.$$

Again, if we assume that c depends continuously on Y, (linearity is obvious) then using Riesz representation theorem we conclude that there exists a vector $\eta \in E^*$ such that $c = \langle \eta, Y\rangle$. Hence,

$$(2.95)\quad \langle \frac{\partial Y}{\partial X_h}, Z\rangle = c\langle\psi,Z\rangle = \langle\eta,Y\rangle\cdot\langle h,Z\rangle .$$

Again, using the Lax - Milgram theorem, we can rewrite this product

$$(2.96)\quad \langle\psi,Z\rangle = \langle A\eta,Z\rangle \cdot \langle Y,Bh\rangle$$

where A,B are linear operators

$B: E \to E^*$

$A: E^* \to E.$

Since this is true for arbitrary $Z \in E^*$, we have equality

$$\frac{\partial Y}{\partial X_h} = \langle Y,Bh\rangle A\eta = \langle Y\xi\rangle A\eta = \langle Y,\xi\rangle\mu, \text{where } \mu = A\eta.$$

Hence, the operator $\frac{\partial}{\partial X_h}$ can be represented as a tensor product

$$(2.97) \quad \frac{\partial}{\partial X_h} = (\mu \otimes \xi), \ \mu \in E, \xi \in E^*$$

$$\frac{\partial}{\partial X_h} Y = \langle Y,\xi\rangle\mu.$$

Similar argument gives us a representation

$$\frac{\partial Y}{\partial X_h} = \langle h,\xi\rangle\mu \ ,$$

where ξ is of the form $\xi = CY$, and C is a linear operator, $C: \ E \to E^*$. In this entire argument the domain of the operators A,B,C is assumed dense in a sufficiently small neighborhood of a region considered for the respective vectors η, h, Y (in E, E*, E, respectively), and the existence of adjoint operators allows the desired manupulations of the inner products $\langle , \rangle$. We can generalize the concept of a gradient by observing that our assumption of continuity with respect to h, allows to define grad $Y = Y_X = (\mu \otimes \xi)$, since

$$\frac{\partial Y}{\partial X_h} = \langle \xi,h\rangle\mu$$

(Recall the definition of a gradient of a functional.) In particular, if there exists a functional $V(X,x)$, $x \in E$, $X \in E$, such that $Y = \text{grad } V(X,x) \in E^*$ (or $Y = V_x$), then we can denote by

$$V_{xX} = (\mu \otimes \xi) = Y_X.$$

If we assume that $V = V(X,x)$, $X,x \in E$, is differentiable, then a theorem of Vainberg states that in the real Hilbert spaces

$$(2.98) \quad V_{Xx} = V_{xX}.$$

However, if real structure is abandoned we may postulate $V_{xX} = -V_{Xx}$. None of this implies $(\xi \otimes \mu) = (\mu \otimes \xi)$! (See [53], Chapter 2.) In fact, commutativity of the "second derivatives" is a necessary condition for the existence of the "potential functional" V, and cannot be assumed arbitrary.

The idea of second derivative is easily generalized to the case when V: $E \times E^* \to \mathbb{R}$ is a functional depending continuously on

$x_1, x_2, \ldots, x_k \in E, X_{k+1}, X_{k+2}, \ldots X_n \in E^*,$

$E = E_1 \otimes E_2 \otimes \ldots \otimes E_k, E^* = E^*_{k+1} \otimes E^*_{k+2} \otimes \ldots \otimes E^*_n,$ and $X_1 \in E_1 \ldots X_n \in E^*_n$, $E_1, E_2, \ldots, E_k$ being subspaces of E, $E^*_{k+1} \ldots E^*_n$ of E^*, vectors $V_{X_1} \ldots, V_{X_n}$ can be regarded as generalizations of the concepts of partial differentials. Consider a functional $V(X^i, Y_j)$ (not necessarily linear) whose values in $\mathbb{R}$ depend on $X^i \in E$, $Y_j \in E^*$. Then, provided all the derivatives shown exist, one constructs the following diagram:

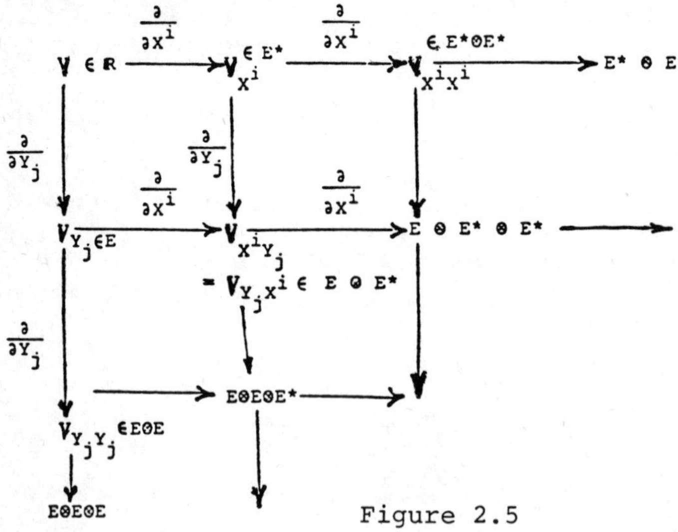

Figure 2.5

It shows the increasingly higher order tensor products obtained by performing repeatedly the gradient operation.

Positive (positive definite) second order tensor product.

Since $(\xi \otimes \eta)$, $\xi \in E$, $\eta \in E$, or $(\mu \otimes \nu)$, $\mu \in E^*$, $\nu \in E^*$) can be regarded as maps from E* into E* (or from E to E) we can define the concept of positive property of such map by the usual definition:

$(\xi \otimes \eta)$ $\xi,\eta \in E$ is positive if

(2.99) $\langle(\xi \otimes \eta)\, Y, Y\rangle \geq 0$ for all $Y \in E^*$

Similarly $(\mu \otimes \nu)$, $\mu \in E^*$, $\nu \in E^*$ is positive if $\langle(\mu \otimes \nu)X,X\rangle \geq 0$ for all $X \in E$. $(\xi \otimes \eta)$ is called positive definite if $\langle(\xi\otimes\eta)Y,Y\rangle > C\|Y\|^2 \geq 0$ for all $Y \in E^*$, and some $C \in R_1$. $\langle(\xi \otimes \eta)Y,Y\rangle = 0$ implies $Y = \emptyset \in E^*$. Similar definition is given for $(\mu \otimes \nu)$, $\mu \in E^*$, $\nu \in E^*$.

Vainberg's lemma and some of its consequences.

Let E_1 be a subspace of a Banach space E, and V a functional $V: E \to R$, the value of V depending on a vector $X \in E_1$, and on $Y \in E_2$ where $E_1 \otimes E_2 \subset E$, $E_1 \cap E_2 = \emptyset$. Suppose that V is defined in some neighborhood of $X_0 \in E_1$. Then a necessary condition for X_0 to be an extremal point of V, when V is restricted to E_1, is $V_X = \emptyset$, $V|_{E_1}$. ($|$ means "restricted to"). A sufficient condition for a min. (max) of $V|_{E_1}$ is that $V_X = \emptyset$, V_{XX} is positive definite (negative definite).

2.29. Examples of application and relation to the Legendre "splitting" transformation.

Consider the behavior of the (non-linear) functional

$$V(X,P) = \langle AX,X\rangle \cdot \langle X,X\rangle^{-1} + \psi(P),\ X \in H_1,\ P \in H_2,$$

where H_1, H_2 are Hilbert spaces, $A: H_1 \to H_1$ is a linear map.

Let P be fixed (in H_2). Find sufficient conditions for a local minimum of V(in H_1). We compute the gradient of V in H_1:

$$V_X = 2\langle X,X\rangle^{-1} \cdot (AX - V(X)\cdot X) .$$

A necessary condition for an extremum of V to occur at $X_0 \in H_1$ is: $AX - V(X)\cdot X = \emptyset$ when $X = X_0$, that is $AX_0 = \Lambda X_0$, where Λ is a number given by

$\Lambda = V(X_0)$, which means that X_0 is an eigenvector, and $V(X_0)$ the corresponding eigenvalue of A.

In what follows let us assume that the multiplicity of Λ_0 is one, and Λ_0 is the lowest eigenvalue of A. The second derivative is computed as follows:

$$(\varphi \otimes \psi)h = V_{XX}h = \lim_{t \to 0} \left(\frac{V_X(X+th) - V_X(X)}{t} \right)$$

$$= 2t^{-1} \{\langle X+th, X+th\rangle^{-1} (AX + tAh - V(X+th)[X+th])$$

$$- \langle X,X\rangle^{-1} (AX - V(X)\cdot X)\}$$

$$= \lim_{t \to 0} 2t^{-1} \left\{ \frac{AX + tAh - V(X+th)(X+th)}{\langle X,X\rangle + t\langle h,X\rangle + t^2\langle h,h\rangle} - \frac{AX - V(X)X}{\langle X,X\rangle} \right\}$$

Having assumed the continuity of V, we obtain as $t \to 0$

$$V_{XX}h = \frac{2(Ah - V(X)h)}{\langle X,X\rangle} .$$

Hence,

$$V_{XX} = 2\left(\frac{A - V(X)I}{\langle X,X\rangle} \right) .$$

At the point X_0 where $V_X(X_0) = \emptyset$, we have

$$V_{XX}\ (X=X_0) = 2\ \frac{A - V(X_0)I}{\langle X_0,X_0\rangle} ,$$

where I is the identity map. Therefore, we have

$$V_{XX}(X=X_0) = 2(A - \Lambda_0 I) \cdot \langle X_0,X_0\rangle^{-1} .$$

We conclude that the extremum of V at X_0 occurs in some (possibly small) neighborhood of X_0: $N_{X_0} \subset H$, only if for every vector $\xi \epsilon N_{X_0}$ it is true that

$$\langle A\xi,\xi\rangle - \Lambda_0\langle\xi,\xi\rangle - \langle V_X(\xi),\xi\rangle > 0 \begin{cases} \xi \neq X_0 \\ \xi \neq 0 \end{cases},$$

$$\text{or } \langle A\xi\ \xi\rangle - \Lambda_0\langle\xi,\xi\rangle - \langle V_X(\xi),\xi\rangle < 0 .$$

In particular, if A is positive definite, completely continuous, and Λ_0 is the lowest eigenvalue, then the Rayleigh quotient V attains its minimum value at X_0. Specifically, we have the equality

$$V(X_0) = \Lambda_0 (\text{since } \langle V_X(X_0),\ X_0\rangle = 0).$$

It is clear that in a sufficiently small neighborhood of X_0, the operator V_{XX} is positive definite and a local minimum takes place for V(X) at X_0. A global theory is much harder, and it is unreasonable to expect that the signs of first and second abstract derivative in some neighborhood should imply anything about the global behavior of a given functional.

An Example: Thin plate theory.

The plate satisfies the linear equation

$$(2.100)\quad \nabla^2(D(X,y)\ \nabla^2 w(x,y)) - (1-\nu)\Diamond^4(D,w) = q(x,y) \text{ (i.e. the Lagrange-Germain equation)}$$

in a region $\Omega \subset R^2$.

$$\text{Here } \Diamond^4(A,B) = \frac{\partial^2 A}{\partial x^2}\ \frac{\partial^2 B}{\partial y^2} - 2\ \frac{\partial^2 A}{\partial x\partial y}\ \frac{\partial^2 B}{\partial x\partial y} + \frac{\partial^2 B}{\partial x^2}\ \frac{\partial^2 A}{\partial y^2} .$$

and ∇^2 is the Laplacian.

The boundary conditions are $w \equiv \frac{\partial w}{\partial n} \equiv 0$ on $\partial\Omega$. The boundary $\partial\Omega$ of Ω is smooth, except for a finite number of external corners. (i.e. corners like the one shown below are not permitted.)

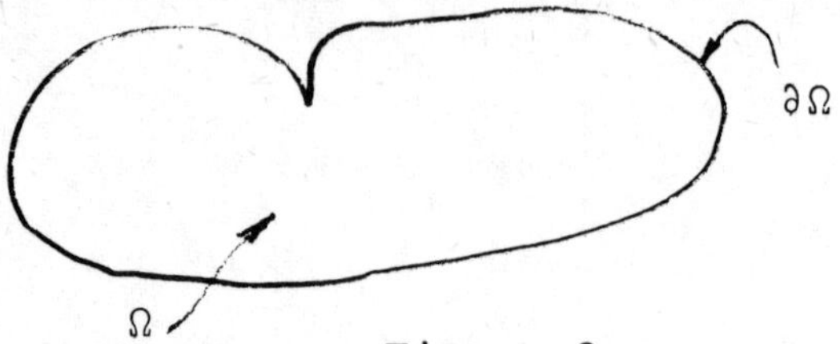

Figure 2

This can be recognized as the Euler-Lagrange equation for the functional

$$(2.101)\quad \Phi = \int_{\Omega} \{D(\nabla^2 w)^2 - D(1-\nu)(\diamondsuit^4(w,w))\}\, dxdy$$

It is not very convenient (or very practical in most engineering applications) for one to assume the existence of all derivatives appearing in the formal Euler-Lagrange equations (2.101) for the extremum of the functional $\Phi(w)$. Physically, we require only twice differentiability of w, and the L_2 property of the derivatives ($w \in H^2(\Omega)$). We introduce the following map

$$N: \; H_4^2(\Omega) \to H_4^2(\Omega),$$

where N is represented by the following positive definite matrix:

(2.102)

$$N = \begin{bmatrix} D(x,y) & \nu D(x,y) & 0 & 0 \\ \nu D(x,y) & D(x,y) & 0 & 0 \\ 0 & 0 & (1-\nu)D(x,y) & 0 \\ 0 & 0 & 0 & (1-\nu)D(x,y) \end{bmatrix}.$$

We assume for physical reasons, $D(x,y) > 0$ in Ω, and $0 < \nu < \frac{1}{2}$. T is a 2×4 matrix of first order differential operators, T* is the transpose of T:

(2.103)
$$T = \begin{bmatrix} \frac{\partial}{\partial x} & 0 \\ 0 & \frac{\partial}{\partial y} \\ \frac{\partial}{\partial y} & 0 \\ 0 & \frac{\partial}{\partial x} \end{bmatrix}.$$

Then we introduce the Legendre transform variables:

(2.104) $M = NT \text{ grad } w = (-NT \text{ grad})w = Aw.$

This is a well-known relation between the moments

$$\underset{\sim}{M} = (M_{xx}, M_{xy}, -M_{yx}, M_{yy})$$

and the displacement $w(x,y)$.

$$A: \ H^2 \ (\Omega) \to L_2 \ (\Omega)$$

At this point in our discussion, it is more convenient to introduce the "modified" moments defined by the formula:

(2.105)
$$\mathcal{M} = (-(N^{\frac{1}{2}}) \ T \cdot \text{grad}) \ w,$$

where we take the positive square root $N^{\frac{1}{2}}$ of the operator N, as follows

$$N^{1/2} = \begin{bmatrix} \sqrt{1-\beta^2}\ \sqrt{D} & \beta\sqrt{D} & 0 & 0 \\ \beta\sqrt{D} & \sqrt{1-\beta^2}\ \sqrt{D} & 0 & 0 \\ 0 & 0 & \sqrt{(1-\nu)D} & 0 \\ 0 & 0 & 0 & \sqrt{(1-\nu)D} \end{bmatrix}$$

and

where β denotes $\left(\frac{1}{2} + \frac{\sqrt{1-\nu^2}}{2}\right)^{\frac{1}{2}}$,

It can be checked that $N^{\frac{1}{2}} N^{\frac{1}{2}} = N$, and $N^{\frac{1}{2}}$ is again a positive definite (and invertible) matrix, and fortunately $N^{\frac{1}{2}}$ is symmetric. (We can easily introduce a splitting of N which is not symmetric.)

The operator $A: H^2(\Omega) \to L_2(\Omega)$ has a formal adjoint $A^*\ L_2(\Omega) \to D^*(\Omega)$, and we have the Legendre duality $Aw = M$, $A^*M = q$ (2.107), where the derivatives now become distributional derivatives. $L_2(\Omega)$ is regarded as a subset of $D^*(\Omega)$, where $D(\Omega) = C_0^\infty(\Omega)$,

$$A^* = \operatorname{div}\,(T^*N^{\frac{1}{2}}),$$

i.e. $A^*M = \operatorname{div}\,(T^*(N^{\frac{1}{2}}M)) = \operatorname{div}\,(T^*\tilde{M}) = q(x,y)$.

The distributional character of $q(x,y)$ being supported by the usual interpretation of admitting point loads, or point moments in engineering practice.

Hence, we have the following mapping diagram

$$w \xrightarrow{\ A\ } M \xrightarrow{\ A^*\ } q,$$

where $A^*Aw = q$ is the original differential equation of static plate deflection:

(2.107) $\quad \nabla^2(D\nabla^2 w) - (1-\nu)\Diamond^4(D,w) = q.$

We can observe that the set of equations:

$$\begin{cases} Aw = M \\ A^*M = q \end{cases}$$

represents two variational principles. We introduce the Hamiltonian

$$\mathfrak{W} = \langle q,w\rangle + \tfrac{1}{2}\{M,M\},$$

where the inner product $\{,\}$ is given by

$$\{A,B\} = \int_\Omega (A_1B_1 + A_2B_2 + A_3B_3 + A_4B_4)\,dxdy,$$

while $\langle,\rangle$ is the usual L_2 product:

$$\langle a,b\rangle = \int_\Omega (a \cdot b)\,dx\,dy.$$

The dual variational principles given for the static plate problem have been given in

[29] by the author, and may be summarized in the observation

$$W_M = M = Aw \,, \tag{2.108}$$
$$W_w = q = A^*M, \tag{2.109}$$

where W_M denotes the gradient of W restricted to the Hilbert space H_M (with the product $\{\cdot,\cdot\}$), while W_w is the gradient of W restricted to H_w (with the product $<,>$.) The examination of second Fréchet derivatives i.e. of the tensor products $(W_M)_M$, $(W_w)_w$ reveals that the corresponding Lagrangian functional

(2.110) $L = \{w, M\} - W$

attains a minimum over the admissible $w \in H_w(\Omega)$ (with M regarded as fixed) and a minumum over the admissible $M \in H_M(\Omega)$ (with w regarded as fixed) at the critical point corresponding to the solution of a system of equations (2.106), (2.107). Combining this system into the form $AA^*w = q$ gives a single equation (2.100) which is the basic static deflection equation of thin plate theory.

The dual variational principles given here originally discovered by the author in [28].
At this point we wish to make a more detailed examination of the equation (2.107) and find alternate variational formulation for this equation.

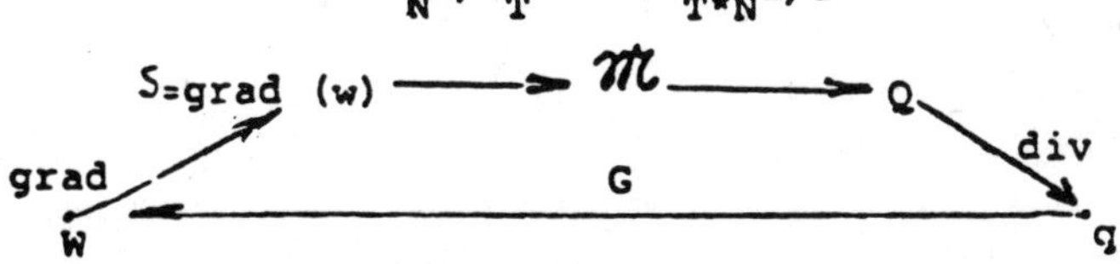

Figure 2.6

representing an abstract form of Legendre transformation.
Let us concentrate on the following portion of the diagram (2.6).

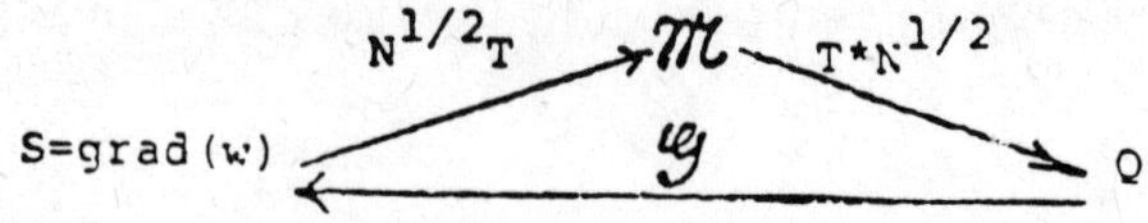

Figure 2.7

$\mathcal{G}$ and G represent the inverse operators. G is the Green's function, or influence function in engineering terminology: w=G*q, where * is the convolution operator (or convolution product (or convolution integral). $\mathcal{G}$ is the corresponding map defined by $\mathcal{G} * Q = \nabla w = S$. It is a map which is to be determined, but whose existence is not hard to prove, if we confirm the existence of the Green function $G(x,y)$. (We do not write $G(x,\xi,y,\eta)$ since the convolution product takes care of introducing the "translated" variables ξ,η.)

The dual equations corresponding to the diagram (2.7) are

$$(2.111^a)\quad N^{\frac{1}{2}}\,TS = M = W_M\ ,$$

$$(2.111^b)\quad T^*N^{\frac{1}{2}}M = Q = W_S,$$

except that W is a functional defined on a pair of different Hilbert spaces: H_M (which is the same as before) and H_S which is $L_2^{(2)}(\Omega)$ only if point moments are excluded, and point loads are allowed as the only loads that are not represented by square integrable functions. Note: For a classification of admissible (distributional) loads of thin plate theory, see author's paper [35].

If point moments are allowed, the space H_S cannot be embedded in $L_2(\Omega)$ and is not a Hilbert space, but only an inner product space which can be at best regarded as a Rigged Hilbert space in the terminology of Gelfand and Shilov (see [20] volume 4, chapter 1, section 4). For the purpose of this discussion we identify H_S with the space $L_2^{(2)}(\Omega)$ having the inner product

$$(A,B)^{(2)} = \int_\Omega [A_1(x,y)B_1(x,y + A_2(x,y)B_2(x,y)]dxdy.$$

The components of the vector Q are recognized as the shears

$$(2.112^a) \quad Q_x = \frac{\partial M_{xx}}{\partial x} + \frac{\partial M_{xy}}{\partial y}$$

$$(2.112^b) \quad Q_y = \frac{\partial M_{yx}}{\partial x} + \frac{\partial M_{yy}}{\partial y}$$

The generalized Hamilton's canonical equations are

$$(2.113^a) \quad N^{\frac{1}{2}} TS = M = W_M ,$$

$$(2.113^b) \quad T^*N^{\frac{1}{2}} = Q = W_S ,$$

$$(2.113^c) \quad L = \tfrac{1}{2} \{M,M\} + (S,Q) = \{N^{\frac{1}{2}}TS, \} - W .$$

The Euler-Lagrange equations

$$L_M \big|_{H_M} = \emptyset ,$$

$$L_S \big|_{H_S} = \emptyset ,$$

are exactly the equations (213^a), (2.113^b) expressed as restrictions of appropriate Fréchet derivatives of L to the spaces H_M and H_S respectively.

The tensor products $W_{MM}, W_{MS}, W_{SM}, W_{SS}$ can be arranged in a 2 x 2 matrix form of operators:

$$(2.114) \quad \begin{bmatrix} I & N^{\frac{1}{2}}T \\ T^*N^{\frac{1}{2}} & T^*NT \end{bmatrix} = \begin{bmatrix} W_{MM} & W_{MS} \\ W_{SM} & W_{SS} \end{bmatrix} .$$

I denotes the identify operator. $W_{MS} = (W_{SM})^*$ is a necessary condition for integrability of the differential system (2.113^a), (2.113^b), while the positive definite nature of I: $H \to H$ and T*NT: $H_S \to H_S$ assure the existence of a double minimum. That is, L attains a minimum in H_w for a fixed S, if the equation (2.113^a) is satisfied, and a minimum in H_S for a fixed w if the equation (2.113^b)

is satisfied. These variational principles suggest numerical techniques of the type introduced by Greenspan in [143] for computation of approximate solutions, and these principles may be simpler than the known principles suggested by equations (2.113^a), (2.113^b). The advantage of obtaining a symmetric form of corresponding operators, and of not having the matrix N appearing in only one set of equations are considerable in actual computation. The positive definite nature of the matrix N allowed us to restate our problem in terms of $\mathcal{M}$ rather than M. Physical limitations on such representation are clear: $D(x,y)>0$ for all $x,y \in \Omega$, (we do not have holes in Ω). If we do, let us relabel what the region Ω really is!), and $\frac{1}{2} > \nu > 0$ which is readily recognized by any engineer as the expected behavior of a physical solid, namely that the solid does not expand volumetrically under pressure.

It is interesting to note how close are the mathematical and the physical limitations imposed on such problems.

The lack of simple connectivity(i.e. the existence of holes) causes complications in proving the existence of solutions to boundary value problems. (In fact, solutions may not exist.) The operators $N^{\frac{1}{2}}$, N^{-1}, $N^{\frac{1}{2}}T$, etc. are not defined if the Poisson ratio ν does not obey the suitable restrictions that coincide with the obvious physical laws of behavior of elastic solids, and our variational principles could not be formulated if the inequality $0<\nu<\frac{1}{2}$ is violated.

Such close affinity of physical and mathematical constraints is common in all areas of applied mathematics and of modeling of physical processes. Perhaps, it should serve as a check on reliability of mathematical models for "real life".

<u>A historical note</u>. The Bernoulli - Euler analysis aroused great interest in behavior of elastic structures and in extending Euler's analysis to vibrating systems and to higher spatial dimensions, i.e. to the theory of plates and simple shells. After E.F. Chladni demonstrated in 1802 in his Paris lectures the vibration patterns of thin plates (the Chladni figures) there was a considerable

activity-encouraged by Napoleon Bonaparte-to develop a mathematical theory explaining Chladni's figures. Sophie Germain attempted a generalization of Bernoulli's relation between the radius of curvature and the bending moment applied to a beam. Her intuition and physical insights were basically correct. The attempt to treat a plate as a "wide beam" and relate the bending moments (they were not called by that name) in mutually perpendicular directions was a first serious effort to produce a reasonable mathematical model. Her inability to manipulate partial derivatives and multiple integrals produced a later criticism of her entire contribution to the theory of elasticity of thin plates. (See the work of L.L. Bucciarelli and Nancy Dworsky: Sophie Germain, An essay in the history of the theory of elasticity [23].) J. Lagrange "cleaned up" the mathematical details and eventually the Lagrange-Germain mathematical model of thin plate deflection phenomena emerged around 1816 after Germain was awarded "prix extraordinaire" by the French Academy of Sciences, but not before alengthy debate involving Lagrange, Navier, Legendre, and Poisson and a host of minor characters. Disputes of the time included molecular theories of Poisson, metaphysics, positivism, and the nature of mathematics. In the opinion of Bucciarelli and Dworsky [23], one of the handicaps of Sophie Germain and some of her contemporaries was inability to understand the variational approach of Lagrange to the rational theory of elasticity.

2.30 A problem in fluid flow theory.

To give a classical example of application, we shall consider an extremely simple case of fluid flow, subject to numerous assumptions.

We look at this example for the reasons of its simplicity, and because it contains the basic ingredients of the much more complex cases. It also gives an indication of difficulties which become apparent when no velocity potential exists, and when mechanical energy is dissipated by friction and converted into heat.

In the very simple case a slow, irrotational, inviscid flow, satisfies Bernoulli's law:

$$p + \rho \frac{\|v\|^2}{2} + h = P, \text{ in } \Omega \subseteq R^2 (\text{or } R^3), \tag{2.115}$$

We introduce the generalized momentum $\underset{\sim}{M} = \rho \cdot \underset{\sim}{V}$, and observe that the existence of velocity potenial Φ such that $\nabla\Phi=\rho \underset{\sim}{V}$ implies the existence of a Hamiltonian system

$$\begin{cases} \nabla\Phi = \rho \cdot \underset{\sim}{V} = P_{\underset{\sim}{M}}, & (2.116) \\ \nabla\cdot \underset{\sim}{M} = 0 = P_{\Phi} \quad (\nabla\equiv \text{grad},\ \nabla\cdot \equiv \text{div}) . & (2.117) \end{cases}$$

A variational principle is easily formulated provided P_{MM} is either a positive or a negative operator. Assuming that on the boundary $\partial\Omega$ of the region Ω the momentum $\underset{\sim}{M}$ satisfies

$$\langle \underset{\sim}{n}\cdot\underset{\sim}{M}\rangle - m_n(x) = 0,\ x \in \partial\,\Omega,$$

we formulate the Hamiltonian

$$H(\underset{\sim}{M},\Phi) = \int_{\Omega} P\ d\underset{\sim}{x} + \int_{\partial\Omega} m_n(x)\Phi\ d\underset{\sim}{s} - \int_{\partial\Omega} \langle\Phi\underset{\sim}{n},\underset{\sim}{M}\rangle\ d\underset{\sim}{s}. \tag{2.118}$$

A rigid boundary implies $m_n(x)\equiv 0$ on $\partial\Omega$, hence $\langle\underset{\sim}{n},\underset{\sim}{M}\rangle = 0$ on $\partial\Omega$. These boundary conditions are natural and do not contribute to the value of the Lagrangian action integral.

A more difficult case of a two-dimensional flow, that is neither isothermal, nor isentropic is discussed by Sewell in [81], [80].

Defining the total energy

$$H = \tfrac{1}{2}\ \|\underset{\sim}{V}\|^2 + E\ (S,\rho)+p/\rho \tag{2.119}$$

and expressing pressure by a functional relation $p = p(V,S,H)$, he derives the relations

$$\underset{\sim}{V} = \text{grad}\ \Phi + \alpha\ \text{grad}\ H + \beta\ \text{grad}\ S,\quad \frac{\partial p}{\partial \underset{\sim}{V}} = \rho \underset{\sim}{V},\quad \frac{\partial p}{\partial H} = \rho,$$

$\dfrac{\partial p}{\partial S} = -\rho T$, where the constants α,β serve the purpose of obtaining correct physical dimension, E is the internal energy, S - entropy, T - absolute temperature, V - velocity.

As before $\text{div}\ (\underset{\sim}{q}) = 0$, $\underset{\sim}{q} = \rho \underset{\sim}{V}$.

The author also assumes that

$$\alpha\langle\underset{\sim}{q},\ \text{grad}\ H\rangle = \beta\langle q,\ \text{grad}\ S\rangle = 0$$

(constant entropy and constant energy along streamlines of flow).

Then denoting symbolically $X = \begin{bmatrix} \underset{\sim}{q} \\ H \\ S \end{bmatrix}$

and $Y = \begin{bmatrix} \Phi \\ \alpha \tilde{q} \\ \beta \tilde{q} \end{bmatrix}$,

he shows that the system of equations describing behavior of the fluid flow can be put into the canonical form

$$\left.\begin{aligned} W_x &= AY, \\ W_y &= A^*X. \end{aligned}\right\}$$

where

(2.120)
$$A = \begin{bmatrix} -\text{div}, & 0, & 0 \\ 0, & \text{grad}, & 0 \\ 0, & 0, & \text{grad} \end{bmatrix}$$

(2.121)
$$A^* = \begin{bmatrix} \text{grad}, & 0, & 0 \\ 0 & -\text{div} & 0 \\ 0 & 0 & -\text{div} \end{bmatrix},$$

and

(2.122) $W = \int_\Omega (\langle \tilde{q}, \tilde{V}\rangle + p(\tilde{V}, H, S))\, d\tilde{x}$,

The corresponding Lagrangian functional is

$L = \langle A^*X, Y\rangle - W.$

The variational principles of [80],[81],of Sewell are equivalent to the statement that L has a critical point. The local convexity or concavity of the originally stated functional have not been discussed in [80]. However, there is no difficulty in formulating necessary conditions for the existence of a local extremum, subject to the usual assumptions of Fréchet differentiability. Later papers of Sewell proceed in a similar manner utilizing physically motivated arguments. (See for example [134], [135] .) A paper of Moreau [144] has considered a similar dual variational formulation for a flow with cavitation. Some older results in fluid flow theory such as Bateman's variational principle or Lin's variational principle also could be restated by the use of the dual formalism.

2.31 An example: Piezo-elasticity

The necessity of incorporating electromagnetic effects into the continuum mechanics of solids is obvious if one studies acoustic devices such as microphones relying on changes in mechanical properties of crystals caused by changes in the electromagnetic field.

The anisotropic properties of crystals have been utilized in the construction of such devices. Recent hydro-acoustic studies revealed the need for better understanding of the electric properties of materials and the need for better theory explaining nonlinear phenomena arising in the coupled systems modeling piezoelasticity.

Static behavior of piezoelastic plates.

We can formulate an analogue of the linear deflection theory for thin plates by including electrostatic potential in the Lagrange-Germaine derivation of the classical formula (2.100) Let W(x,y) denote displacement, V(x,y) electrostatic potential. We introduce linear operators L_1, L_2, L_3 acting on Sobolëv space $H_0^1(\Omega)$, $H_0^2(\Omega)$, $H_0^2(\Omega)$, respectively. Ω denotes a bounded, closed region of $\mathbb{R}^2$, $H_0^m(\Omega)$ is the closure of C_0^∞ functions in the Sobolëv norm induced by the inner product

$$\int_\Omega \sum_{|\alpha|\le m} \left(\frac{\partial u}{\partial x^{\alpha_1} \partial x^{\alpha_2}} \quad \frac{\partial v}{\partial x^{\alpha_1} \partial x^{\alpha_2}}\right) dx_1\, dx_2 \;.$$

We define

$$L_1 = D_{44} \frac{\partial^2}{\partial x_1{}^2} + 2\, D_{45} \frac{\partial^2}{\partial x_1 \partial x_2} + D_{55} \frac{\partial^2}{\partial x^2} \,,$$

where $D_{ij} = E_{ij}\ h^3/[12\ (1-\nu_{ij}{}^2)] \cdot \gamma_{ij}$.

Here h(x,y) denotes the thickness

ν_{ij} - the Poisson ratio constants

E_{ij} - Young's modulus constants

γ_{ij} - a constant depending on the piezo-elastic and dialectric properties of the material to simplify this relation we can introduce a single anisotropic constant matrix A_{ij} such that

$$D_{ij} = A_{ij}\, h^3/12.$$

The factor 1/12 is retained for the sake of convenience.

The operators L_2, L_3 are given, respectively, by relations

$$L_2 = D_{14}\frac{\partial^3}{\partial x_1^3} + (D_{15} + 2\,D_{34})\frac{\partial^3}{\partial x_1^2 \partial x_2} + (D_{24} + 2\,D_{35})\frac{\partial^3}{\partial x_1 \partial x_2^2} + D_{25}\frac{\partial^3}{\partial x_2^3} \qquad (2.123^a)$$

$$L_3 = D_{11}\frac{\partial^4}{\partial x_1^4} + 4\,D_{13}\frac{\partial^4}{\partial x^3 \partial x_2} + (2D_{12} + 4\,D_{33})\frac{\partial^4}{\partial x_1^2 \partial x_2^2} + 4\,D_{23}\frac{\partial^4}{\partial x_1 \partial x_2^3} + D_{22}\frac{\partial^4}{\partial x_2^4}\,. \qquad (2.123^b)$$

The equilibrium equations are

$$\lambda_1(w,v) = L_3\, w - 2h^{-1}\, L_2\, V = q \qquad (2.124^a)$$

$$\lambda_2(w,v) = L_2\, w + 2h^{-1}\, L_1\, V = 0\,. \qquad (2.124^b)$$

Equations (2.124^a), (2.124^b) are derived, using Kirchhoff's hypothesis, by I. Vekovishcheva in [107]. Also see R.D. Mindlin [33], [32], and the work of H.F. Tiersten [31].

We wish to confirm that equations (2.124^a), (2.124^b) are the Euler-Lagrange equations for some functional $\Phi(w,v,q)$ and to exhibit the corresponding splitting of the operator $\begin{Bmatrix} \lambda_1 \\ \lambda_2 \end{Bmatrix}$ given in these equations by introducing the Legendre transformation. The equations (124^a), (125^b) can be written as

$$\begin{bmatrix} L_3 & -2h^{-2}\, L_2 \\ L_2 & +2h^{-2}\, L_1 \end{bmatrix} \cdot \begin{bmatrix} w \\ v \end{bmatrix} = \begin{bmatrix} q \\ 0 \end{bmatrix}$$

We observe that L_1, L_3 are self-adjoint operators, while $L_2 = -L_2{}^*$ if appropriate zero boundary conditions are assumed.

The Lagrangian product is given by

$$L(w_1, w_2, V_1, V_2) = \langle L_3 w_1, w_2\rangle + 2h^{-1} \langle L_2{}^* V_1, w_2\rangle$$

$$+\langle L_2 w_1, V_2\rangle + 2h^{-1}\langle L_1 V_1, V_2\rangle - \langle q, w_2\rangle$$

The system is not conservative since the operator L_2 is not self-adjoint. The decomposition is similar to the duality in classical elasticity or to factoring through intermediate Hilbert spaces but this is not immediately obvious. Fortunately, Tiersten's formulas may be reformulated in a symmetric form if we simplify some assumptions and substitute equations of classical elasticity for Sophie Germaine plate hypothesis. The Gurtin's approach to this problem is discussed later.

2.32 Possible applications of Noble's duality to the basic theories of solid materials.

The basic engineering thermodynamics laws applicable to solid materials can be summarized in the equations (2.125), (2.126) and the inequalities (2.127), (2.128) below. The solid occupies a region $\Omega \subset \mathbf{R}^3$, in which the following relations are valid.

(2.125) $\mathbb{E} = \frac{1}{2} (\rho_0/\rho)\tau_{ij} \dot{C}^{ij} - h^i{}_{,i} + \dot{Q}$

(first law of thermodynamics) ,

(2.126^a) $\psi = \psi(C^{ij}, \theta, q^\alpha)$,

(2.126^b) $\tau_{ij} = \tau_{ij} (C_{mn}, \theta, q^\alpha)$.

(the principle of material indifference)

(2.127) $\theta \geq 0$ (Absolute temperature is non-negative)

$\theta\dot{S} \geq 0$ (Positive rate of dissipation).

(2.128) $h^i\theta_{,i} \leq 0$.

(Heat conduction in the direction of negative temperature gradient.) The symbols used here have the following meaning. $\mathcal{E}$ is the internal energy, C_{ij} is the Cauchy-Green strain tensor, ρ is the mass density $\rho(0) = \rho_0$, θ the temperature, S entropy, ψ the free energy, h^i the heat flow, Q the heat supply per unit of mass. We assume the existence of entropy.

Dots denote differentiation with respect to time, commas-covariant derivatives. The summation notation is used unless otherwise stated. q^α are additional independent variables, called internal variables, which are not necessarily observable, i.e. the implicit variable theorem may be not applicable to relations (2.126).

It has been shown (see for example [49]) that the stress distribution consistent with the first and second law of themodynamics satisfies the equation

$$(2.129)\qquad \frac{1}{2}\,\tau_{ij} - \left(\frac{\rho}{\rho_0}\right)\frac{\partial\psi}{\partial C^{ij}} = 0,$$

where $\psi = \psi(C^{ij}, \theta, q^\alpha)$.

That is, assuming no constitutive equations, and choosing τ_{ij}, C^{ij} (and independently of each other) the correct choice will produce a critical point of the functional:

$$(2.130)\qquad \left(\frac{1}{2}\,W - \frac{\rho}{\rho_0}\,\psi\right) = \Phi(C^{ij},\ldots),$$

where $W(C^{ij})$ satisfies the relation

$$(2.131)\qquad W_{C^{ij}} = 2\,\tau_{ij},$$

i.e. τ_{ij} is the gradient of $\frac{1}{2}W$ in the appropriate product of Sobolev spaces, which are subsets of $L_2(\Omega)\times L_2(\Omega)$. Recalling that the entropy S satisfies the relation

$$(2.132)\qquad -S = \frac{\partial\psi}{\partial\theta},$$

and that the temperature θ is always positive,

the first law of thermodynamics can be written as:

$$(2.133)\quad -\dot{S} = \frac{d}{dt}\left(\frac{\partial\psi}{\partial\theta}\right) = \frac{1}{\theta}\,\dot{\Omega} - h^{i}_{,i} - \frac{\partial\psi}{\partial q^{\alpha}}\,\dot{q}^{\alpha}$$

We note that it makes no reference to constitutive equations for the material. We shall attempt to follow at this point a fascinating idea of Ilyushin [22] who suggested that the material has no concept of our idea of time, and that processes within a solid should be parametrized with respect to a "material time", which was later defined by Rivlin by the relation $d\tau = (dC^{ij}\, dC^{ij})^{\frac{1}{2}}$ ([44]). Also see [139], [141]. In particular, the internal variables $\underset{\sim}{q}$ should only depend on τ, and preferably in a simple manner. We assume

$$(2.134)\quad \tau = \tau(C^{ij}(t),\ q^{\alpha}(t),\ \psi(t), t),$$

and

$$(2.135)\quad \frac{d\tau}{dt} = \frac{\partial\tau}{\partial C^{ij}}\,\dot{C}^{ij} + \frac{\partial\tau}{\partial q^{\alpha}}\,\dot{q} + \frac{\partial\tau}{\partial\psi}\,\dot{\psi} + \frac{\partial\tau}{\partial t}$$

We hypothesize that τ is differentiable and

$$\frac{d\tau}{dt} > 0,$$

which is a sensible physical assumption, since otherwise the material would react to future physical condition, or could, in a manner of speaking predict the future. The other assumption frequently made here is the existence of a steady state condition. The assumption that $\dot{\underset{\sim}{C}} \equiv 0$ is possible-independently of q^{α}, S, θ or t - or that $\dot{q}^{\alpha} \equiv 0$, etc. implies that each term could be zero, hence that $\frac{d\tau}{dt} = 0$ is possible. We state also that each term of the sum (2.135) is non-negative. This is a very basic and a non-trivial assumption. For the time being we shall try to avoid it, assuming $\frac{d\tau}{dt} > 0$ for all $t \in (-\infty, +\infty)$.

Now since $\frac{d\tau}{dt} > 0$, we can use implicit function theorem and express t as a function of τ:

$$(2.136)\quad t = \Phi(\tau),\ C^{ij}(t) = \hat{C}^{ij}(\tau),\ q^{\alpha}(t) = \hat{q}^{\alpha}(\tau),$$

$$S(t) = \hat{S}(\tau), \text{ etc.}$$

It also follows from $\frac{d\tau}{dt} > 0$ that

$$(2.137)\quad -\frac{\partial\psi}{\partial q^{\alpha}} \cdot \frac{dq^{\alpha}}{d\tau} \geq 0 \qquad \alpha = 1,2,\dots,m.$$

It is a standard argument (see Valanis [141]) that there is a set of constitutive equations of the form

$$\frac{dq^{\alpha}}{d\tau} = f_{\alpha}(C^{ij}, q^{\beta}, \psi),$$

otherwise the inequality (2.137) can be violated.

Following this discussion the following relations could be hypothesized:

$$(2.138)\quad \frac{1}{2}\cdot\sigma_{k\ell} = \frac{\partial\psi}{\partial C^{k\ell}} = \alpha_{ijk\ell}(\tau) * L_1 C^{ij}(\tau)$$

and

$$(2.139)\quad -S = \frac{\partial\psi}{\partial\theta} = \beta_{ij}(\tau) * L_2 C^{ij} + \gamma(\tau) * L_3\,\theta,$$

where $\sigma_{k\ell}$ is one of the internal variables

$(k,\ell = 1,2,3)$ and $*$ is the convolution operation:

$$f * g = \int_{\tau_0}^{\tau} f(\tau-\tau')g(\tau')\,d\tau', \qquad \text{and}$$

L_1, L_2, L_3 are linear differential operators,

which are convolutions (see Mikusiński [115]). These relations reduce to usual assumptions of materials with memory in the material time parametrization, if the forms of the operators L_1, L_2, L_3 are specifically given as in [145], or [103]

For purposes of variational formulation we shall leave the constitutive relations in the form (2.138) and (2.139).

It is apparent that in the "material time"

parametrization (2.138) is Hooke's law if $\alpha_{ijk\ell}(\tau)$ is independent of τ, while $(*L_1)$ is the identity operator, i.e. $*L_1\ (x) = x$.

Of course, the equations (2.138), (2.139) have to be consistent with the first and second laws of thermodynamics.

$$(2.140)\quad -\dot{S} \geq 0 = L_3\theta\ \frac{\partial\gamma}{\partial\tau} + L_2\ C^{ij}\ \frac{\partial\beta_{ij}}{\partial\tau} \geq 0$$

$$(2.141)\quad h_{i,i}(\tau) = \theta L_2 C^{ij} * \frac{d\beta}{dt} + L_3\theta * \frac{d\gamma}{dt} - \frac{\partial\psi}{\partial\sigma_{ij}}\ L_1\ C^{ij} * \frac{d\alpha}{dt} + \frac{dQ}{d\tau}\ ,$$

$$(2.142)\quad -\frac{\partial\psi}{\partial\sigma_{ij}}\quad L_1\ C^{ij} * \frac{d\alpha}{dt} \geq 0;$$

i,j not summed in (2.141), (2.142).

If we put θ = const = θ_0, $Q \equiv$ const., $h_{i,i} \equiv 0$, i.e. ignore the thermal effects, we obtain mechanical laws

$$(2.143)\quad \theta_0 \cdot L_2(C^{ij} * \beta) - \frac{\partial\psi}{\partial\sigma_{ij}} \cdot \bar{L}_1(C^{ij} * \alpha) = 0$$

$$(2.144)\quad \theta_0 \cdot \bar{L}_2(C^{ij} * \beta) \leq 0.$$

We can only hypothesize at this point that the constitutive equations must be of such form that the equations (2.138), (2.139) or (2.141), (2.142) represent a critical point of some functionals Φ_1, Φ_2, and that

$$A^* q^\alpha = f_\alpha(C^{ij}, q^\beta, \psi) = \Phi_{3_{p_\alpha}}\ ,$$

$$A\ p_\alpha = \Phi_\alpha(C^{ij}, q^\beta, \psi) = \Phi_{3_{\dot{q}\alpha}}\ ,$$

$$A^* = -\frac{d}{d\tau}\ ,\quad A^*:\quad H_2 \to H_1\ ,$$

$$A = \frac{d}{d\tau}\ ,\quad A:\quad H_1 \to H_2\ ,$$

where we expect f_α, Φ_α to be convolutions. (2.129) is of the required form

$$\left(\frac{\rho_0}{\rho}\right) \tau_{ij} = \Phi_{4_{C^{ij}}}, \quad \text{if we identify } \Phi_4 = \psi,$$

corresponding to the variational principle

$B(\sigma^{ij}) = \Phi_{4_{C^{ij}}}$. Here B is a linear operator, which is mapping σ^{ij} space into τ_{ij} space, with properties of these spaces still undetermined.

A simplified versions of such equations of state without memory have been also suggested in Russian literature

$\tau_{ij} = \omega_{ij\alpha\beta}\, \varepsilon^{\alpha\beta}$, where $\varepsilon^{\alpha\beta}$ is the linear strain component and

$$\tau^{ij} = \frac{1}{2} \cdot \left(\frac{\partial}{\partial C^{ji}} + \frac{\partial}{\partial C^{ij}}\right) \Phi(\underset{\sim}{C})),$$

leading to variational principle of Hu, corresponding to a critical point of the functional:

$$I_1(\tau,\varepsilon,u) = \int_\Omega [\tfrac{1}{2}\, \omega^{ij\alpha\beta}\, \varepsilon_{\alpha\beta}\varepsilon_{ij} - \tau^{ij}(\varepsilon_{ji} - \nabla_i u_j)]d\underset{\sim}{v}$$

$$- \int_\Omega (p^j u_j\, ds - v \cdot \tau^{ij} u_j) d\underset{\sim}{s}.$$

(see O.M. Guz [100]).

For computational purposes it does not suffice to know that certain equations of solid mechanics represent a critical point of a functional. Iterative techniques which have been eminently successful in such computations always had the additional information that this functional attains either a maximum or a minimum. To make certain that this occurs we need the generalization of the concept of second and higher order derivatives. This can be done in the manner of section 2.27, or by using differential geometry.

It is not clear at this point how to formulate "a universal variational principle" which would incorporate the equations of state, if they exists, and we shall stop conjecturing at this point and return to an area where variational principles are easily established.

A link between thermodynamics of solids and classical mechanics has been suggested in a monograph of Truesdell and Bharatha [148].

Another attractive idea has been advocated by several authors, particularly K. Valanis suggesting that constitutive equations should include internal variables. This parallels some concepts of microelasticity championed by C. Eringen, and seems particularly attractive, since it allows an almost exact duplication of variational arguments of classical mechanics with variables defined pointwise replaced by variables which represent the averages of more complex and generally random processes.

For an outline of internal variable approach to thermodvnamics of solids see for example [141] [96] or [97]. It is impossible to present the outline of this theory within the scope of this monograph. Instead, we shall make a few heuristic observations and deduce a Lagrangian formulation by formal methods.

Other extensions and possible consequences of our arguments shall be left as open questions.

Internal Variables.

In a series of papers Gurtin [145],[96], Lubliner [120], Valanis [139], [140], [141] have considered the averaging effects of the internal structure of a deformed solid by introducing internal variables. Roughly speaking these variables represent statistical averages of some molecular displacements.

It should be clear that a microsystem can not obey laws of classical mechanics in arbitrarily small, or infinitesimal regions. If physical quantities such as internal energy density, entropy, stress, material density are to be physically significant in a medium which is made of molecules, these quantities are either undefined at the molecular level, or at least are badly discontinuous. In the case of polymeric materials K. Valanis identified the internal variables as actual displacements along a typical very large molecule. Hence, in the study of polymers the internal variables q_α could be identified with average displacements, and the internal forces with the energy gradient $\partial \varepsilon / \partial q_\alpha$.

These internal variables are then adjoined to the observable thermodynamic variables.

In the treatment of internal variables we could follow the procedure originally introduced by Duhem and Hadamard by defining the specific internal energy function $\mathcal{E}$ at each point of the continuum. Also see L.A.Segel [119] .

Some fundamental difficulties in the problems of calculus of variations and control theory.

The difficulties of minimization of functionals are well known and provide the most interesting reading in any historical account of the subject. (The easy and beautiful ideas of Euler-Lagrange and Dirichlet led to extremely difficult problems of analysis because of the lack of closure in the classes of solutions considered in 18th and 19th century.) Various counterexamples and paradoxes troubled Lebesgue, de Bois Raymond, Weierstrass and Hilbert. For a historical account see for example Chapter 1 of [12], or Young's account in [146].

One way of overcoming the difficulty of not having a sufficiently "rich" class of (parametrized) curves has been suggested by Young in [146]. It adapts the idea of Laurent Schwartz of constructing a dual space and embedding the original space (of admissible curves, in this instance) into its dual. Consider, in the way of illustrating this idea, the problem of minimizing a functional $\int \ell\,(x,x',t)\,dt$, $t_1 \le t \le t_2$, where x(t) varies along a parametrized arc Γ of finite length. In this context ℓ and Γ are elements of dual spaces. Let the function ℓ belong to some normed space B then $\int_\Gamma \ell\,dt \equiv \langle \ell, \Gamma \rangle$ clearly defines a linear functional in B*. Defining the operation $\oplus$ as union of arcs, we have $\langle \ell, (\Gamma_1 \oplus \Gamma_2)\rangle = \langle \ell, \Gamma_1\rangle + \langle \ell, \Gamma_2\rangle$ and $\langle \ell, c\Gamma\rangle =$

$c\langle \ell, \Gamma\rangle = \langle c\ell, \Gamma\rangle$ for any real number c. We confirm that this functional is indeed bilinear and that it is bounded. Hence Γ is an element of the dual space B*.

Following the ideas of Laurent Schwartz we can define weak convergence in B*. Such weak limits are called generalized arcs. The limitations of this approach are almost identical with those of the theory of distributions.

Concluding remarks.

The specific purpose of this part was to present extensions of Legendre transform to a class of general problems in the dynamics of solids. The author has attempted to link these theories with the Ilyushin - Rivlin - Valanis approach [44] , [52], [51]. This aspect is still in preliminary research stages. See [22],[138-141].

For additional refinements of the now classical dual variational principle postulation, we refer readers to the articles of A.M. Arthurs, such as [5] , or his monograph [4] , and to the monograph of Robinson [45] . This topic will be continued in the next chapter.

Other physical applications have been given in numerous papers of A.M. Arthurs, M.J. Sewell, P. Robinson, and others. See [84], [135], [134], [64], [81], [47], [8] , [80] .

We only mention in closing that there have been discovered applications to quantum mechanics, ideal plasticity, classical theory of electrical networks(not necessarily linear),to economics and to various areas of optimization theory.

Generalizations are suggested in [28], [30] [133], [62], [8], [84]. Relation with the hypercircle method of Synge [82] has been established in papers of Arthurs and Duggan. See [149], [102].

References for Chapter 2

[1] N.J. Akhiezer, The calculus of variations, Blaisdell Publ. Co., New York, 1962.

[2] N.I. Akhiezer and I.M. Glazman, Theory of linear operators in Hilbert Space, vol. I & II, Ungar Publishing Company, New York, 1963, translated from Russian by M. Nestell.

[3] A. Alexiewicz and W. Orlicz, On differentials in Banach Spaces, Ann. Soc. Polon. Math. 25, (1952), p. 95-99.

[4] A.M. Arthurs, Complementary Variational Principles, Clarendon Press, Oxford University, Oxford, 1971.

[5] A.M. Arthurs, A note on Komkov's class of boundary value problems and associated variational principles, J. Math. Analysis & Applications, 33, (1971), p. 402-407.

[6] Jacob Bernoulli, Acta Eruditorum, (see in particular pages 268-269), Leipzig, 1696 .

[7] G.A. Bliss, Lectures on Calculus of Variations, University of Chicago Press, Phoenix Science Series, Chicago, 1946.

[8] W. Collins, Dual Extremum principles and Hilbert space decomposition-in Duality in problems of mechanics of deformable bodies, Ossolineum, Warsaw, 1979.

[9] L. El'sgolc, Differential equations and the calculus of variations, Mir, Moscow, 1973, second English edition.

[10] Leonhardi Euleri Opera Omnia, vol. X and XI, C. Truesdell, editor. Soc. for Nat. Sciences of Switzerland, Basel, Switzerland, 1960.

[11] H. Flanders, Exterior Forms, Academic Press, New York, 1968.

[12] A.R. Forsyth, Calculus of variations, Dover Publications, New York, 1960.

[13] M. Frechét, La notion de differentielle dans l'analyse generale. Ann. Soc. de l'Ecole Norm. Super., 42, (1925), p. 293-323.

[14] M. Frechét, Sur la notion differentielle, Jour. de Math., 16, (1937), p. 233-250.

[15] K.O. Friedrichs, On the boundary value problems in the theory of elasitcity and Korn's inequality, Ann. of Math., 48, (1947), p. 441-471.

[16] R. Gateaux, Sur les functionelles continues et les functionelle analitiques, Comptes Rendues, 157, (1913), p. 325-327.

[17] M.K. Gavurin, Analytic methods for the study of non-linear functional transformations, Uchebnyje Zapiski Leningradskovo Universiteta (Scientific Notes of Leningrad University) Math. Series, 19, (1950), p. 59-154.

[18] I.M. Gel'fand, Abstrakte Functionen und Lineare Operatoren, Matem. Sbornik, 4, #46, (1938), p. 235-283.

[19] I.M. Gel'fand and G. Ye. Shilov, Generalized Functions vol. I & II, Academic Press, New York, 1964.

[20] I.M. Gel'fand and N. Ya. Vilenkin, Generalized Functions, vol. IV, Academic Press, New York, 1964.

[21] L. Hörmander, Linear Partial Differential Operators, Springer Verlag, (Grundlagen der Math. Wissenshaften Series), Berlin, 1963.

[22] A.A. Ilyushin, On the relation between stresses and small deformations in the mechanics of continuous media, Prikl. Math. Mekh. (P.M.M.), 18, (1954), p. 641-647.

[23] L.L. Bucciarelli and Nancy Dworski, Sophie Germain, an essay in the history of elasticity, Reidel Publishing Co., Amsterdam, Holland, 1984.

[24] L.V. Kantorovich and G.P. Akilov, Functional Analysis in Normed Spaces, Pergamon Press, Oxford, 1964.

[25] T. Kato, On some approximate methods concerning the operators T*T, Math. Ann. 126, (1953), p. 253-257.

[26] T. Kato, Perturbation Theory for Linear Operators, Springer Verlag, Berlin and New York, Die Grundlehren der mathematischen Wissenschaften, vol. 132, (1966).

[27] J. Kolomy, On differentiability of mappings in functional spaces, Comment. Math. Univ. Carolinas, 8, (1967), p. 315-329.

[28] V. Komkov, Application of Rall's Theorem to classical elastodynamics, J. Math. Anal. Applic., vol. 14, #3, (1966), p. 511-521.

[29] V. Komkov, A note on the vibration of thin inhomogeneous plates, ZAMM, #408, (1968), p. 11-16.

[30] V. Komkov, On variational formulation of problems in the classical continuum mechanics of solids, International J. of Eng. Science, vol. 6, (1968), p. 695-720.

[31] H.F. Tiersten, Natural boundary and initial conditions from a modification of Hamiltons principle, J. Math. Phys, 1968, 9, p. 1445-1451.

[32] R.D. Mindlin, Waves and vibrations in isotropic elastic plates, in Structural Mechanics, New York, N.Y., 1960, p.209

[33] R.D. Mindlin, High frequency vibrations of piezo-elastic crystal plates, International Journal of Solids and Structures, vol. 10, #4, (1974), p. 343-359.

[34] V. Komkov, Optimal control theory for the damping of vibrations of simple elastic systems, Mir, Moscow, 1976. A corrected and modified edition of Springer Verlag Lecture Notes of 1972.

[35] V. Komkov, The class of admissible loads of the linear plate and beam theory, Inst. Lomb. di Science e Lettere, Acad. di Science e Lett., Classe A, 105, (1971), p. 329-335.

[36] S.G. Mikhlin, Mathematical Physics - an advanced course, North Holland Publishing Co., Amsterdam, (1970), (Translated from Russian).

[37] J.V. Neumann, Über adjungierte functional Operatoren, Ann. Math., 33, (1932), p. 294-310.

[38] B. Noble, Complementary variational principles for boundary value problems, I, Basic Principles, Report #473, Math. Res. Center, University of Wisconsin, Madison, Wisc. (1964).

[39] B. Noble, Complementary variational principles II, Non-linear networks, Report #643, Math. Res. Center, Univ. of Wisconsin, Madison, Wisc., (1966).

[40] M.Z. Nashed, The Role of Differentials, published in Non-linear Fuctional Analysis and Applications. Editor L.B. Rall - Proceedings of an advanced seminar @ Math. Res. Center, Madison, Wisc., Academic Press, New York, (1971).

[41] J.T. Oden and J.N. Reddy, Variational methods in theoretical mechanics, Springer Verlag, Berlin (1976).

[42] L.B. Rall, Variational Methods for Non-linear integral equations, Symposium on Non-linear Integral Equations, P.M. Anselone,editor, University of Wisconsin Press, Madison, Wisc., (1964).

[43] L.B. Rall, On complementary variational principles, J. Math. Anal. Applic., 14, (1966), p. 174-182.

[44] R.S. Rivlin, Non-linear viscoelastic solids, SIAM Review, 7, (1965), p. 323.

[45] P.D. Robinson, Complementary Variational Principles, in Non-linear Functional Analysis and Applications. L.B. Rall editor - proceedings of an advanced seminar

at M.R.C. Univ. of Wisconsin, Academic Press, (1971). See Nashed's article, reference[40] in the same issue.

[46] E.H. Rothe, Gradient Mappings, Bull. Amer. Math. Soc., 59, (1953), p.5-19.

[47] M.J. Sewell, On dual approximation principles and optimization in continuum mechanics, Phil. Trans. Royal Society, Ser. A265 (1969), p. 319-351.

[48] S.L. Sobolëv, Applications of Functional Analysis in Mathematical Physics, Amer. Math. Soc. Translations, vol. 7, Rhode Island, (1963). (Particularly pages 33-45).

[49] C. Truesdell and R. Toupin, Handbuch der Physik, III, Springer Verlag. Berlin, (1960). (The classical field theory)

[50] C. Truesdell and W. Noll. The non-linear field theories of mechanics, Handbuch der Physik, 3, Springer Verlag, Berlin, (1965).

[51] K.C. Valanis, Thermodynamics of Continuous Media, (notes) University of Iowa (1973), Springer Verlag (1974).

[52] K.C. Valanis, A unified theory of thermo-mechanical behavior of viscoelastic materials, Proceedings - Mechanical Behavior of Materials under Dynamic Loads, V.S. Lindholm editor, Springer Verlag, New York, (1968).

[53] M.M. Vainberg, Variational Method for the Investigation of Non-linear Operators, Holden Day, San Francisco (1963), (Translated from Russian).

[54] M.M. Vainberg and R.R. Kačurovski, On the variational theory of non-linear operators and equations. Dokl. Akad. Nauk S.S.S.R., 129, (1959), p. 1199-1202.

[55] M.M.Vainberg, On differentials and gradients of mappings, Uspekhi Matem. Nauk, 7, #49, (1952), p. 139-143.

[56] M.M. Vainberg, Some questions of differential calculus in linear spaces, Uspekhi Matem. Nauk, 7, #50, (1952), p. 55-102.

[57] A.N. Vasil'ev and A.K.Kazanskiĭ, Higher order Legendre transformations, Akad. Nauk S.S.S.R., J. Theoret. and Math. Physics, vol. 14, #3, March 1973, p. 289-304.

[58] Fraeijs de Vreubeke, Theory of elasticity, Springer, Berlin, 1979.

[59] K.Washizu, A note on the conditions of compatibility, J. Math. Phys. 36, (1958), p. 306-312.

[60] M. Zorn, Derivatives and Fréchet differentials, Bull. Amer. Math. Soc., 52, (1946), p. 133-137.

[61] I. Kuti, On the proximal variational principles in linear elastodynamics, Acta Technica Acad. Scient. Hungaricae, 93, p. 101-114, (1981).

[62] I. Herrera and J. Bielak, A simplified version of Gurtin's variational principles, Arch. Rat. Mech. Anal. 53, p. 131-149, (1974).

[63] E. Tonti, On the variational formulation for linear initial value problems, Annali di Matematica pura ed applicata 95, p. 331-360, (1973),

[64] M.J. Sewell, The governing equations and extremum principles of elasticity and plasticity generated from a single functional. Part I, J. Struct. Mech. 2, p. 1-32, (1973).

[65] C. Lanczos, Variational Principles of Mechanics, Univ. of Toronto Press, Toronto, 1949.

[66] L.S. Polyak, Variational Principles of Mechanics (in Russian), Fizmatgiz, Moscow, 1960.

[67] L.D. Landau and E.M. Lifshitz,Mechanics (volume I in the series: A Course in Theoretical Physics, Pergamon Press, New York, 1960.

[68] L.D. Landau and E.M. Lifshitz, ibid., volume 6, 1962.

[69] G. Duvaut and J.L. Lions, Les inequations en mechanique et en physique, Dunod, Paris, 1972.

[70] S. MacLane, Hamiltonian Męchanics and Geometry, Amer. Math. Monthly, vol. 77, (1970), p. 570-586.

[71] L.S. Pontryagin, V.G. Boltyanski, R.V. Gamkrelidze and E.F. Mishchenko, The Mathematical Theory of Optimal Processes, Interscience Publishing Co., New York, 1962.

[72] C. Caratheodory, Calculus of Variations and Partial Differential Equations of the First Order, Holden Day Inc. San Francisco, 1965.

[73] A. Korn, Über einige Ungleichungen, welche in der Theorie der elastischen und elektrischen Schwingungen eine Rolle spiellen. Bull. Internat. Cracovie Akad. Umiedet., Classe des sciences mathematiques et naturelles, 1909, S. 705-724.

[74] H.L. Langhaar, Energy methods in applied mechanics, New York: Wiley, 1962.

[75] S. Nemat-Nasser,ed., Variational methods in the mechanics of solids, Pergamon Press, 1980.

[76] V. Komkov, Optimal Control theory for damping of vibrations of simple elastic systems, Springer Verlag, Lecture Notes in Mathematics, #256, Berlin, 1973.

[77] V. Komkov, Optimal controls of vibrating thin plates, SIAM J. of Control, vol. 8, #2, 1970, pp. 273-304.

[78] V. Komkov, Formulation of Pontryagin's maximality principle in a problem of structural mechanics, International Journal of Control, 1972, vol. 17, #3, pp. 455-563.

[79] R. van Doren, Derivation of the Lagrange equations for a non-holonomic Chetaev system from a modified Pontryagin's maximum principle, ZAMP, vol. 28, 1977, p. 729-734.

[80] M.J. Sewell, On dual approximation principles and optimization in continuum mechanics, Philos. Trans. Roy. Soc. of London, Ser. A, 1969, vol. 265, p. 319-351.

[81] M.J. Sewell, On reciprocal variational principle for perfect fluids, J. Math. Mech., 1963, vol. 12, p. 495-504.

[82] J.L. Synge, The hypercircle in mathematical physics, Cambridge Univ. Press, 1957.

[83] Avner Friedman, Variational Principles and Free Boundary Problems, Wiley-Interscience, New York, 1982.

[84] A.M.Arthurs, Complementary variational principles and error bounds for biharmonic boundary value problem, Nuovo Cimento, B, II, 17, 1973, p. 105-112.

[85] E.R. Barnes, Necessary and sufficient optimality conditions for a class of distributed parameter control systems, SIAM J. Control, Vol. 9, No. 1, 1971, pp.62-82.

[86] R. Becker, Introduction to theoretical mechanics, McGraw-Hill, New York, 1954.

[87] E.J. Beltrami, An algorithmic approach to non-linear analysis and optimization, Academic Press, New York, 1970.

[88] E.K. Blum, The calculus of variations, functional analysis, and optimal control problems, Topics in Optimization, G. Leitman, ed., Vol. 31, Mathmematics in Science and Engineering Series, Academic Press, New York, 1967, pp. 415-461.

[89] Y.C. Bruhat, Problems and solutions in mathematical physics, Holden Day, San Framcisco, 1967.

[90] A.G. Butkovskii and A.Y. Lerner, "Optimal controls with distributed parameter," Dokl. Akad. Nauk S.S.S.R., No. 134, 1960, pp. 778-781.

[91] R.P. Gaida, Ukrainian Crystallographic Journal, (UDK) 530, 12 (1976) p. 23-28. Also see: R.P. Gaida, Acta Physica Polonica, B5, #613, (1974).

[92] A.G. Butkovskii, "The maximum principle for optimum systems with distributed parameters," Avtomatika, No. 10, Vol. 22, 1961, pp. 1288-1301.

[93] A.G. Butkovskii and L.N. Poltavskii, "Optimal control of a distributed oscillatory system," English translation -- Automation and Remote control, 26, 1965, pp. 1835-1848.

[94] A.G. Butkovskii, A.I. Egorov, and A.K. Lurie, "Optimal control of distributed systems," SIAM J. on Control, Vol. 6, No. 3, 1968, pp. 437-476.

[95] L. Cesari, "Existence theorems for weak and usual optimal solutions in Lagrange problems with unilateral constraints," Parts I and II, Trans. Am. Math. Soc. No. 124, 1966, pp. 369-412, 413-430.

[96] B.D. Coleman and M. Gurtin, Thermodynamics of internal variables, J. Chem. Phys., 47, 597, 1967.

[97] B.D. Coleman, J. Markovitz, and W. Noll, Viscometric flows of non-Newtonian fluids, Springer Verlag, New York, 1966.

[98] M.K. Gavurin, On the construction of differential and integral calculus in Banach spaces, Dikl. Akad. Nauk S.S.S.R. (U.S.S.R.), 22, No. 9, 1939, pp. 547-551.

[99] H. Goldstein, Classical Mechanics, Addison Wesley, Reading, Massachusetts, 1950.

[100] O.M. Guz, On variational principles of Washizu and Hu, Dokl. Ukrain. Akad, Nauk, 539, 3, 1972, pp. 1013-1015.

[101] E.J. Haug, Engineering design Handbook, Headquarters U.S. Army Material Command, AMCP-706-192 Publication, July 1973.

[102] R.C. Duggan, On dual extremal principles and hypercircle for a class of partial differential equations, Jour. Inst. Math. Applic. 26, (1980), p. 259-267.

[103] I. Herrera and J. Bielak, Dual variational principles for diffusion equations, Quart. Appl. Math., 85, 1976.

[104] N.A. Ivanov, On Gateaux and Fréchet derivatives, Uspekhi Matem. Nauk, 10, Vyp. 2, No. 64, 1955, pp. 161-166.

[105] C. Truesdell, Absolute temperatures as a consequence of Carnot's General Axiom, Archive for History of Exact Sciences 20, 1979, 357-380.

[106] V. Komkov, "A classification of boundary conditions in optimal control theory of elastic systems. Proceedings of the IFAC Proceeding, Banff, Canada, 1972.

[107] I.A. Vekovishcheva, Two-dimensional problem of electroelasticity for piezoelastic plates, Prikl. Mekhanika, Vol. 11, #2, (1975), p. 85-89.

[108] V. Komkov, "Another look at dual variational principles". J. Math. Analysis and Appl., Vol. 63, #2, (1978), p. 319-323.

[109] E. Cartan, Leçons sur les Invariants Intégraux, (English and Russian translations are available) Hermann et Cie., Paris, 1934.

[110] A.E.H. Love, A treatise on the mathematical theory of elasticity, fourth ed., Dover Publ., New York, 1944.

[111] J. Lubliner, On the structure of rate equations for materials with internal variables, Acta Mechanica, 17, 109, 1973.

[112] E.H. Mansfield, "The bending and stretching of plates," McMillan Co., New York, 1964.

[113] S.G. Mikhlin, Mathematical physics -- An advanced course, North Holland Publishing Co., Amsterdam, Translated from Russian, 1970.

[114] S.G. Mikhlin, Linear equations of mathematical physics, Holt, Rinehart, and Winston, New York, 1967.

[115] J. Mikusiński, The Theory of operators, Polska, Akad. Nauk, Monografje Matematyczne, Warsaw, 1951.

[116] L.S.D. Morley, Skew plates and structures, McMillan Co., New York, 1963.

[117] P.M. Morse and H. Feshback, Methods of theoretical physics, McGraw-Hill, New York, 1953, Vol. 1 and 2.

[118] N.I. Muskhelishvili, Some basic problems of the mathematical theory of elasticity, Noordhoff, Groningen, Holland, 1953.

[119] L.A. Segel, An Introduction to continuum theory, Lectures in Appl. Math, Vol. 16, 1977, American Mathematics Society.

[120] J. Lubliner, On The Structure of rate equations for materials with internal variables, Acta Mechanica, 17, 109, 1973.

[121] N. Anderson and A.M. Arthurs, Variational principles for Maxwell equations. Int. J. Electr. 51, (1981), p. 71-77.

[122] R.S. Sandhu and K.S. Pister, Variational principles for boundary value and initial value problems in continuum mechanics, Internat. J. Solids & Structures, 7 (1971), 639-654.

[123] P.D. Robinson, Complementary variational principles, in "Nonlinear Functional Analysis and Applications" (L.B. Rall, Ed.), pp. 507-576, Academic Press, New York, 1971.

[124] J. Mikusiński, Lectures on the constructive theory of distribution, Univ. of Fla, 1969.

[125] L.S.D. Morley, Skew plates and structures, McMillan Co, New York, 1963.

[126] W. Flügge, Stresses in Shells, Springer Verlag, Berlin, 1960.

[127] R.C. Buck, Advanced Calculus, McGraw Hill, New York, 1961.

[128] M.Z. Nashed, "Differentiability and related properties of non-linear operators: Some aspects of the role of differentials in non-linear functional analysis", in L.B. Rall, ed., Non-linear functional analysis applications, Academic Press, New York, 1971, pp. 103-309.

[129] B. Noble and M.J. Sewell, On dual extremum principles in applied mathematics, J. Inst. Maths. Applics., 9, 1972, pp. 123-193.

[130] W. Prager, Introduction to mechanics of continua, Ginn and Co., Boston, 1961.

[131] L.B. Rall, Variational methods for non-linear integral equations, symposium of non-linear integral equations, P.M. Anselone, ed., Univ. of Wisconsin Press, Madison, Wisconsin, 1964.

[132] E. Reissner, "On bending of elastic plates, Quart. Appl. Math. No. 5, 1947, pp. 395-401.

[133] R.S. Sandhu and K.S. Pister, Variational methods in Continuous mechanics, (in international conference on variational methods in engineering), Southhampton Univ. England, 1972, pp. 1.13-1.25.

[134] M.J. Sewell, The governing equations and extremum principles of elasticity and plasticity generated from a single functional, part I, J. Struct. Mech., 2, 1973, pp. 1-32.

[135] M.J. Sewell, The governing equations and extremum principles of elasticity and plasticity generated from a single functional, part II, J. Struct. Mech., 2, 1973, pp. 135-158.

[136] I.S. Sokolnikoff, Mathematical theory of elasticity, McGraw-Hill, New York, 1956.

[137] S.P. Timoshenko, History of strength of materials, McGraw-Hill, New York, 1954.

[138] K.C. Valanis, Irreversible thermodynamics with internal inertia, Principle of stationary total dissipation, Arch. of Mechanics, 24, No. 5, 948, 1972.

[139] K.C. Valanis, Thermodynamics of internal variables in the presence of internal forces, report G123-DME-76-006, Division of materials engineering, Univ. of Iowa.

[140] K.C. Valanis, Irreversability and the existence of entropy, J. Non-linear Mech., 6, 338, 1971.

[141] K.C. Valanis, Twelve lectures in thermodynamics of continuous media, International institute of mechanical sciences, Udine, Italy, 1971.

[142] M.M. Vaînberg, Variational methods for the investigation of non-linear operators, Holden Day, San Francisco, Translated from Russian, 1963.

[143] D. Greenspan, Numerical solutions of non-linear equations, Holden Day, San Francisco, 1966. (See report no. 466, Mathematics Research Center (MRC), Univ. of Wisconsin, Madison, Wisconsin, 1964).

[144] J.J. Moreau, Fonctions convexes duales et points prominaux dans un espace Hilbertien, C.R. Acad. Sci., Paris, vol. 255, (1962), p. 2897-2899.

[145] M.E. Gurtin, Variational principles for linear initial value problems, Quart. Appl. Math. 22, 1964, pp. 252-256.

[146] L.C. Young, The calculus of variations and optimal control theory, Saunders, New York, 1969.

[147] C. Truesdell, Improved estimates of the efficiencies of irreversible heat engines, Annali di Mathematica Pura ed Applicata (4) 108, 1976, pp. 305-323.

[148] C. Truesdell and S. Bharatha, The concepts and logic of classical thermodynamics as a theory of heat engines, rigorously constructed upon the foundation laid by S. Carnot and F. Reech, Springer Verlag, Berlin, New York, 1977.

[149] A.M. Arthurs, On variational principles and the hypercircle for boundary value problems, Proc. Roy. Irish Acad., Vol. 77, Section A, (1977), p. 75-83.

[150] W. Yourgrau and S. Mandelstam, Variational principles in dynamics and quantum theory, Pitman and Sons, London, 1956. (Reprinted in 1968, Philadelphia).

[151] O.C. Zienkiewicz, The finite elements method, Third edition, revised, McGraw-Hill, New York, 1977.

[152] A Kneser, Lehrbauch der Variationsrechnung, Vieweg Verlag, Brunswick, 1925.

[153] H. Rund, The Hamilton-Jacobi theory in the calculus of variations, Van Nostrand-Reinhold Publishing Co., Princeton, N.J., 1966.

[154] R. Baumeister, Generalized Hamilton-Jacobi theories, J. Math. Physics, Vol. 19, (1978), p. 2377-2387.

Related references on finite element approach to variational principles in continuum mechanics

[155] J.H. Argyris and S. Kelsey, Energy Theorems and Structural Analysis. Aircraft Eng.,Oct. 54 - May 1955 (Reprinted by Butterworth, 1960).

[156] G. Strang and G.J. Fix, An Analysis of the Finite Element Method, Prentice Hall, 1973.

[157] E.R. Arantes e Oliveira, Completeness and Convergence in the Finite Element Method. Proc. 2nd Conf. Math. Meth. Struct. Mech. Wright-Patterson AFB, Dayton, Ohio, 1968.

[158] P.A. Raviart and J.M. Thomas, Cours d'Analyse Numérique-Eléments Finis. Université de Paris VI, 1973.

[159] J.T.Oden, Finite Element of non-linear Continua., McGraw Hill, N.Y. 1972.

[160] B. Fraeijs de Veubeke, Sur certaines inégalités fondamentales et leur généralisation dans la théorie des bornes supérieures et inférieures en élasticité. Revue Universelle des Mines, XVII, No. 5, 1961.

[161] R. Clough and E.L. Wilson, Dynamic Finite Element Analysis of Arbitrary Thin Shells. IUTAM Symposium on High Speed Computation of Elastic Structures. Fraeijs de Veubeke edition, Lèige, Belgium, 1970.

[162] R.H. Gallagher, Finite Element Representations for Thin Shell Instability Analysis. IUTAM Symposium on Buckling of Structures. Harvard University Press, Mass., June 1974.

[163] J. Robinson, Integrated Theory of Finite Element Methods, J. Wiley and Sons, London 1973.

Chapter 3

SOME KNOWN VARIATIONAL PRINCIPLES IN ELASTICITY

Introduction to variational formulation of the laws of mechanics for elastic systems

3.0. Some comments

In the 1970-s and 80-s a significant change of style took place in variational treatment of continuum mechanics.

A heuristic treatment of partial differential equations describing the behavior of a continuum, expansions in infinite series, largely heuristic perturbation techniques was replaced by rigorous arguments making a full use of recent developments in functional analysis, operator theory and to a somewhat lesser extent in modern algebra and differential geometry.

The names of Vaĭnberg, Lions, Stampacchia, Oden and Reddy, Lions and Magenes, Duvaut and Lions are associated with the development of functional analytic tehcniques that were found useful in continuum mechanics. (See for example [30], [31], [32], [33], [34], [29].)

3.1. Review of basic equations and known variational principles in the general theory of elasticity.

Let us list some existing principles in linear elasticity.

"Classical" linear elasticity is generally introduced by listing equations (3.1), (3.2), (3.3) stated below. The nine components of strain are given in Cartesian coordinates by the linear formula

(3.1) $\varepsilon_{ij} = \frac{1}{2}(u_{i,j} + u_{j,i}), \ i,j = 1,2,3.$

This may be written symbolically as

$A\, \underset{\sim}{u} = \underset{\sim}{\varepsilon}$.

The Hooke's law relates strain and stress components

(3.2) $C^{kjk\ell} \varepsilon_{ij} = \bar{\sigma}^{k\ell}$ (summation convention is employed),

or $\underset{\approx}{C} \cdot \underset{\sim}{\varepsilon} = \underset{\sim}{\sigma}$.

The linear equilibrium equations are

(3.3) $\sigma^{ij}{}_{,i} = f^j$,

where f^j, j = 1,2,3 are components of the body forces ,

or (3.3^a) $T \underset{\sim}{\sigma} = \underset{\sim}{f}$.

3.1^a Variational principles

1. The minimum potential energy principle.

The potential energy is given by a quadratic approximation formula

(3.4) $U(\underset{\sim}{u}) = \int_{\Omega} (\tfrac{1}{2} u_{i,j} \cdot C^{ijk\ell} u_{k,\ell} - f^i u_i) \, d\underset{\sim}{x}$

$- \int_{\partial\Omega} \tilde{u}_i \cdot \tilde{T}^i \, ds$,

where $\tilde{u}_i$, $\tilde{T}^i$ are, respectively, displacements and tractions assigned on parts of the boundary $\partial\Omega$.

The equations (3.1) and (3.2) may be regarded as constraints assigned to the variational problem of extremizing $U(\underset{\sim}{u})$. The equilibrium equations (3.3) and the remaining matching conditions on $\partial\Omega$ can be derived as the Euler-Lagrange conditions for extremizing the potential energy $U(\underset{\sim}{u})$ given by (3.4).

If $f^i \equiv 0$ and natural conditions are assigned on the boundary $\partial\Omega$, the potential energy is equal to the strain energy.

Thus, we have an equivalent principle of minimum strain energy.

3.1^b The complementary energy variational principle.

(3.5) $J(\underset{\sim}{\sigma}) = -\tfrac{1}{2} \int_{\Omega} (C^{ijk\ell})^{-1} \underset{\sim}{\sigma}^{ij} \underset{\sim}{\sigma}^{k\ell} \, d\underset{\sim}{x}$

$+ \int_{\partial\Omega_1} n_i \sigma^{ij} \tilde{u}_j \, ds$, where n is a unit vector normal to the boundary $\partial\Omega$.

The constraints are the equations (3.2), (3.3) and boundary conditions

$$n_i \cdot \underset{\sim}{\sigma}^{ij} = \tilde{T}^j \quad \text{on } \partial\Omega_1 \subset \partial\Omega.$$

The Euler equations are:

$$\tfrac{1}{2}(u_{i,j} + u_{j,i}) = (C^{ijk\ell})^{-1} \sigma^{ij}, \text{ in the region } \Omega,$$

and include the matching of displacements

$$\lim_{\underset{\sim}{x} \to \partial\Omega} u_i = \tilde{u}_i \text{ on } \partial\Omega_2, \text{ with } \partial\Omega = \partial\Omega_1 \cup \partial\Omega_2.$$

<u>3.1^c</u> <u>A "simple" constitutive variational principle</u>.

The functional

$$(3.6)\quad L(\underset{\sim}{\varepsilon}) = \int_\Omega \tfrac{1}{2}[\varepsilon_{ij} \cdot C^{ijk\ell} \varepsilon_{k\ell} - \varepsilon_{ij} \mu^{ij}] \, d\underset{\sim}{x}$$

assumes an extremal value subject to constraints

$$\left.\begin{aligned} \tfrac{1}{2}(u_{i,j} + u_{j,i}) &= \varepsilon_{ij} \\ \mu^{ij}{}_{,j} + f^i &= 0 \end{aligned}\right\} \text{ in } \Omega$$

$$\sigma^{ij} - \mu^{ij} = 0 \quad \text{on } \partial\Omega,$$

$$\left.\begin{aligned} u_i &= \tilde{u}_i \text{ on } \partial\Omega_1, \\ n_j \cdot \sigma^{ij} &= \tilde{T}^i \text{ on } \partial\Omega_2. \end{aligned}\right\}$$

This results in Euler's equation

$$\sigma^{ij} = C^{ijk\ell} \varepsilon_{ij}.$$

<u>3.1^d</u> <u>The Reissner-Hellinger principles</u>.

The extremum of

$$(3.7)\quad L_\sigma(\underset{\sim}{u}, \underset{\sim}{\sigma}) = \int_\Omega \{\tfrac{1}{2}(u_{i,j} + u_{j,i}) \sigma^{ij} - \tfrac{1}{2}(C^{ijk\ell})^{-1} \sigma^{ij} \sigma^{k\ell} - u_i f^i\} \, d\underset{\sim}{x}$$

$$- \int_{\partial\Omega_1} n_i \cdot \sigma^{ij} (u_i - \tilde{u}_i)\, ds - \int_{\partial\Omega_2} (u_i \cdot \tilde{T}^i)\, ds$$

with the constraint condition

$\varepsilon_{ij} = (C^{ijk\ell})^{-1}\, \sigma^{k\ell}$ (in Ω) results in Euler-Lagrange equations

$$\left.\begin{aligned} &\tfrac{1}{2}(u_{i,j} + u_{j,i}) = (C^{ijk\ell})^{-1} \cdot \sigma^{k\ell} \\ &\sigma^{ij}{}_{,j} + f^i = 0, \end{aligned}\right\} \text{ in } \Omega$$

with

$u_i = \tilde{u}_i$ on $\partial\Omega_1$,

$n_i \cdot \sigma^{ij} = \tilde{T}^i$ on $\partial\Omega_2$.

The extremum of

(3.8) $$L(\underset{\sim}{u}, \underset{\sim}{\varepsilon}) = \int_\Omega \{\tfrac{1}{2}(u_{i,j} + u_{j,i}) C^{ijk\ell} \varepsilon_{k\ell} - \tfrac{1}{2} \varepsilon_{ij} C^{ijk\ell} \varepsilon_{k\ell} - f^i u_i\}\, d\underset{\sim}{x}$$

$$- \int_{\partial\Omega_1} n_i \cdot \sigma^{ij} (u_j - \tilde{u}_j)\, ds - \int_{\partial\Omega_2} u_i \tilde{T}^i\, ds,$$

with constraint

$$\sigma^{ij} = C^{ijk\ell} \varepsilon_{k\ell}$$

results in Euler-Lagrange equations

$$\left.\begin{aligned} &\tfrac{1}{2}(u_{i,j} + u_{j,i}) = \varepsilon_{ij} \\ &f^i + C^{ijk\ell} \varepsilon_{k\ell,j} = 0 \end{aligned}\right\} \text{ in } \Omega$$

with

$u_i = \tilde{u}_i$ on $\partial\Omega_1$

$n_j \cdot C^{ijk\ell} \varepsilon_{k\ell} = \tilde{T}^i$ on $\partial\Omega_2$.

3.1^e The Hu-Washizu variational principle

This is one of the most general principles of linear elasticity.

The functional that attains an extremum is

$$(3.9)\quad L(\underset{\sim}{u}, \underset{\sim}{\varepsilon}, \underset{\sim}{\sigma}) = \int_\Omega \{\tfrac{1}{2}\, \varepsilon_{ij}\, C^{ijk\ell}\, \varepsilon_{k\ell} + [\tfrac{1}{2}(u_{i,j} + u_{j,i}) - \varepsilon_{ij}]\, \sigma^{ij} - f^i u_i\}\, d\underset{\sim}{x}$$

$$- \{\int_{\partial\Omega_1} u_i \cdot \sigma^{ij}(u_i - \tilde{u}_i)\, ds$$

$$+ \int_{\partial\Omega_2} (u_i\, \tilde{T}^i)\, ds\}\ ,$$

where $\tilde{u}_i$ are displacements assigned on $\partial\Omega_1$ part of the boundary and $\tilde{T}^i$ are the tractions assigned on the part $\partial\Omega_2$ of the boundary. No constraints are assigned.

The Euler-Lagrange equations are

$$\left.\begin{aligned} \varepsilon_{ij} &= \tfrac{1}{2}(u_{i,j} + u_{j,i}) \\ \sigma^{ij} &= C^{ijk\ell}\, \varepsilon_{k\ell} \\ \sigma^{ij}{}_{,j} + f^i &= 0 \end{aligned}\right\} \text{in } \Omega\ ,$$

$$u_i = \tilde{u}_i \qquad \text{on } \partial\Omega_1$$

$$T^i = \tilde{T}^i \qquad \text{on } \partial\Omega_2$$

3.2 A simplified notation for classical elasticity.

Since the stress and strain tensors are assumed to be symmetric in the elementary theory in which dislocations and point sources are not considered, we could replace the nine component stress tensor $\underset{\sim}{\sigma} = \sigma_{ij}$, $i,j = 1,2,3$, and the strain tensor $\underset{\sim}{\varepsilon}$ by corresponding six component vectors

$$\underset{\sim}{\sigma} = \{\sigma_{11}, \sigma_{22}, \sigma_{33}, \sigma_{12}, \sigma_{23}, \sigma_{13}\}$$

$$\equiv\{\sigma_{xx}\ \sigma_{yy}\ \sigma_{zz}\ \sigma_{xy}\ \sigma_{yz}\ \sigma_{xz}\}$$

in a Cartesian x,y,z coordinate system.

Similarly, $\underset{\sim}{\varepsilon}$ may denote the six component vector

$$\underset{\sim}{\varepsilon} = \{\varepsilon_{11}\ \varepsilon_{22}\ \varepsilon_{33}\ \varepsilon_{12}\ \varepsilon_{23}\ \varepsilon_{13}\}.$$

(we have assumed that $\sigma_{ij} = \sigma_{ji}$, $\varepsilon_{ij} = \varepsilon_{ji}$.)

$\underset{\sim}{u} = \{u_1, u_2, u_3\}$ denotes the displacement vector in a fixed coordinate system.

In an isotropic elastic medium Hooke's law takes the form

(3.10) $\underset{\sim}{\sigma} = \underset{\approx}{C} \cdot \underset{\sim}{\varepsilon}$

where $\underset{\approx}{C}$ is the Lammé matrix

$$(3.10^a)\quad \underset{\approx}{C} = \begin{bmatrix} (\lambda+2\mu) & \lambda & \lambda & 0 & 0 & 0 \\ \lambda & (\lambda+2\mu) & \lambda & 0 & 0 & 0 \\ \lambda & \lambda & (\lambda+2\mu) & 0 & 0 & 0 \\ 0 & 0 & 0 & \mu & 0 & 0 \\ 0 & 0 & 0 & 0 & \mu & 0 \\ 0 & 0 & 0 & 0 & 0 & \mu \end{bmatrix},$$

λ, μ are Lamme´ coefficients which can be expressed in terms of the Young's modulus and Poisson's ratio.

The operators A,A*,B, are given in the matrix forms convenient for the 6-component vector representation of $\underset{\sim}{\varepsilon}$ and $\underset{\sim}{\sigma}$, as follows:

$$(3.11)\quad A = \begin{bmatrix} \partial/\partial x & 0 & 0 \\ 0 & \partial/\partial y & 0 \\ 0 & 0 & \partial/\partial z \\ \partial/\partial y & \partial/\partial x & 0 \\ 0 & \partial/\partial z & \partial/\partial y \\ \partial/\partial z & 0 & \partial/\partial x \end{bmatrix}$$

$$(3.11^{a})\quad A^* = \begin{bmatrix} \partial/\partial x & 0 & 0 & \partial/\partial y & 0 & \partial/\partial z \\ 0 & \partial/\partial y & 0 & \partial/\partial x & \partial/\partial z & 0 \\ 0 & 0 & \partial/\partial z & 0 & \partial/\partial y & \partial/\partial x \end{bmatrix}$$

(3.12) B =

$$\begin{bmatrix} \partial^2/\partial y^2 & \partial^2/\partial x^2 & 0 & -\partial^2/\partial x\partial y & 0 & 0 \\ 0 & \partial^2/\partial z^2 & \partial^2/\partial y^2 & 0 & -\partial^2/\partial y\partial t & 0 \\ \partial^2/\partial t^2 & 0 & \partial^2/\partial x^2 & 0 & 0 & -\partial^2/\partial x\partial t \\ \partial^2/\partial y\partial z & 0 & 0 & -\frac{1}{2}\frac{\partial^2}{\partial x\partial z} & -\frac{1}{2}\frac{\partial^2}{\partial x^2} & -\frac{1}{2}\frac{\partial^2}{\partial x\partial y} \\ 0 & \partial^2/\partial x\partial z & 0 & -\frac{1}{2}\frac{\partial^2}{\partial y\partial z} & -\frac{1}{2}\frac{\partial^2}{\partial x\partial y} & -\frac{1}{2}\frac{\partial^2}{\partial y^2} \\ 0 & 0 & \partial^2/\partial x\partial y & \frac{1}{2}\frac{\partial^2}{\partial z^2} & -\frac{1}{2}\frac{\partial^2}{\partial x\partial z} & -\frac{1}{2}\frac{\partial^2}{\partial y\partial z} \end{bmatrix}$$

B* is the transpose of B.
In this notation the basic equations of linear elasticity take the same form as (3.1), (3.2), (3.3) :

$$(3.13)\quad \begin{cases} A\underset{\sim}{u} = \underset{\sim}{\varepsilon} & \\ A^*\underset{\sim}{\sigma} + \underset{\sim}{f} = 0 & \text{in a region } \Omega \subset \mathbb{R}^3 \\ \underset{\sim}{\sigma} = \underset{\approx}{C}\,\underset{\sim}{\varepsilon} = \underset{\approx}{C}\,A\underset{\sim}{u} & \end{cases}$$

While equations (3.13) are identical with (3.1)-(3.3) in appearance they are in fact basically different. We have satisfied all formal relations simplifying all equations to a vector calculus form in a six-dimensional space, while sacrificing some important transformation laws related to tensor properties of the stresses and strains.

These transformation laws are of great importance in the study of several nonlinear phenomena, or of large deflections, and also in the theory of shells and in the study of elastic stability.

The form (3.13) generally suffices in "classical" linear elasticity restricted to Cartesian coordinate representation.

We also introduce boundary operator B_n:

$$(3.14)\quad B_n = \begin{bmatrix} \cos(n,x) & 0 & 0 \\ 0 & \cos(n,y) & 0 \\ 0 & 0 & \cos(n,z) \\ \cos(n,y) & \cos(n,x) & 0 \\ 0 & \cos(n,z) & \cos(n,y) \\ \cos(n,z) & 0 & \cos(n,x) \end{bmatrix},$$

where

$$(3.15^a)\quad B_n \sigma = \underset{\sim}{\gamma}_n \text{ on } \partial\Omega_2$$

$$(3.15^b)\quad \underset{\sim}{u} = \underset{\sim}{\tilde{u}} \quad \text{on } \partial\Omega_1,$$

are boundary conditions that assign pressure γ_n normal to the boundary on the $\partial\Omega_2$ part of the bounday, while $\tilde{u}$ are the deflections assigned to the part $\partial\Omega_1$ of the boundary.

3.3 Thermoelastic principles.

$T(x_1, x_2, x_3, t)$ will denote temperature distribution, $j(x_1, x_2, x_3, t)$ - the heat flux, ρ-density, c_v - the specific heat at constant volume, k- heat conductivity.

The so called "restricted" law of energy conservation is

$$(3.16)\quad \rho\, c_v\, T_i = \partial j_i/\partial x_i$$

while the Fourier law of heat conduction relates the gradient of temperature to the components of heat flux.

$$(3.17)\quad j_i = k\,\frac{\partial T}{\partial x_i}.$$

If the temperature is assigned as a function of time on the boundary $\partial\Omega$ of a bounded region $\Omega \subset \mathbb{R}^3$ then the distribution of temperature in the region Ω (supposedly a smooth function of x and t)

$$T = T(x,t), \quad x \in \Omega,$$

can be derived as the solution of the Euler-Lagrange equation for the bilinear Prigogine-Glansdorff functional

$$(3.18)\quad \Phi = \tfrac{1}{2}\int_{\Omega}[k(\tau)\tau_o^2\,(\frac{\partial T^{-1}}{\partial x_i})^2 - \rho(\tau)c_v(\tau)\dot{\tau}T^{-1}]\,d\underset{\sim}{x},$$

where $\cdot \equiv \frac{d}{dt}$, i.e. $\cdot$ is the total derivative. The Euler-Lagrange equation is the law of energy conservation. t is regarded as a parameter.

It is a more or less a standard approach to heuristic treatment of such problems to identify the dual variables such as τ and T after setting to zero the first variation of the functional $\Phi(\tau,T)$. Duality is not introduced in a strict sense of duality between topological spaces.

It is easy to incorporate temperature effects into classical elasticity if one assumes that the density, conductivity and specific heats are constant (independent of time or temperature) and the medium is fully isotropic. We shall assume Cartesian representation, ignoring sub-and super-scripts. The basic equations of thermoelastcity are the equation of elastic equilibrium (Newton's second law).

$$(3.19)\quad \text{i)}\quad \rho\,\ddot{\underset{\sim}{u}}_i = \sum_j \frac{\partial}{\partial x_j}\sigma_{ij} + f_i$$

where f_i are component of body forces

The Energy balance:

$$\text{ii)}\quad \rho\,\dot{e} + \sum_{ij}\frac{\partial j_i}{\partial x_i} = \sum_{ij}\sigma_{ij}\,\dot{\varepsilon}_{ij}$$

Fourier's law of heat conduction

$$\text{iii)}\quad j_i = k\,\frac{\partial T}{\partial x_i}$$

The Duhamel linear law of thermoelasticity

$$\text{iv)}\quad \varepsilon_{ij} = (C_{ijk})^{-1}\,\sigma_{k\ell} + \gamma_{ij}\,(T - T_o)$$

As before ε_{ij} , σ_{ij} are components of strain and stress, respectively, and e is internal energy density

$$\varepsilon_{ij} = \tfrac{1}{2} \left(\frac{\partial u_i}{\partial x_j} + \frac{\partial u_j}{\partial x_i}\right)$$

The Legendre transformation introduces entropy $\bar{s}$

$$(3.20^a) \quad \bar{s}\,(T^{-1}) = s(e,\varepsilon_{ij}) - T^{-1}e$$

The rate of entropy production is

$$(3.20^b) \quad \rho\,\dot{\bar{s}} = \sum_i j_i \frac{\partial T^{-1}}{\partial x_i} = \sum_{i,j} L_{ij} \frac{\partial T^{-1}}{\partial x_i} \cdot \frac{\partial T^{-1}}{\partial x_j} ,$$

where L_{ij} is the heat conduction tensor.

3.4 Piezoelastic phenomena.

An interaction between elastic, thermal and magnetic phenomena in an essentially anisotropic elastic medium is considered in a rational theory relating the stress components σ_{ij}, strain components ε_{ij}, temperature θ, entropy s, components of the electric field E_i, and magnetic field H_i.

An elastic vibration of a piezoelastic crystal may be described in a Hamiltonian formalism in which the Hamiltonian is the modified Gibbs free energy function. For example R.D. Mindlin introduces the following form of Gibbs' free energy $G = G(E,\varepsilon,\theta)$. Mindlin [37] defines the free energy:

$$G = \tfrac{1}{2} \langle C_{ijk\ell}\ \varepsilon_{ij}, \varepsilon_{k\ell} \rangle - \langle \varepsilon^{s\theta}_{ij}\ E_i, E_j \rangle$$

$$- \tfrac{1}{2} \langle \alpha\theta, \theta \rangle - \langle e^{\theta}_{ijk}\ E_i, \varepsilon_{jk} \rangle \quad \langle \rho^{s}_{i}\ \theta, E_i \rangle$$

$$- \langle \gamma^{E}_{ij}\ \theta \rangle, \text{ where}$$

$C_{ijk\ell}$, λ_{ij} are moduli of elasticity and thermoelastic coefficients, respectively. $\varepsilon^{s,\theta}_{ij}$, p^{s}_{i} are the dielectric and piezoelectric coefficients, respectively, e^{θ}_{ijk} are the piezoelectric constants,

c_v is the specific heat, $\alpha = c_v T_o^{-1}$.

The equations of state are

$$\left.\begin{aligned}
\sigma_{ij} &= C_{ijk\ell} \cdot \varepsilon_{k\ell} - e^{\theta}_{kij} \cdot E_k - \lambda^{E}_{ij}\,\theta = \frac{\partial G}{\partial \varepsilon_{ij}} \\
D_i &= e_{ik\ell}\,\varepsilon_{k\ell} - \varepsilon^{s,\theta}_{ij}\,E_j + p^{s}_{i}\,\theta = -\frac{\partial G}{\partial E_i} \\
S &= \lambda^{E}_{ij}\,\varepsilon_{ij} + p^{s}_{i}\,E_i + \alpha\theta = -\frac{\partial G}{\partial \theta}
\end{aligned}\right\}$$

$\theta = T - T_o$, where T_o is the initial (reference) temperature, E_i are components of the electrical field

$\varepsilon_{ij} = \frac{1}{2}(u_{i,j} + u_{j,i})$ is the linear strain tensor,

S is the entropy.

3.5 Variational problems in engineering statics.

Several authors studying boundary value problems of engineering statics in the 1970-s developed theoretical techniques based on functional analysis.

This is not restricted to problems of statics and, in fact, functional analytic techniques have been even more aggressively applied to related problems of buckling, or general loss of stability, to steady state vibration of elastic systems to control and optimization of elastic systems. A strong incentive to pursue this line of investigation was provided by an entry of computer oriented numerical techniques in analysis and optimization and by the influence of finite element analysis of complex elastic systems. (See Aubin [29], the book of Ciarlet [41],Ciarlet and Destuynder [42] .)

A variational approach consists of replacing a system of partial differential equations with some boundary conditions by a variational statement or perhaps an equivalent statement concerning some energy forms. This is not a new technique. Recent advances in functional analysis, or perhaps the awareness of its applications allowed us to derive rigorous bounds, to prove existence and uniqueness theorems that are essential to implementation of variational methods and to justify numerical techniques based on finite element approximations.

In order to take a full advantage of the variational approach it is necessary to introduce appropriate function spaces for "admissible displacement fields" and "admissible force fields".

Traditionally, engineering texts introduced only the classical solutions to boundary value problems of engineering mechanics. No discussion of admissible classes of functions was offered and most manipulations would now be regarded as purely heuristic. It is not the purpose of this monograph to repeat classical arguments.

On the other hand, this is not a textbook on functional analysis.

Quick review of relevant elementary concepts of functional analysis are given in appendices A and B.

3.5[a] Static bending of beams

Let us assume the small deflection linear model of Bernoulli and Euler.

A(x) will denote the cross-sectional area I(x) moment of inertia about the neutral axis of bending. Other symbols are the same as in Chapter 1, section 1. We assume that $A(x) \in L^\infty[0,\ell]$, where $L^\infty[a,b]$ is the space of functions that are Lebesgue measurable and bounded on $[0,\ell]$.

$$\|A\|_\infty = \inf(M \in R) \text{ such that}$$

$$|A(x)| < M.$$

For the sake of simplicity we may also assume that the geometry of the cross-section does not vary, i.e. $I(x) = C[A(x)]^\alpha$, where C is a constant and α is a positive number, $\alpha > 1$. Frequently, we could take $\alpha = 2$. (That is the case for homogeneous materials and for rectangular or circular cross-sections. For multiply layered laminated beams $\alpha = 1$ may be the correct exponent.)

Thus, we could accept the Euler-Bernoulli relation in the form :

(3.21) $(EC[A(x)]^2 w_{xx})_{xx} = q(x)$,

where the applied load q(x) is an element of the Sobolëv space $H^{-2}[0,\ell]$ and w(x) is regarded as a weak solution of (3.21), that is

$$w(x) \in H_0^2 [0,\ell].$$

Note: w(x) is <u>not</u> assumed to be four times differentiable.
Equation (3.21) is rewritten in the weak form

$$\int_0^\ell \{[\mathfrak{A}^*_{(E,C,A(x))} \cdot \phi(x)]\, w(x) - q(x)\phi(x)\}\, dx = 0,$$

where $\Phi(x)$ is a test function that is an arbitrary $H^2[0,\ell]$ function, while the operator $\mathfrak{A}^*$ is the adjoint of the operator $\mathfrak{A}$ defined by the relation $\mathfrak{A} y \equiv (ECA^2\, y_{xx})_{xx}$.

If natural boundary conditions are assigned to w(x) then $\mathfrak{A} = \mathfrak{A}^*$.

The corresponding quadratic form may be derived by integration by parts.

(3.22) $$\int_0^\ell [(ECA^2\, w_{xx})_{xx}\, \phi(x)]\, dx$$

$$= [(ECA^2\, w_{xx})_x \phi - ECA^2\, w_{xx}\, \phi_x]_0^\ell$$

$$+ \int_0^\ell (ECA^2\, w_{xx}\, \phi_{xx})\, dx = \int_0^\ell (\underset{\sim}{q}\, \phi)\, dx.$$

(3.23) $\int_0^\ell (ECA^2\, w_{xx}\, \phi_{xx})\, dx$ satisfies all requirement of the inner product with $A \in L_\infty[0,\ell]$, E>0, C>0.

We define the energy product $\{w,\phi\}$ of w and ϕ by the formula (3.23), and we define the energy norm of w(x) to be $|||\, w\, ||| = \{w,w\}^{\frac{1}{2}}$.

It is physically justified that admissible displacement functions are not necessarily four times differentiable. The bending moment function $\mathcal{M}(x) = ECA^2(x)\, w_{xx}$ does not have to be

a continuous function in the domain of w(x).
It suffices if w(x) is an element of the Sobolev space $H^2_0\ [0,\ell]$, i.e. w has compact support and $\int_0^\ell (w^2 + w_x^2 + w_{xx}^2)\ dx < \infty$.

Since E, C are positive constants, $A^2(x) \in L_\infty$, and obviously A^2 is positive it is a trivial conclusion that $\mathcal{M}(x) \in H_0^2\ [0,\ell]$.

Note: In the usual introduction to the theory of Sobolëv spaces, the space H_0^2 is constructed as a space of all function that are limits of $C^\infty[0,\ell]$ functions whose support is compact (is contained in a closed, bounded subset of $\mathbb{R}$ in the one-dimensional case) and whose Sobolëv H_0^2 norm is bounded.

3.6 Comments about complementary energy and the Legendre transformation

Let us consider static problem of the form

$$V_o = \min_{(\underset{\sim}{\varepsilon},\underset{\sim}{u})} V_1(\underset{\sim}{\varepsilon},\underset{\sim}{u}),$$

for a homogeneous elastic medium
where $\underset{\sim}{\varepsilon} = A\underset{\sim}{u} \in \Omega$, $C^{\frac{1}{2}}\underset{\sim}{\varepsilon} \in L_2\ (\Omega)$,

$$\underset{\sim}{u} = \underset{\sim}{u}_o \text{ on } \partial\Omega.$$

$$V_1(\underset{\sim}{\varepsilon},\underset{\sim}{u}) = \tfrac{1}{2} \langle C\underset{\sim}{\varepsilon}, \underset{\sim}{\varepsilon}\rangle_\Omega - \langle u, f_\rho\rangle_\Omega - \langle u, \Phi_\Gamma\rangle_{\partial\Omega},$$

where

$$\langle \underset{\sim}{\varepsilon}, \underset{\sim}{\varepsilon}\rangle_\Omega = \int_\Omega (C\underset{\sim}{\varepsilon}\cdot\underset{\sim}{\varepsilon}^T)\ d\underset{\sim}{x}\ \text{, etc...}$$

$\underset{\sim}{f}_\rho$ is the vector of body forces,

Φ_Γ are the boundary tractions.

The classical statement of minimum potential energy assumes the form:
Find a minimum of $V_1\ (\underset{\sim}{\varepsilon},\underset{\sim}{u})$ with constraint conditions

$$\underset{\sim}{\varepsilon} = A\underset{\sim}{u} \in \Omega$$
$$\underset{\sim}{u} = \underset{\sim}{u}_0 \text{ on } \partial\Omega.$$

Let λ_1, λ_2 be Lagrangian multipliers converting this minimization problem to an unconstrained problem of minimizing

$$\tilde{V}_1(\underset{\sim}{\varepsilon}, \underset{\sim}{u}, \underset{\sim}{\lambda}_1, \underset{\sim}{\lambda}_2) = V_1(\underset{\sim}{\varepsilon}, \underset{\sim}{u})$$

$$+ \langle \underset{\sim}{\lambda}_1, (A\underset{\sim}{u} - \underset{\sim}{\varepsilon})\rangle_\Omega + \langle \underset{\sim}{\lambda}_2, (\underset{\sim}{u} - u_0)\rangle_{\partial\Omega}$$

Eliminating λ_1, λ_2, or setting the Fréchet derivatives of $\tilde{V}_1$ with respect to $\underset{\sim}{\varepsilon}$ and $\underset{\sim}{u}$ equal to zero one can derive

$$\begin{cases} \underset{\sim}{\lambda}_1 = C\,\underset{\sim}{\varepsilon} & \text{in } \Omega \\ \underset{\sim}{\lambda}_2 = B_n^T\,\underset{\sim}{\lambda}_1 & \text{on } \partial\Omega_1 , \end{cases}$$

where

$$B_n = \begin{bmatrix} \cos(n,x) & 0 & 0 \\ 0 & \cos(n,y) & 0 \\ 0 & 0 & \cos(n,z) \\ \cos(n,y) & \cos(n,x) & 0 \\ 0 & \cos(n,z) & \cos(n,y) \\ \cos(n,z) & 0 & \cos(n,x) \end{bmatrix} .$$

We recover one of the "canonical" relations

$$\underset{\sim}{\lambda}_1 = \frac{\partial V_1}{\partial \underset{\sim}{\varepsilon}} = \underset{\sim}{\sigma} \quad \text{in } \Omega .$$

A Legendre transformation is given by

$$H(\underset{\sim}{\sigma}, \underset{\sim}{u}) = \langle \underset{\sim}{\sigma}, \underset{\sim}{\varepsilon}\rangle_\Omega - V_1(\underset{\sim}{\varepsilon}, \underset{\sim}{u}) =$$

$$\langle \underset{\sim}{\sigma}, C^{-1}\cdot \underset{\sim}{\sigma}\rangle_\Omega - V_1(\underset{\sim}{\sigma}, \underset{\sim}{u}) =$$

$$\tfrac{1}{2}\langle C^{-1}\underset{\sim}{\sigma}, \underset{\sim}{\sigma}\rangle_\Omega + \langle \underset{\sim}{u}_\rho, \underset{\sim}{f}\rangle_\Omega , \text{ or}$$

$$\underset{\sim}{\sigma} = \frac{\partial V_1}{\partial \underset{\sim}{\varepsilon}} , \quad \underset{\sim}{\varepsilon} = \frac{\partial H}{\partial \underset{\sim}{\sigma}} ,$$

$$H = \langle \underset{\sim}{\sigma}, \underset{\sim}{\varepsilon}\rangle - V_1$$

$$(\underset{\sim}{\sigma} = \underset{\approx}{C}\cdot\underset{\sim}{\varepsilon}) .$$

The Hamilton canonical equations become

$$\left. \begin{array}{c} A^T \underset{\sim}{\sigma} = -\dfrac{\partial H}{\partial \underset{\sim}{u}} \\[2mm] A\underset{\sim}{u} = \dfrac{\partial H}{\partial \underset{\sim}{\sigma}} \end{array} \right\} \text{ in } \Omega.$$

Some easy manipulations show that this pair of Hamilton's canonical equations gives exactly the Reissner variational principle, i.e. the necessary conditions for the minimum of $V_1(\underset{\sim}{\varepsilon},\underset{\sim}{u})$, or

$$V_2(\underset{\sim}{\sigma},\underset{\sim}{u}) = \{<A\underset{\sim}{u}, \underset{\sim}{\sigma}>_\Omega - H\}$$

$$- <\underset{\sim}{u},\underset{\sim}{\phi}_\Gamma>_{\partial\Omega} - <(\underset{\sim}{u}-\underset{\sim}{u}_0), B_n \cdot \underset{\sim}{\sigma}>_{\partial\Omega_2},$$

where ϕ_Γ are forces assigned to $\partial\Omega$, part of the boundary and $\underset{\sim}{u}_0$ - displacements to $\partial\Omega_2$, $\partial\Omega_1 \cup \partial\Omega_2 = \partial\Omega$.

(3.24) An alternative formulation

3.7 The Friedrichs' version of Legendre transformation

Let H be the Hamiltonian functional for a problem in 3-dimensional elasticity.

$$H(\underset{\sim}{\varepsilon},\underset{\sim}{u}) = <\underset{\sim}{\sigma},\underset{\sim}{\varepsilon}> - V_1(\underset{\sim}{\varepsilon},\underset{\sim}{u})$$

We introduce new variables

$$\underset{\sim}{f} = \dfrac{\partial H}{\partial \underset{\sim}{u}}, \qquad \underset{\sim}{\sigma} = \dfrac{\partial H}{\partial \underset{\sim}{\varepsilon}},$$

and the function

$$\psi(\underset{\sim}{\sigma},\underset{\sim}{f}) = <\underset{\sim}{\sigma},\underset{\sim}{\varepsilon}> + <\underset{\sim}{f}_\rho,\underset{\sim}{u}> - H(\underset{\sim}{\varepsilon},\underset{\sim}{u})$$

or

$$\tilde{\psi}(\underset{\sim}{\sigma}) = \tfrac{1}{2}<C^{-1} \cdot \underset{\sim}{\sigma},\underset{\sim}{\sigma}>_\Omega.$$

The variational problem is formulated as follows:
Find a critical point of the functional

$$V_2(\underset{\sim}{\sigma},\underset{\sim}{u}) = \tilde{\psi}(\underset{\sim}{\sigma}) + <\underset{\sim}{u},B_n\underset{\sim}{\sigma}>_{\partial\Omega_1} - <\underset{\sim}{u},\underset{\sim}{\phi}_\Gamma>_{\partial\Omega_1} +$$

$$<\underset{\sim}{u}-\underset{\sim}{u}_0, B_n\underset{\sim}{\sigma}>_{\partial\Omega_2},$$

with the constrain condition

$$A^T \underset{\sim}{\sigma} = -\underset{\sim}{f}_\rho \quad \text{in } \Omega.$$

If all boundary conditions are restated as constraints, we derive the stationary condition due to Castigliano

$$V_3(\underset{\sim}{\sigma}) = -\tfrac{1}{2}\langle C^{-1}\underset{\sim}{\sigma}, \underset{\sim}{\sigma}\rangle_\Omega + \langle \underset{\sim}{u}, B_n \underset{\sim}{\sigma}\rangle_{\partial\Omega_2}$$

Some important inequalities.

All variational principles can be restated as variational inequalities. There are close relations between some classical inequalities one learns in differential equations or functional analysis courses and the energy inequalities derived in physics or in mechanics courses. For example, the Prager-Schields mutual complience inequality (or principle) turns out to be the Cauchy-Schwartz inequality with respect to the Friedrich's norm.

In [30] G. Fichera gives a proof that the energy product satisfies the following inequalities

(3.24) $\{f,\phi\} \leq k \,\|f\|_{H^2}\,\|\phi\|_{H^2}$ for some constant

$k > 0$, if some bounds are imposed(e.g $A^2>,0, C>0$ and $E>0$, for the beam equation)

and that

(3.25) $\{f,f\} = |||f|||^2 > \gamma \,\|f\|^2_{H^2}$

for some $\gamma>0$, provided that A is bounded away from zero for an elastic beam, that is, if

$A^2(x) > A_0 > 0$ for all $x \in [0,\ell]$.

Let us consider an abstract form of the beam equation:

$$\langle q,\phi\rangle = \langle (EI(x)\, w_{xx})_{xx}, \phi\rangle = \langle Aw,\phi\rangle =$$

$$\langle EI(x) w_{xx}, \phi_{xx}\rangle + \text{boundary terms} = \{w,\phi\} +$$

boundary terms. (3.26)

If one of the "natural" support conditions is assumed (free support, built in end, free end), the boundary terms vanish. The inequality (3.25) implies that the beam operator A is strongly elliptic.

The important property of elliptic equations is the smoothness of solutions. Normally, one can expect that the smallness of the $L_2(=H^0)$ norm of

$$([\int_a^b u^2dx]^{\frac{1}{2}})$$

does not influence the magnitude of H^1_0 norm $([\int_a^b (u^2 + u_x^2)\,dx]^{\frac{1}{2}})$. The saw-tooth function is a clear counterexample, and obviously, $\int_0^1 (u_x)^2 dx$ could be very large, while $\int_0^1 u^2 dx$ is small.

However, one can estimate the L_2 norm of $\underset{\sim}{u}(\underset{\sim}{x})$ if norms of derivatives of $\underset{\sim}{u}$ are known. For example, for any scalar function f(x) continuously differentiable on $[0,\pi]$, with $f(0) = f(\pi) = 0$, the Wirtinger inequality is true:

$$\int_0^\ell (y')^2dx \geq (\frac{\Pi}{\ell})^2 \int_0^\ell (y^2)\,dx.$$

Exact equality is true only for $y = C\sin x$. More general inequalities (partial differential operator ∇ replacing $\frac{d}{dx}$) were derived by Poincaré, Levi, and others. These may be found in the book of D. Mitrinovič [5] and also in the books of Hardy and Littlewood [3], Rabczuk [8], Lakshmikantham and Leela, [4], Szarski [9]. The ellipticity of the operator A does allow to assert the "wrong way" inequalities of the Gårding type (see [9]). It is easy to see that if a function $f(x)\in C^1[0,1]$, $f(0) = 0$, has its derivative bounded by some constant: $(|f'(x)| \leq C)$ then on the interval [0, 1]

$\max|f(x)| \leq C$. If $f(\bar{x}) > C$ for some $\bar{x}$, satisfying the statement of our problem, that is :

$0 < \bar{x} \leq 1$, the question: "how did f(x) get there?" cannot be answered. Therefore the knowledge of bounds on higher order derivatives suffices to produce some bounds on the values of any function having a sufficient degree of smoothness.

The Gårding inequality.

Let Lu denote the formal operator

$$Lu = \sum_{|\alpha| \le 2m} a_\alpha D^\alpha u, \qquad (3.27)$$

which can be rewritten in the divergence form

$$Lu = \sum_{0 \le |r|, |s| \le m} (-1)^r D^{|r|}(a_{rs}(x) D^s u), \qquad (3.27^a)$$

provided $a_{rs}(x)$ are sufficiently smooth. The formal ajoint of L is

$$L^* = \sum_{0 \le |r|, |s| \le m} (-1)^{|r|} D^r (a_{rs}(x) D^s \cdot) \qquad (3.27^b)$$

Let $B(u,v)$ be the bilinear form

$$B(u,v) = \sum_{0 \le |r|, |s| \le m} (D^r v(a_{rs} D^s u)). \qquad (3.28)$$

This known as the Lagrange bilinear form associated with the operator L. We repeat that the coefficients a_{rs} need not be differentiable m-times, that is we do not require existence of classical derivatives.

Gårding inequality states that if

a) L is strongly elliptic with a modulus of elipticity independent of x in Ω,

b) The coefficient a^{rs} are bounded. There exists some constant C_1, such that $|a^{rs}| < C_1$.

c) The highest order coefficients have a bounded modulus of continuity. There exists C_2 such that

$$|a^{rs}(\underset{\sim}{x}_1) - a^{rs}(\underset{\sim}{x}_2)| < C_2 \cdot |x_1 - x_2|, \text{ for}$$

$|r|, |s| = m$ and for $\underset{\sim}{x}_1, \underset{\sim}{x}_2 \in \Omega$, and C_2 is small in a small neighborhood by the origin,

then there exist constants C_3, C_4 such that

$$\|\underset{\sim}{u}\|_m^2 \le C_3 B(u,u) + C_4 \|\underset{\sim}{u}\|_0^2, \text{ where}$$

$\|\underset{\sim}{u}\|_0^2$ denotes the usual $L_2(\Omega)$ norm of $\underset{\sim}{u}(\underset{\sim}{x})$

That is, strong ellipticity permits us to estimate high order derivative of solutions in terms of the lower order derivatives. The "wrong way" inequalities are very important in estimating values of functionals. Strong "ellipticity" is essential in this estimate and in other estimates, for example the Rellich type inequalities for elliptic systems. (The "fundamental Rellich inequality" is a predecessor of the Sobolëv imbedding lemma.)

Some inequalities of the thin plate theory.

Constitutive equations (Hooke's law).

Associated with any displacement $\underset{\sim}{u} = (u_1, u_2, u_3)$ in $\mathbb{R}^3$ is its linear strain tensor

$$\varepsilon_{ij}(\underset{\sim}{u}) = \tfrac{1}{2}(\partial_j u_i + \partial_i u_j) \qquad i,j = 1,2,3.$$

and the corresponding stress tensor

(3.29) $$\sigma_{ij}(\underset{\sim}{u}) = C_{ijk\ell}\varepsilon_{k\ell}\varepsilon_{k\ell}(\underset{\sim}{u}) .$$

The fourth-order tensor $C_{ijk\ell}$ satisfies

(3.30) $$C_{ijk\ell} = C_{jik\ell} = C_{ij\ell k} = C_{k\ell ij} ;$$

We assume that the elastic energy

$$\langle C_{ijk\ell}\, \varepsilon_{ij}, \varepsilon_{k\ell} \rangle$$

is a positive definite functional.

We shall always assume that the planes of elastic symmetry are horizontal. This means that

$$C_{\alpha\beta\gamma 3} = 0, \quad C_{\alpha 333} = 0.$$

Finally, we define a positive definite fourth-order tensor

(3.31) $$\tilde{B}_{\alpha\beta\gamma\delta} = C_{\alpha\beta\gamma\delta} - \frac{C_{\alpha\beta 33}C_{\gamma\delta 33}}{C_{3333}} .$$

3.5[b] Plate geometry

The plate geometry is determined by

i) a smoothly bounded domain Ω in the $x_1 - x_2$ plane representing the midplane,

ii) a real parameter a, $0 < a < \infty$, determining the length scale of the thickness variation, and

iii) a bounded function $h(\eta) \geq 0$, defined for any $\eta \in \mathbb{R}^2$ and periodic in η_α with periods T_α, $\alpha = 1,2$.

The three-dimensional region occupied by the plate is $D = \Omega \times [-h,+h]$, $= \{x : x \in \Omega, |x_3|<h\}$

We assume throughout that D is a connected, $C^{2,\alpha}$ domain, for some Hölder exponent $\alpha > 0$. The function h may nonetheless have discontinuities- <u>i.e.</u> parts of ∂D may be vertical, and h may vanish on a set of positive measure - <u>i.e.</u> our plates may have holes.

$\partial_0 D$ denotes the outer edge of the plate in the $(x_1 - x_2)$ plane

3.5c <u>Loads and equations of equilibrium</u>

The following discussion applies for $h \equiv$ constant. Otherwise it is more natural to work with the load per unit projected surface area on the $x_1 - x_2$ plane.

The "usual" equations of equilibrium for thin plates of almost uniform thickness are derived in an engineering course by summing up the forces and the moments in the manner illustrated on figure 3.1

Several authors observed that routine optimization procedures in plate designs seem to lead to discontinuous variations of the plate thickness, that is to the formation of ribs. This phenomenon was observed by K.T. Cheng, by Cheng and Olhoff, N.V. Banichuk, G.J. Semitses, E.F. Masur and others. This is hardly a surprising result. Practicing engineers seem to have realized it for centuries. Specifically, one could observe a typical design (dating to middle 19th Century) of cast iron manholes decorating most of our streets, to conclude that whoever designed it knew instinctively the best optimal design for a given weight, without the advantages of functional analytic arguments.

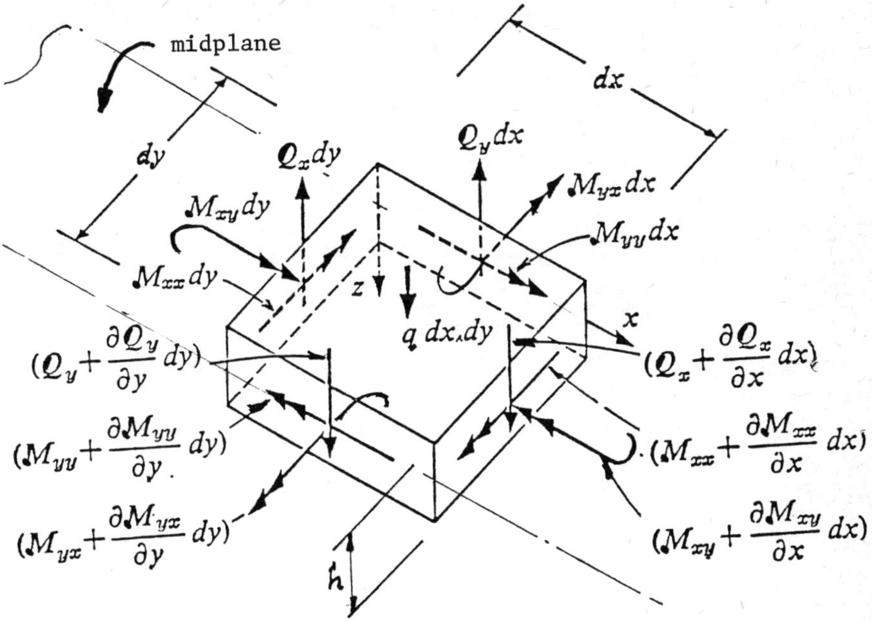

Figure 3.1

3.5[d] Integral estimates

We derive certain integral inequalities for displacement vector $\underline{u} \in H^1(D)$.

$$D = \{\underset{\sim}{x} : \underset{\sim}{x} \in \Omega, |x_3| < h$$

A method suggested by Kohn and Vogelius is to decompose D into $\mathcal{O}(\varepsilon^{-2})$ subdomains Q_i, each with diameter of order M, and to apply Korn's inequality on each subdomain.

Let $\mathcal{R}$ be the space of rigid motions,

$$\underset{\sim}{\gamma} \in \mathcal{R} \iff \gamma_i(\underset{\sim}{x}) = c_{ij}x_j + d_i, \text{ for some } \underset{\sim}{d} \in \mathbb{R}^3 \text{ and some skew-symmetric matrix } \underset{\approx}{C}.$$

$\nabla\underline{u}$ denotes the tensor $\partial_j u_i$, and $\varepsilon(\underline{u})$ denotes the (symmetric) strain tensor $\frac{1}{2}(\partial_j u_i + \partial_i u_j)$.

Lemma For any $u \in H^1(Q)$ there exists $\gamma \in \mathcal{R}$ such that

$$\int_Q |\nabla(\underset{\sim}{u}-\underset{\sim}{\gamma})|^2 \, d\underline{x} \le C \int_Q |\varepsilon(\underset{\sim}{u})|^2 \, d\underline{x} \, . \quad (3.32)$$

and

$$\int_Q |\underset{\sim}{u}-\underset{\sim}{\gamma}|^2 d\underline{x} \le C \int_Q |\varepsilon(\underset{\sim}{u})|^2 \, d\underline{x} \; ; \quad (3.33)$$

where the constant C depends only on Q.

Rescaling (3.32) and (3.33) yields the following results.

Lemma For any $\underset{\sim}{u} \in H^1_{\ell oc}(D)$ and any pair of integers 1,2 or 3 (k,ℓ) there exists $\gamma^{k\ell} \in \mathcal{R}$ such that

$$\int_{D_{k\ell}} |\nabla(u-\gamma^{k\ell})|^2 \, d\underline{x} \le C \int_{D_{k\ell}} |\varepsilon|^2 \, d\underline{x}$$

and

$$\int_{D_{k\ell}} |u-\gamma^{k\ell}|^2 d\underline{x} \le C \mu^2 \int_{D_{k\ell}} |\varepsilon(u)|^2 \, d\underline{x} \, .$$

The constant C depends only on h.

Corrolary

Let $$\underset{\sim}{\gamma}^{k\ell}(\underset{\sim}{x}) = \underset{\approx}{c}^{k\ell} \underset{\sim}{x} + \underset{\sim}{d}^{k\ell}, \quad (3.34)$$

where $\underset{\sim}{d}^{k\ell} \in R^3$ and $\underset{\sim}{c}^{k\ell}$ is a skew-symmetric matrix. Clearly

$$\mu^4 |c^{k+1,\ell} - \underline{c}^{k\ell}|^2 + \mu^2 |d^{k+1,\ell} - d^{k\ell}|^2$$

$$\le C \int |\gamma^{k+1,\ell}(\underline{x})|^2 dx_2 dx_3$$

$$\le C \int (|\underset{\sim}{\gamma}^{k+1,\ell} - \underset{\sim}{u}|^2 + |\underset{\sim}{u}-\underset{\sim}{\gamma}^{k\ell}|^2) dx_2 dx_3 ,$$

where the integrals are taken over the interface between $D_{k\ell}$ and $D_{k+1,\ell}$.

On any Lipschitz domain one has the trace estimate

$$\int_{\partial Q} |\underset{\sim}{w}|^2 ds \leq C \int_Q (|\nabla \underset{\sim}{w}|^2 + |\underset{\sim}{w}|^2) dx \qquad (3.35)$$

for any $w \in H^1(Q)$. Rescaling and combining the result with (3.34) we obtain

$$\mu^4 |c^{k+1,\ell} - c^{k\ell}|^2 + \mu^2 |d^{k+1,\ell} - d^{k\ell}|^2$$

$$\leq C\Big(\mu \int_{D_{k+1,\ell}} |\nabla(\underline{u} - \underline{\gamma}^{k+1,\ell})|^2 dx + \qquad (3.36)$$

$$\mu^{-1} \int_{D_{k+1,\ell}} |u - \gamma^{k+1,\ell}|^2 dx$$

$$+ \mu \int_{D_{k\ell}} |\nabla(u - \gamma^{k\ell})|^2 dx + \mu^{-1} \int_{D_{k\ell}} |u - \gamma^{k\ell}|^2 d\underline{x}\Big).$$

(See Kohn and Vogelius [35].
This implies

$$\mu^2 |c^{k+1,\ell} - c^{k\ell}|^2 + |d^{k+1,\ell} - d^{k\ell}|^2 \leq$$

$$C\mu^{-1} \int_{D_{k+1,\ell}} \int_{D_{k\ell}} |\underset{\sim}{\varepsilon}|^2 d\underline{x}.$$

As a consequence of such inequalities of Poincaré's and of Korn's type we are allowed to average properties on non-uniform plates concluding that local displacement of the plate cannot exceed in absolute value the displacement derived from the "averaged" plate by more than some power of the diameter of the subdivision. For example, if the plate is subdivided into regions $D_{k\ell}$ of size μ, and the plate has no holes, the following estimate is given in [35].

$$\{ \int_\Omega |w - w^\mu|^2 dx \}^{\frac{1}{2}} < C \mu^{\frac{1}{2}} \qquad (3.37)$$

Such estimates are of importance in estimating energy terms for plates of rapidly varying cross-section, such as plates with ribs, or with re-inforcing strips.

3. 8 . Vibration of elastic plates satisfying the Euler-Bernoulli hypotheses.

As usual, we assume Hoke's Law, and the usual linear hypothesis of the thin plate theory, which in turn imply the correctness of Duhamel's Principle. The plate occupies a compact, connected region $\bar{\Omega}$, which is a subset of the Euclidean plane $\mathbb{R}^2$. The boundary of $\bar{\Omega}$ is a smooth Jordan curve. It will be denoted by $\partial\Omega$. The interior of $\bar{\Omega}$ will be denoted by Ω. x, y are cartesian coordinates. The deflection function W(x,y,t) obeys the differential equation

(3.38)

$$[D(x,y)\,\nabla^2(\nabla^2 W) + 2\frac{\partial D(x,y)}{\partial x}\frac{\partial}{\partial x}(\nabla^2 W) + 2\frac{\partial D(x,y)}{\partial y}\frac{\partial}{\partial y}(\nabla^2 W) + \nabla^2(D(x,y))\nabla^2 W] -$$

$$-(1-\nu)\left[\frac{\partial^2 W}{\partial x^2}\frac{\partial^2 D(x,y)}{\partial y^2} - 2\frac{\partial^2 W}{\partial x\,\partial y}\frac{\partial^2 D(x,y)}{\partial x\,\partial y} + \frac{\partial^2 W}{\partial y^2}\frac{\partial^2 D(x,y)}{\partial x^2}\right] = -\frac{\gamma\, h(x,y)}{g}\frac{\partial^2 W}{\partial t^2} + p(x,y)\,,$$

where D(x,y) and p(x,y) are known functions, and we assume that W(x,y) is of the class C^2 in Ω, D(x,y) is of the class C^2 in Ω. The operator ∇^2 is the Laplacian: $\nabla^2 \equiv \partial^2/\partial x^2 + \partial^2/\partial y^2$.

On $\partial\Omega$ $W(x,y,t)$ obeys some physically motivated conditions which may be of the form:

(3.39ᵃ)

$$W = 0,\quad M_n = D\left\{\frac{\partial^2 W}{\partial n^2} + \nu\left(\frac{\partial^2 W}{\partial s^2} + \frac{\partial\psi}{\partial s}\frac{\partial W}{\partial n}\right)\right\} = 0 \quad ,$$

where ψ is the angle which the boundary forms with the direction of the x axis, n denotes the direction of the outer normal to the boundary, while $\partial/\partial s$ denotes differentiation along the arc length of the boundary $\partial\Omega$, and

(3.39ᵇ)

$$W = 0\,,\quad \frac{\partial W}{\partial n} = 0 \text{ on some subarc } \Gamma_2 \text{ of } \partial\Omega\,,$$

(3.39c) $M_n = 0,$

$$D\left[\frac{\partial}{\partial n}\nabla^2 W + (1-\nu)\frac{\partial}{\partial s}\left(\frac{\partial^2 W}{\partial n\,\partial s} - \frac{\partial\psi}{\partial s}\frac{\partial W}{\partial s}\right)\right] + \frac{\partial D}{\partial n}\left[\frac{\partial^2 W}{\partial n^2} + \nu\left(\frac{\partial^2 W}{\partial s^2} + \frac{\partial\psi}{\partial s}\frac{\partial W}{\partial n}\right)\right] +$$

$$+ 2(1-\nu)\frac{\partial D}{\partial s}\left(\frac{\partial^2 W}{\partial n\,\partial s} - \frac{\partial\psi}{\partial s}\frac{\partial W}{\partial s}\right) = 0 \quad \text{on some subarc } \Gamma_3 \text{ of } \partial\Omega,$$

where M_n stands for the same expression as in (3.39a). Γ_1, Γ_2, Γ_3, have at most the end points in common, and $\bar{\Gamma}_1 \cup \bar{\Gamma}_2 \cup \Gamma_3 = \partial\Omega$ (the bar over Γ indicates the closure operation).

The physical meaning of the symbols used is:

γ - specific weight of the material
$h(x,y)$ - the thickness of the plate ($h(x,y)>0$ in Ω)
E - Young's Modulus ($E>0$),
ν - Poisson's Ratio (By a physical argument ν must obey $0 \leqq \nu \leqq \frac{1}{2}$) .

$D(x,y)$-rigidity$\left(= \frac{Eh^3(x,y)}{12(1-\nu^2)}\right)$,

g - earth's gravitational constant,
$p(x,y)$ - the applied distribution load, $p(x,y)$ is assumed to be a square integrable function in Ω.

The well-known linearalized relationships between the moments M_{ij} and the derivates of $W(x,y)$ are:

(3.40)
$$\begin{cases} M_{xx} = -D(x,y)\left(\frac{\partial^2 W}{\partial x^2} + \nu\frac{\partial^2 W}{\partial y^2}\right), \\ -M_{xy} = M_{yx} = -(1-\nu)D(x,y)\frac{\partial^2 W}{\partial x\,\partial y}, \\ M_{yy} = -D(x,y)\left(\frac{\partial^2 W}{\partial y^2} + \nu\frac{\partial^2 W}{\partial x^2}\right). \end{cases}$$

We shall denote by **M** the vector

$$M = \begin{bmatrix} M_{xx} \\ M_{xy} \\ M_{yx} \\ M_{yy} \end{bmatrix}.$$

Assuming that each point of the plate oscillates with the same frequency ω, we can write the equations of equilibrium in the following form:

(3.41)
$$\left(\frac{\partial^2 M_{xx}}{\partial x^2} - \frac{\partial^2 M_{xy}}{\partial x\,\partial y} + \frac{\partial^2 M_{yx}}{\partial y\,\partial x} + \frac{\partial^2 M_{yy}}{\partial y^2}\right) = \gamma\frac{\omega^2 h(x,y)}{g} W(x,y,t) + p(x,y) =$$
$$= \omega^2 m(x,y)\, W(x,y,t) + p(x,y) \qquad (x, y \in \Omega, t > 0),$$

where m(x,y) is the mass distribution function of the plate. We introduce the differential operator:

$$(3.42)\qquad T = \begin{bmatrix} \frac{\partial}{\partial x} & 0 \\ 0 & \frac{\partial}{\partial y} \\ \frac{\partial}{\partial y} & 0 \\ 0 & \frac{\partial}{\partial x} \end{bmatrix}$$

and the symmetric matrix N:

$$(3.43)\qquad N = \begin{bmatrix} D(x, y) & \nu D(x, y) & 0 & 0 \\ \nu D(x, y) & D(x, y) & 0 & 0 \\ & 0 & -(1-\nu) D(x, y) & 0 \\ 0 & 0 & 0 & (1-\nu) D(x, y) \end{bmatrix}.$$

Because of the physical limitations imposed on ν, N is nonsingular. The equations (3.40) can be written in the operator form:

$$(3.44)\ -M = N\,T\,\mathrm{grad}\,W, \qquad (3.45)\ M = M_0(x,y)e^{j\omega t} + M_1(x,y).$$

Equation (3.44) is rewritten as:

$$(3.46)\ -N^{-1}M = T\cdot\mathrm{grad}\,W,$$

$$(3.47)\qquad W(x,y,t) = W_0(x,y)e^{j\omega t} + W_1(x,y).$$

$$(3.48)\qquad V(x, y, t) = \sqrt{m(x, y)}\,\frac{\partial W(x, y, t)}{\partial t} = j\,\omega\,\sqrt{m(x, y)}\,W_0(x, y)\,e^{j\omega t}.$$

For the time dependent part of M(x,y,t) we have

$$(3.49)\qquad -N^{-1}M_0(x, y) = T\,\mathrm{grad}\left(\frac{1}{j\,\omega\,\sqrt{m(x, y)}}\,V_0(x, y)\right),$$

where $V_0(x,y)$ is defined by the relationship:

$$(3.50)\qquad V(x, y, t) = V_0(x, y)\,e^{j\omega t}.$$

Let the operator B denote

$$(3.51)\qquad B \overset{\mathrm{def}}{=} \left(\sqrt{m(x, y)}\,j\,\omega\right)^{-1},$$

and let the operator $\mathfrak{A}$ stand for

$$(3.52)\qquad \mathfrak{A} = T\cdot\mathrm{grad}\,.$$

We introduce Hilbert Spaces H_M and H_W. The elements of H_M are four-dimensional vectors, whose components are real valued functions with the domain Ω, and square integrable in Ω. The inner product in H_M is defined for A,B, in H_M:

$$\langle A, B\rangle_{(\Omega)} = \int_\Omega (A_1 B_1 + A_2 B_2 + A_3 B_3 + A_4 B_4)\, d\sigma \tag{3.53}$$

(σ is the usual two dimensional Euclidean measure) The elements of H_W are real valued functions, whose domain is Ω, square integrable in Ω. The inner product is defined for ϕ, ψ in H_W:

$$(\varphi, \psi)_{(\Omega)} = \int_\Omega (\varphi\, \psi)\, d\sigma\,. \tag{3.54}$$

We consider an operator S which maps some dense subset of H_W into H_M. Therefore,the adjoint operator S^* is uniquely defined.

We now restrict our discussion to the static case:

$$W \equiv W_1(x, y)\,, \qquad M \equiv M_1(x, y)\,, \qquad \omega = 0\,. \tag{3.55}$$

The components of $M_1(x,y)$ belong by assumption to the Sobolëv class W_2^2 that is they possess weak second derivatives in Ω, which are square integrable functions in Ω. Consequently,we are dealing with dense subsets of L_2 spaces. If the admissible displacements W_1 are assumed to be four times continuously differentiable in Ω, and their fourth derivatives are assumed to be square integrable functions in Ω,then the admissible displacements $W_1(x,y)$ form again a dense subset of an L_2 space.

We define the following bilinear functional:

$$\mathscr{W}_1(W_1, M_1) = \frac{1}{2}(p(x, y), W_1(x, y))_{(\Omega)} = \frac{1}{2}(\mathfrak{A}^* M_1(x, y), W_1(x, y))_{(\Omega)} \tag{3.56}$$

Integrating by parts,we can easily show that if the only conditions imposed on the boundary concerning the displacements are : $W_1(x, y) = 0$ on $\partial\Omega$, and the only boundary condition imposed on

$M_1(x,y)$ is: $M_1(x,y) = 0$ on $\partial\Omega$, then the operators $\mathfrak{A}$ and $\mathfrak{A}^*$ defined by relationships (3.52) and ( ) are adjoints of each other. For this special case ($W_1 = 0$ on $\partial\Omega$, $M_1 = 0$ on $\partial\Omega$) we have:

$$(3.57)\quad \mathscr{W}_1(W_1, M_1) = \frac{1}{2}(\mathfrak{A}^* M_1(x, y), W_1(x, y))_\Omega = \frac{1}{2}\langle M_1(x, y), \mathfrak{A}\, W_1(x, y)\rangle_\Omega = \frac{1}{2}\langle - N^{-1} M_1(x, y), M_1(x, y)\rangle_\Omega .$$

We assume that the functional $\mathscr{W}(W_1, M_1)$ has first and second Fréchet derivatives with respect to M_1 and W_1. It follows from (3.56) and from (3.46) since ($\mathfrak{A}^* M_1$ depends on W_1) that:

$$(3.58)\quad \frac{\partial \mathscr{W}_1}{\partial W_1} = p(x, y) = \mathfrak{A}^* M_1(x, y) .$$

Hence, we have a "canonical" system of equations

$$(3.59)\quad \frac{\partial \mathscr{W}}{\partial \theta} = T\,U, \qquad \frac{\partial \mathscr{W}}{\partial U} = T^*\,\theta,$$

identical with the Hamiltonian systems discussed previously in chapters 1 and 2. $\mathscr{W}$ is the Hamiltonian. Let $\mathfrak{L}$ denote the corresponding Lagrangian. If the signs of the quantities

$\langle -(\partial^2\mathscr{W}/\partial U^2)\, U_0, U\rangle_{(\Omega)}$ and $\langle -(\partial^2\mathscr{W}/\partial\theta^2)\,\theta_0, \theta\rangle_{(\Omega)}$ differ in neighborhood of a critical point $\binom{U_0}{\theta_0}$ then h_0 is the minimax point of $\mathfrak{L}$. To apply this theorem to the system (3.59) we check on the signs of

$((\partial^2\mathscr{W}_1/\partial W_1^2)\, \widetilde{W}_1, W_1)_{(\Omega)}$ and $\langle(\partial^2\mathscr{W}_1/\partial M_1^2)\, \widetilde{M}_1, M_1\rangle_{(\Omega)}$ where

$\widetilde{W}_1, \widetilde{M}_1$ are solutions of the system (3.59)

$$\left(-\frac{\partial^2\mathscr{W}_1}{\partial W_1^2}\widetilde{W}_1, W_1\right) = (\mathfrak{A}^* N\,\mathfrak{A}\,\widetilde{W}_1, W_1)$$

where W_1 is restricted to some neighborhood of

$\widetilde{W}_1$, i.e. $\widetilde{W}_1 = W_1 + \varepsilon\,\xi$, where $\|\xi\| < 1$, $\|\mathfrak{A}\,\xi\| < 1$ and ε is some real number. Then

$$\left(\frac{\partial^2\mathscr{W}}{\partial W_1^2}\widetilde{W}_1, W_1\right)_\Omega = \langle N\,\mathfrak{A}\,\widetilde{W}_1, \mathfrak{A}\,W_1\rangle + \varepsilon\,\langle N\,\mathfrak{A}\,W_1, \varepsilon\,\mathfrak{A}\,\xi\rangle_\Omega$$

which is clearly positive for a sufficiently small ε, provided $W_1(x,y)$ is not identically equal to zero in Ω. Similarly, we obtain the confirmation of the positiveness of $\langle(\partial^2 \mathcal{W}_1/\partial M_1^2)\, \tilde{M}_1, M_1\rangle_{(\Omega)}$ in

some neighborhood of $(\tilde{W}_1, \tilde{M}_1)$, provided $\mathcal{M}$ is not identically equal to zero in Ω.

It follows from the positiveness of these products in some neighborhood of the point

$\begin{pmatrix}\tilde{W}_1\\ \tilde{M}_1\end{pmatrix}$ that $\mathcal{L}$ assumes a minimum over all choices

of $\tilde{M}$ with $\tilde{W}$ fixed and over all choices of $\tilde{W}$ with $\tilde{M}$ fixed. This is not the most desirable result. A similar discussion follows if $M = M_0(x,y)e^{jwt}$, where we define the inner produce in H_M to be:

$$\int_0^{t=\frac{2\pi}{\omega}} \int_\Omega [(A_1\, B_1) + (A_2\, B_2) + (A_3\, B_3) + (A_4\, B_4)]\, dt\, d\sigma$$

and the product in H_W to be

$$\int_0^{t=\frac{2\pi}{\omega}} \int_\Omega (\varphi\, \psi)\, dt\, d\sigma$$

and we define the functional $\mathcal{W}_0$ by the formula

$$\mathcal{W}_0 = \frac{1}{2}\langle - N^{-1} M_0, M_0\rangle_\Omega = \frac{1}{2}\langle A\, V_0, M_0\rangle_\Omega = \frac{1}{2}(V_0, V_0)_\Omega \,. \qquad (3.60)$$

Fréchet derivatives of $\mathcal{W}_0$ are:

$$\frac{\partial \mathcal{W}_0}{\partial M_0} = - N^{-1} M_0 = A\, V_0\,, \qquad \frac{\partial \mathcal{W}_0}{\partial V_0} = V_0 = A^* M_0\,. \qquad (3.61)$$

Formula (3.60) expressed Lagrange's equality of the mean values of kinetic and potential energy, and formulas (3.61) again reduce the system to the generalized canonical form.

The formulas (3.48) and (3.61) are superimposed by defining $W = W_1 + W_0$ and by combining equations (3.56) with (3.61). The equations (3.59), (3.61) and the equations (3.56) and (3.57) are of interest for their own sake, since they reveal the generalized canonical formulation of the physical problem. The boundary conditions discussed so far were restricted to the case $W = 0$ on $\partial\Omega$. The variational principle introduced in this work is of little help in numerical solution of the equation (3.38) with this boundary condition, since it is reduced easily to the familiar statement of Hamilton's Principle and known numerical techniques.

3.9 Basic ideas of the linear shell theory.

3.91 A general discussion.

We accept the assumptions of Kirchhoff-Love theory (see [44]) and the Novozhilov principle. The neutral middle surface has an atlas of local orthogonal coordinates denoted by $\mathbf{e}_1, \mathbf{e}_2, \mathbf{e}_3$, where $\mathbf{e}_1$, $\mathbf{e}_2$ are tangential to the middle surface.

The forces acting on the middle surface are illustrated on figure 3.2 in a conventional engineering "free body" manner.

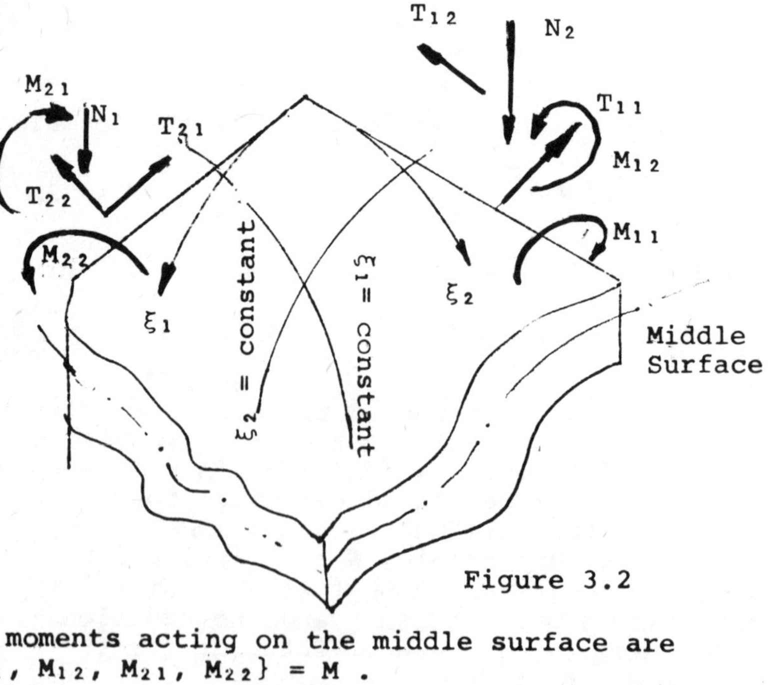

Figure 3.2

The moments acting on the middle surface are $\{M_{11}, M_{12}, M_{21}, M_{22}\} = \underset{\sim}{M}$.

The tractions are T_{11}, T_{12}, T_{21}, T_{22}. Let R_1, R_2 denote the principal radii of curvative. Let u_1 denote the displacement in the direction of the coordinate ξ_1, u_2 in the direction of ξ_2, w in the direction perpendicular to the middle surface, that is perpendicular to both ξ_1 and ξ_2. Let h_1, h_2, h_3 denote the Lammé parameters:

$h_i = \sqrt{g_{ii}}$, where the (distance) metric is defined by the quadratic form

$$ds^2 = \sum_{i=1}^{2} g_{ii}(d\xi^i)^2,$$ with the local coordinates

ξ_1, ξ_2, ξ_3, (ξ_3 = w) forming a mutually orthogonal system. We can introduce tangential load S given by

$$S = T_{12} - M_{21}/R_2 = T_{21} - M_{12}/R_1$$

and the torque

$$\tau = \tfrac{1}{2}(M_{12} - M_{21}),$$

where R_1, R_2 are radia of curvature of the shell surface along coordinates ξ_1, ξ_2, respectively. We can represent in an abstract form the relation

$$\underset{\sim}{\varepsilon} = A_1 \underset{\sim}{u} \tag{3.62}$$

where A_1 is the matrix

$$A = \begin{bmatrix} a_{11} & a_{12} & a_{13} \\ \vdots & & \vdots \\ a_{61} & a_{62} & a_{63} \end{bmatrix},$$

$$a_{11} = \frac{1}{h_1}\frac{\partial}{\partial\xi_1} \quad a_{12} = \frac{1}{h_1 h_2}\frac{\partial h_1}{\partial\xi_2}, \quad a_{13} = \frac{1}{R_1}$$

$$a_{21} = \frac{1}{h_1 h_2}\frac{\partial h_2}{\partial\xi_1}, \quad a_{22} = \frac{1}{h_2}\frac{\partial}{\partial\xi_2}, \quad \ldots \quad .$$

$\underset{\sim}{u} = (u_1, u_2, w)$ where u_i is displacement in the

direction of ξ_i coordinate, i = 1,2.
Moments $(M_{11}, S, M_{22}, T_{11}, \mathbf{S}, T_{22}) = \underset{\sim}{\mu}$ are related to body forces by a linear relation

$$B\,\underset{\sim}{\mu} = -f_\rho\ , \tag{3.63}$$

where elements of B are:

$$b_{11} = \left(\frac{1}{h_1 h_2}\, \frac{\partial}{\partial \xi_1} \cdot (h_2\ \cdot) \right)$$

$$b_{12} = -\frac{1}{h_1 h_2}\, \frac{\partial h_2}{\partial \xi_1}$$

$$b_{13} = \frac{1}{h_1^2 h_2}\, \frac{\partial}{\partial \xi_2} \cdot h_1^2,\quad b_{14} = \frac{1}{h_1 h_2}\, \frac{1}{R_1}\, \frac{\partial}{\partial \xi_1}\, h_2, \ldots$$

(See Novozhilov [45], or Goldenveizer [44] for details and derivations.) The linear elasticity hypothesis becomes:

$$\underset{\sim}{\mu} = \underset{\approx}{D} \cdot \underset{\sim}{\varepsilon}\ , \tag{3.64}$$

where D is the Novozhilov matrix

$$\boldsymbol{D} = \begin{pmatrix} \frac{Eh}{1-\nu^2} & \frac{\nu Eh}{1-\nu^2} & 0 & 0 & 0 & 0 \\ \frac{\nu Eh}{1-\nu^2} & \frac{Eh}{1-\nu^2} & 0 & 0 & 0 & 0 \\ 0 & 0 & \frac{Eh}{2(1+\nu)} & 0 & 0 & 0 \\ 0 & 0 & 0 & \frac{Eh^3}{12(1-\nu^2)} & \frac{\nu Eh^3}{12(1-\nu^2)} & 0 \\ 0 & 0 & 0 & \frac{\nu Eh^3}{12(1-\nu^2)} & \frac{Eh^3}{12(1-\nu^2)} & 0 \\ 0 & 0 & 0 & 0 & 0 & \frac{Eh^3}{12(1+\nu)} \end{pmatrix}$$

The following equations may be assembled.

$$\left.\begin{aligned} \underset{\sim}{\varepsilon} &= A_1\,\underset{\sim}{u} \\ \underset{\sim}{\mu} &= D\ \underset{\sim}{\varepsilon} \\ B\,\underset{\sim}{\mu} &= -\underset{\sim}{f}_\rho \end{aligned}\right\} \in \Omega \tag{3.65}$$

with boundary conditions

$$\mathcal{B}^{(1)}\mu = \mu_0 (\text{or } = 0) \text{ on } \partial\Omega,$$

or $$\mathcal{B}^{(2)} u = u_0 \text{ on } \partial\Omega,$$

where $\mathcal{B}^{(1)}$, $\mathcal{B}^{(2)}$ are the 4x6 and 4x3 boundary projection matrices, respectively

$$\mathcal{B}^{(1)} = \begin{bmatrix} b_{11}^{(1)} \ldots b_{16}^{(1)} \\ \\ b_{41}^{(1)} \ldots b_{46}^{(1)} \end{bmatrix}$$

$$b_{11}^{(1)} = \cos^2 \gamma$$

$$b_{12}^{(1)} = \sin^2 \gamma$$

$$b_{13}^{(1)} = \sin^2 \gamma$$

$$b_{14}^{(1)} = \sin^2 \gamma \cos^2 \gamma \left(\frac{1}{R_1} - \frac{1}{R_2}\right)$$

$\vdots$

$$b_{11}^{(2)} = \cos \gamma$$

$$b_{12}^{(2)} = \sin \gamma = - b_{21}^{(2)}$$

$\ldots$

$$b_{13}^{(2)} = 0 = b_{31}^{(2)}$$

$$b_{23}^{(2)} = 0 = b_{32}^{(2)}$$

$\vdots$

$$b_{14}^{(2)} = - b_{41}^{(2)} = - \frac{1}{R_1} \cdot \cos \gamma$$

$\ldots$

The potential energy may be written in the form

$$V = \tfrac{1}{2} \langle (I_0 \underset{\sim}{\mu}) , \underset{\sim}{\varepsilon} \rangle_\Omega$$

where $$I_0 = \begin{bmatrix} 1 & \cdots & & & & 0 \\ & 1 & & & & \\ & & 1 & & & \\ & & & 1 & & \\ & & & & 1 & \\ 0 & & & & & 2 \end{bmatrix}$$

where the 2 in the last diagonal place is the penalty we pay for converting 3x3 matrices representing 2-nd rank tensors into 6-dimensional vectors.

The Castigliano functional representing the Lagrangian is given by

$$\langle \underset{\sim}{\mu} , \underset{\sim}{u}_0 \rangle_{\partial\Omega} - V.$$

The functional generating the Hu-Washizu variational principle is given by

$$\Pi(\underset{\sim}{\mu}. \underset{\sim}{\varepsilon} .\underset{\sim}{u}) = \tfrac{1}{2} \langle I_0 D\underset{\sim}{\varepsilon}, \underset{\sim}{\varepsilon}\rangle_{\Omega}$$

$$+ \langle A\underset{\sim}{u} - \underset{\sim}{\varepsilon}, I_0 \underset{\sim}{\mu} \rangle_{\Omega} - \langle \underset{\sim}{u}, f_{\rho}\rangle_{\Omega}$$

$$- \langle \underset{\sim}{u}_0, \underset{\sim}{\mu}\rangle_{\partial\Omega_1} - \langle \underset{\sim}{\mu}, \underset{\sim}{u}-\underset{\sim}{u}_0 \rangle_{\partial\Omega_2} .$$

If $k_1 = k_2 = \infty$, and if ξ_1, ξ_2 are the Cartesian coordinates all equations (3. 62) - (3. 65) reduce to the familiar equations of thin plate theory.

The outline offered in this section is certainly sketchy, inadequate and deals only with specific general assumptions. The subject is too extensive to be considered in a more comprehensive manner in a monograph attempting some generality. Hopefully, we indicated availability of the Hamiltonian formalism to the theory of thin shells in a manner similar to the techniques applied to thin plate theory in the section 3.8. There are numerous complications - the most obvious one concern inapplicability of the Cartesian coordinate representation and rather deep arguments of differential geometry that become unavoidable in the corresponding treatment of deformable thin shells. We deliberately bypassed this aspect of variational arguments that must emerge in a more complete discussion of this topic.

For a more conventional treatment of variational principles of Kirchhoff-Love shell theory read [44] , [45], [46] , [47], [48],[52] .

An early critique of the Kirchhoff-Love theory may be found in the Novozhilov-Finkelstein article [49] or the Mushtari-Galimov's treatise [50], and in the article [51]. More modern objections concern the inability of Kirchhoff's hypothesis to incorporate rapid changes in cross-sectional area, that are frequently associated with either optimal, or near-optimal plate and shell designs.

References

[1] den Hartog, J.P., Advanced Strength of Materials (New York; McGraw-Hill), 1952.

[2] Euler, L., Opera Omnia. Seriei secundae (11) Introduction to Vols. X and XI by C. Truesdell, 1960, The rational mechanics of flexible or elastic bodies, pp. 1638–1788.

[3] Hardy, G.H., Littlewood, J.E., and Polya, G., Inequalities. Second edition (Cambridge University Press), 1952.

[4] Lakshmikantham , V. and Leela, S., Differential and integral inequalities, theory and applications, Vol. 1 and 2, Academic Press, New York, 1969.

[5] Mitrinović, D.S., Analytic Inequalities, Grundlagen der Mathematischen Wissenchaften, Volume 165 (Berlin: Springer Verlag), 1970.

[6] Nikolai, E.L., Collected Works on Mechanics (Moscow: Gostekhizdat), pp. 452-53, 1955.

[7] Panovko, Ja. G., and Gubanova, I.I., Stability and Vibrations of Elastic Systems. Third edition (Moscow: Nauka), 1974.

[8] Rabczuk, R., Elements of differential inequalities, PWN, Warsaw, 1976.

[9] Szarski, J., Differential inequalities, PWN, Warsaw, 1967.

[10] C. Truesdell, The rational mechanics of flexible or elastic bodies, seriei secundae, vol. X, sect. altera, Society for Natural Sciences of Switzerland, pp. 345-370, 1960.

[11] S. Timoshenko, Theory of Stability for Elastic Systems (Moscow: Nauka), p. 91, 1955.

[12] S. Timoshenko and J.M. Gere, Theory of Elastic Stability. Second edition (New York: McGraw-Hill), 1952.

[13] G. Frobenius, Über adjungierte lineare Differential Ausdrucke, Jour. fur Math., 85(1878) p. 185-213.

[14] A.M. Arthurs, Complementary Variational Principles, Clarendon Press, Oxford University, Oxford, 1971.

[15] A.M. Arthurs, A note on Komkov's class of boundary value problems and associated variational principles, J. Math. Analysis & Applications, 33, (1971) p. 402-407.

[16] Jacob Bernoulli, Acta Eruditorum, (see in particular pages 268-269), Leipzig, 1696.

[17] G.A. Bliss, Lecture on Calculus of Variations, University of Chicago Press, Phoenix Science Series, Chicago, 1946.

[18] L. El'sgolc Differential equationa and the calculus of variations, Mir. Moscow, 1973, second English edition.

[19] A.R. Forsyth, Calculus of variations, Dover Publications, New York, 1960.

[20] M. Fréchet, La notion de differentielle dans l'analyse generale. Ann. Soc. de l'Ecole Norm. Super., 42, (1925), p. 293-323.

[21] M. Fréchet, Sur la notion differentielle, Jour. de Math., 16, (1937), p. 233-250.

[22] K.O. Friedrichs, On the boundary value problems in the theory of elasiticity and Korn's inequality, Ann. of Math., 48, (1947) p. 441-471.

[23] L.V. Kantorovich and G.P. Akilov, Functional Analysis in Normed Spaces, Pergamon Press, Oxford, 1964.

[24] T. Kato, On some approximate methods concerning the operators T*T, Math. Ann. 126, (1953), p. 253-257.

[25] V. Komkov, Anote on the vibration of thin inhomogenous plates, ZAMM, #408, (1968), p. 11-16.

[26] V. Komkov, The class of admissible loads of the linear plate amd beam theory, Inst. Lomb. di Science e lettere, Acad. di Science e Lett., Classe A, 105, (1971), p. 329-335.

[27] S.G. Mikhlin, Mathematical Physics - an advanced course, North Holland Publishing Co., Amsterdam, (1970), (Translated from Russian).

[28] J.V. Neumann, Über adjungierte functional operatoren, Ann. Math., 33, (1932), p. 294-310.

[29] Aubin, J. P., Approximation of Elliptic Boundary-Value Problems, Wiley-Interscience, New York, 1972.

[30] Fichera, G., Existence theorems in elasticity, Handbuch der Physik, Volume VI / A2, (1972), p347-389.

[31] Oden, J. T., Applied Functional Analysis, Prentice-Hall, Englewood Cliffs, N. J., 1977.

[32] Duvaut, G., and Lions, J. L., Les Inéquations en Mécanique et en Physique, Dunod, Paris, 1972.

[33] Vainberg, M. M., Variational Methods for the Study of Nonlinear Operators, (translated from the 1956 Russian monograph), Holden-Day, San Francisco, 1964.

[34] Oden, J. T., and Reddy, J. N., Variational Methods in Theoretical Mechanics, Springer-Verlag, Heidelberg, 1976.

[35] R.Kohn and M.Vogelius,A new model for thin plates with rapidly varying thickness,Parts 1 and 2, University of Maryland ,Laboratory for Numerical Analysis,Technical Notes, March , 1983.

[36] R.D.Mindlin,Waves and vibrations in isotropic elastic plates,in Structural Mechanics,New York,N.Y.,1960, p.209

[37] R.D.Mindlin,High frequency vibrations of piezo-elastic crystal plates,International Journal of Solids and Structures,vol.10,#4,(1974),p.343-359.

[38] R.D.Mindlin, - ditto - Part II ,vol.10,#6,(1974), p.625-637.

[39] A.S.Kosmodemianskii and V.N.Lozhkin,Generalized plane stressed state of piezoelastic plates,Prikl.Mekhanika, vol.13,#10(1975),p.75-79.

[40] I.A.Vekovishcheva,Two-dimensional problem of electroelasticity for piezo-elastic plates,Prikl.Mekhanika, vol.11,#2,(1975),p.85-89.

[41] P.G Ciarlet, Numerical Analysis of the Finite Element Method for Elliptic Boundary Value Problems, North Holland, Amsterdam, 1977.

[42] P.G. Ciarlet and P. Destuynder, A justification of the 2-dimensional linear plate model, J. Mecanique, <u>18</u>, (1979) p. 315-344.

[43] R.A. Adams, Sobolëv spaces, Academic Press, New York, 1975.

[44] A.L. Goldenveizer, Theory of elastic thin shells (in Russian), Nauka, Moscow, 1976.

[45] V.V. Novozhilov, Theory of thin shells, Oborongiz, Miscow, 1941.

[46] Paul Seide, Small elastic deformations of thin shells, Noordhoff, Leyden, 1975.

[47] A.E.H. Love, A treatise on the mathematical theory of elasticity, Dover Press, 4th edition, New York, 1944.

[48] F. John, Estimates for the derivatives of stresses in a thin shell and the interior shell equations, Comm. Pure and Appl. Math. 18, (1965), p. 235-267.

[49] V.V. Novozhilov and R.M. Finkelstein, On the errors of Kirchhoff's hypotheses in the theory of shells, P.M.M., Vol. 7 (1943) #5.

[50] Kh.M. Mushtari and K.Z. Galimov, Non-linear theory of elastic shells, Tatingosizdat, Kazan, 1957, English translation by the Israel program for scientific translations, Jerusalem, Israel, 1961.

[51] Kh.M. Mushtari, Region of applicability of the approximate shell theory of Kirchhoff and Love, P.M.M. Vol. 11, #5, (1947).

[52] N.P.Abovskii and N.L.Andreev,Variational principles in the theory of elasticity and in the theory of shells, Nauka,Krasnoyarsk,1973 (in Russian).

Chapter 4

VARIATIONAL FORMULATION OF PROBLEMS OF ELASTIC STABILITY. TOPICS IN THE STABILITY OF BEAMS AND PLATES

Variational formulation of the eigenvalue problems in engineering continuum mechanics.

4A. Statics

4.A.0 Introduction

Several authors studying the theory of boundary value problems describing deformation, buckling, and steady state vibration of elastic systems have developed a variational approach based on functional analytic arguments. See Aubin [1]. Generally a system of equations of the classical boundary value type is reduced to a variational, or to an energy formulation. This is not a new technique. A rigorous existence and uniqueness theory can be developed in a formulation that provides a direct link with the engineering theory of finite element analysis. See Aubin [1], or Ciarlet [2].

In order to take full advantage of the powerful variational formulation it is necessary to introduce appropriate function spaces for admissible displacement fields", and admissible force fields.

Traditionally, engineering texts introduced only the classes of relatively smooth functions. Only so called classical solutions were considered, or else no admissible classes of functions were mentioned and the manipulations were of purely heuristic nature. This is not the intention of this author. Several specific examples are presented in this chapter. Appendices A and B provide the necessary quick review of relevant elementary concepts of functional analysis.

4.A.1 Static buckling of columns.

Consider the column of Fig. 4.1, with a normalized axial coordinate x, clamped ends, and variable cross-sectional area $A(x) \epsilon L_\infty$. The area distribution $h(x)$ may be taken as a piecewise smooth function bounded above and below, in order to respect the engineering sensibilities; thus, $0 < h_0 < h(x) < h_1$, or it may be taken in the larger space $L_2[0,\ell]$, or of essentially bounded functions $L_\infty(0,\ell)$. Here $L_\infty(a,b)$ is the space of Lebesgue measurable functions $h(x)$ on the interval $a \leq x \leq b$, such that $\|h\|_\infty \equiv \inf\{M>0: |h(x)| < M$ a.e., in $[a,b]\}$. For an introductory treatment of such function spaces, see, for example, the text of Adams [3]. In some design problems it is presumed that all dimensions of the cross-section vary in the same ratio (that is the essential shape is fixed), so the moment of inertia of the cross-sectional area is $I(x) = \alpha A^2(x)$, where α is a positive constant depending on the shape of the cross-section. The boundary-value problem for displacement $w(x)$ is formally written as

$$(E\alpha h^2(x) w_{xx})_{xx} = q(x) ,$$

$$w(0) = w_x(0) = w(\ell) = w_x(\ell) = 0,$$

where E is Young's modulus, $q(x)$ is distributed load, and subscript notation is used to indicate derivatives.

4.A.1 The buckling phenomena.
Historical origins - Euler's formulation of the problem.

Leonhardi Euleri Opera Omnia contains an excellent exposition by C. Truesdell of Euler's papers and two Lagrange papers concerning basic principles of static equilibrium of elastic bodies. (See [4].)

The early treatise on elastic curves (1743) contains the section "On the strength of columns", and an earlier derivation and classification of the (nine) types of elastic curves.

The elastica problem posed by Daniel Bernoulli involves minimization of the integral

$\int_0^\ell \frac{ds}{r^2}$, where r is the radius of curvature.

In a letter to Euler of October 20th, 1742 Daniel Bernoulli clearly states a variational principle for a uniform elastic bar. Here Bernoulli assumes that the length of the elastic strip is invariant, and that the positions and slopes of the end points are given.

Using infinite series expansion for elliptic integrals Euler determined all possible shapes of the elastica. If one eliminates the elasticity B from Euler's formula, the following formula follows as an easy consequence

$\left(\frac{P}{P_{cr}}\right)^{\frac{1}{2}} = \frac{2}{\pi} K\{\sin(\alpha/2)\}$, where $K\{\alpha\}$ is the

complete elliptic integral of the first kind, while α is the angle between the direction of the load and the tangent to the elastica curve at the quarter period of this curve.

(See [4] pages 205-210)

This observation appears to have been made for the first time in E.L. Nikolai's treatise [6].

From the considerations of the elastica curves Euler derives the celebrated buckling load formula. Let me quote from "Opera Omnia":

> "If the elasticity and thickness of the column remain constant then the load P that it may bear without danger varies inversely as the square of the height", and "Unless the weight P to be supported satisfies the inequality
>
> $P > \pi^2 \frac{B}{\ell^2}$ there is no fear of bending".

The important feature of this early work of Euler and the variational postulation of Daniel Bernoulli is the lack of assumption concerning linearity. The onset of buckling, or "bifurcation of solutions" in the terminology of Lagrange (See [5]) can be predicted accurately using the linear approximation. However, the buckled shape and the post-buckled behavior of an elastic strip deduced from the linear theory must alway be viewed with great suspicion. Let me quote Truesdell: "That many of these results (of Euler) are often attributed to later authors may be due to the fact that Euler's work became more generally known through later publications, in which he gave more verbal description but less mathematical detail. In particular, the later literature, including Euler's own subsequent papers, unfortunately emphasizes the misleading connection between buckling and proper numbers of the linearalized theory, insufficient to predict the magnitude of bending." In his fundamental work of 1743, Euler derives the forms of bent elastic bands directly from the nonlinear differential equation. He does not overcome completely all difficulties inherent in the variational formulation, but he brilliantly observes the analogy between the integral formula $\int \frac{B}{r}\, d(\frac{1}{r}) = \frac{B}{2r^2}$ and the work of the "force of elasticity" by which he means the bending moment.(B is called the elasticity

Here he mentions the static version of the de Maupertuis principle, which is the principle of minimum for the potential energy.

A truly fascinating aspect of Euler's research involves his study of discrete models. His paper "De aequilibrio et motu corporum flexuris elasticis iunctorum", Novi Comm. Acad. Sci Pettrop. (of St. Petersburg), vol. 13 of 1768, presented, in 1763 and 1764, in two parts to the Academy contains basic ideas which have been reintroduced in modern literature in the context of finite element analysis. The elastic effects for small rotations of a twisted bar, or other elastic object, are replaced by those of a torsion cord (or a torsion spring).

This scheme has been recently used by many authors in problems involving optimization of vibrating elastic rods and in schemes determining parametric resonance (see for example [7]).

Unfortunately, mathematical sophistication necessary to derive estimates relating the behavior of such discrete models to the behavior of corresponding continuous systems was two centuries away in the future.

Truesdall remarks sadly that "Further studies of discrete oscillating systems do not seem to have contributed even indirectly to theories of flexible or elastic bodies".

Euler's paper of 1743 contained so many brilliant and new ideas that one would expect an explosion of knowledge to follow shortly. To the contrary, his ideas appear to have been ahead of their time, attracted no attention, and were followed only by his own papers and those of Lagrange many years later.

In his 1757 paper Euler points out applicability of his buckling analysis to the calculation of critical loads for columns of variable stiffness.

He reduces his second order equations to the form which now bears the name of Riccati, then completes his analysis by assuming a specific form of the elasticity function B (i.e. of EI(x)), where E is Young's modulus and I(x) the variable moment of inertia of the cross-section with respect to the neutral axis of bending).

In 1770 Lagrange observed the multiplicity of shapes of buckled columns corresponding to high critical loads, but his analysis of optimal shape was incorrect. That difficult problem had to await almost two centuries before significant progress was made towards its solution.

Before proceeding with some recent developments let us recapitulate the basic facts of the Euler elastica theory.

Let us assign x - axis to coincide with the axis of an undeformed column. We consider a rod whose ends are constrained to lie on the x - axis, but are free to rotate. Let ψ denote the angle between the axis of the rod (in the deformed state) and the x - axis. The rod has a unit length.

The elastica problem is described by the equations.

$$\psi_{xx} + \Lambda \sin \psi = 0, \qquad 0 \leq x \leq 1 , \tag{4.1}$$

$$\psi_x (0) = \psi_x (1) = 0 . \tag{4.1^a}$$

The number Λ is proportional to the applied axial load.

For small angles ψ this problem has been replaced by the linear version

$$(4.2) \quad \psi_{xx} + \Lambda \psi = 0 \quad ,$$

with the same boundary conditions (4.1^a). We should make clear that this is not the version discussed by Euler in his early work of 1743.

Clearly $\psi \equiv 0$ is a solution of $(4.1),(4.1^a)$ and of $(4.2),(4.1^a)$.

Multiplying (4.1) by ψ_x and integrating one easily derives an equation of Ricccati type

$$(4.2^a) \quad \frac{1}{2} (\psi_x)^2 = \Lambda(\cos \psi(x))$$

with $\psi_0 = \psi(0)$.

Since $0 \leq \psi \leq \Pi$, $0 \leq \psi_0 \leq \Pi$, $0<\Lambda$,

we see that $\psi_0 < \psi(x)$, if $\psi \not\equiv 0$ (since the right hand side of (4.2^a) must be positive). Following Euler's ideas we can change the dependent variable

$$\phi(x) = \text{arc} \sin(\frac{1}{k} \sin \frac{\psi}{2}),$$

where $k = \sin(\psi_0/2)$,

introduce the elliptic integral

$$F(k) = \int_0^{\pi/2} (1 - k^2 \sin^2 \phi)^{-\frac{1}{2}} d\phi \; ,$$

and solve the differential equation

$$(4.3) \quad \mu \frac{dx}{d\phi} = (1 - k^2 \sin^2 \phi)^{-\frac{1}{2}}$$

in terms of infinite series.

The interesting points of Euler's theory which were revised or rediscovered by modern authors involve comparison of solutions and eigenvalues of (4.1)and (4.2).

The eigenvalues of (4.3) are $\mu_n = 2n\,F(k)$, $n = 1,2,\ldots$, which are exactly the square roots of the eigenvalues of the linear problem (4.2^a), i.e. (4.2).

This could be deduced by a purely physical

argument. Since the shape of the rod bifurcates from the initial solution $\psi \equiv 0$ the critical loads should be determined by "the small angle theory", i.e. by the linear theory.

Both theories predict a countably infinite number of possible solutions of the corresponding eigenvalue problem. The choice of the buckled shape has to be determined by appropriate energy levels.

The shape of the buckled solution corresponds to the minimum of potential energy. The branching occurs at values of Λ in (4.2) or μ in (4.3) where equal levels of potential energy correspond to at least two different shapes of the buckled rod.

For details of the infinite series expansion I recommend a study of the original paper of Euler or [4].pages 205-215 .

4A.2 Elastic stability of a compressed column Euler's approach (continued)

Leonard Euler studied the stability of a compressed column and discovered in 1743 the phenomenon of buckling. In his treatise on shapes of elastic curves (1743) he analyses the equilibrium state of a column and comes to the conclusion that in the theory of compressed columns one can ignore the bending effects if the load P satisfies the inequality

$$P < P_{cr} = \pi^2 \beta / l^2 \tag{4.4}$$

For columns with constant cross-section β is a constant which Euler called the elasticity of the column and ℓ is the length of the column.

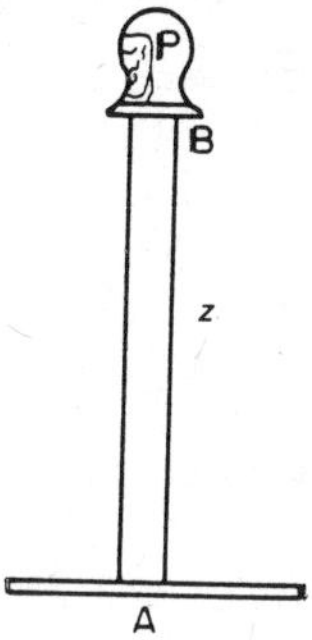

Figure 4.1 Euler's diagram for the buckling problem.

To interpret the meaning of AB on the figure above, see [4]. In the figure z denotes the length of a quarter of the period of the deflection curve. P is the compression load.

In this exposition the Cartesian coordinate system is adopted. The x-axis coincides with the axis of underformed column. The deflection curve for the loaded column is designated by the functional relation y = y(x), $0 \leq x \leq \ell$. The basic relation between the bending moment M(x) acting on a cross-section of a beam or column and the transverse deflection y(x) was first derived by Euler

$$\beta(x)=EI(x),\ EI(x)y''(x)=M(x) ,$$

- trully the fundamental formula. E is a constant depending on the properties of the material (it is called Young's modulus), I(x) is the moment of inertia of the cross-sectional area about the neutral axis of bending (the line whose length is unchanged as the column bends), y"(x) denotes d^2y/dx^2. However, Mariotte and Leibniz realized the importance of M(x) I(x) in their work on properties of elastic beams (1684). For the derivation of the basic relation β=EI see reference [8], pages 143-146, 387-388 and 403-404.

In an article published in 1759 (Hist. Acad. Sci. Berlin, 252-282) Euler derived the basic linear equation By" + py = 0. He made various assumptions (see [4, p. 315] for details), which were absent in his earlier more precise arguments.

Lagrange made an attempt to optimize the design of a compressed column having a variable cross-section, but was not very successful in tackling this difficult problem. He rederived Euler's formula for the critical load, remarking on the class of problems dealing with the 'bifurcation of the equilibrium' (see [4, pp. 352-353]).

Modern engineering textbooks generally follow Euler's arguments. The derivation given in most textbooks of the fundamental relation

$$M(x) = (EI)y''(x) \tag{4.5}$$

between the bending moment M(x) and the transverse displacement y(x)- generally follows Euler's arguments (see, for example, den Hartog [8, Ch. 8, pp. 251-258]). The relation (4.5) can be used to establish the basic differential equation for the transverse displacement of a slender column.

$$(d^2/dx^2)(EI(x)d^2y/dx^2) + Pd^2y/dx^2 = 0 \tag{4.6}$$

Equation (4.6) is the equilibrium equation for the deflected column. Euler did encounter some difficulties in disposing of the boundary conditions which are imposed by physical considerations. His estimates of the critical loads were wrong because of a simple error in arithmetic, but his analysis based on physical intuition was correct. A detailed description of Euler's solution can be found in the monograph of Nikolaĭ [6], and in chapter 1, entitled 'Euler's error', of the monograph of Panovko and Gubanova [9].

Truesdell comments that the paper on linearalized theory (the one to which the authors of reference [9] refer) should not be taken in preference to the treatment which was given by Euler earlier and which was exact. The 'bifurcation of equilibrium' appears there first and is handled correctly.

4.3 The energy approach

An energy approach to the buckling problem

was first introduced by Kirchhoff in 1850. A modernized version of Kirchhoff's argument can be found in Timoshenko's textbook [11].

Basically, he states that the stability or instability of a compressed column depends on the sign of an appropriate energy form.

Let us consider the case of column with a constant cross-section which is simply supported at both ends and is subjected to a point load applied at each end. This configuration is shown in figure 4.2

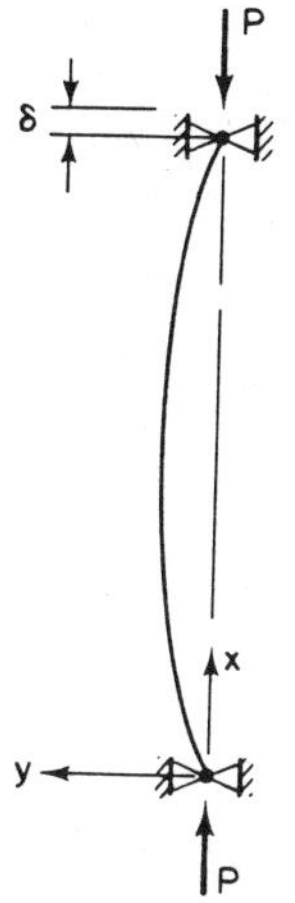

Figure 4.2

The constant load P may change the shape of the column from a straight shape($y\equiv 0$) to some unknown buckled shape $y=\tilde{y}(x)$. This change of shape causes an axial displacement equal to δ.

At this point we need to define the class of admissible shapes $\tilde{y}(x)$. For physical reasons, only continuously differentiable functions can be considered as deflection functions for a beam. Discontinuity in $\tilde{y}'(x)$ could exist only if the beam ceased to obey the hypothesis of the theory of elasticity.

Following the classical arguments of Euler we assume that the total length of the beam remains constant as the beam changes its shape and that the slope is 'small'.

We compute the deflection δ. Clearly, since $y \in C^1[0,1]$

$$
\begin{aligned}
l &= \int_0^{l-\delta} [1+(y')^2]^{1/2}\,dx \\
&= \int_0^{l} [1+(y')^2]^{1/2}\,dx - \int_{l-\delta}^{l} [1+(y')^2]^{1/2}\,dx \qquad (4.7)\\
&\approx \int_0^{l} [1+(y')^2]^{1/2}\,dx - \delta
\end{aligned}
$$

Hence,

$$
\begin{aligned}
\delta &\approx \int_0^{l} [1+(y')^2]^{1/2}\,dx - l \\
&\approx \left[\int_0^{l} dx + o((y'))^4\right] - l + \frac{1}{2}\int_0^{l} (y')^2\,dx \\
&\approx \frac{1}{2}\int_0^{l} (y')^2\,dx
\end{aligned}
$$

The work performed by the force P is

$$W = P\delta = \frac{1}{2}P\int_0^{l} (y')^2\,dx. \qquad (4.8)$$

The strain energy of the beam bent into the shape $y = \tilde{y}(x)$ is given by

$$\frac{1}{2}\int_0^{l} M\,d\theta,$$

which is approximately equal to

$$V(\tilde{y}) = \frac{1}{2}\int_0^{l} EI(x)(\tilde{y}'')^2\,dx \qquad (4.9)$$

(see the relation (4.5) The principle of lowest potential energy implies that the beam will buckle if for any admissible function $\tilde{y}(x) \neq 0$ the inequality

$$V(\tilde{y}) < W(\tilde{y}) \qquad (4.10)$$

is true.

The trivial solution $y \equiv 0$ is stable (the beam will not buckle) if the potential energy function $J(y) = V(y) - W(y)$ attains its minimum in the class of admissible functions when $y \equiv 0$. We shall use this fact and a well known analytic inequality, known as Wirtinger's inequality, to derive directly the critical load formulas.

We comment that the commonly used derivations of Euler's critical load formula use either some expansion of the solution in infinite series, or a rather sophisticated variational principle due to Lord Rayleigh. For example, the proof given in a very popular textbook written by Timoshenko and Gere [11,pp. 88-92] uses expansion of the solution in trigonometric series. The authors equate the work performed by the force P with the increase in the strain energy and finally give P in terms of the Fourier coefficients of the deflection function. Infinitely many values of P appear to solve this problem. By choosing the smallest value of $P \neq 0$, the authors derive Euler's critical load formula.

Other textbooks use Rayleigh's principle to find the smallest eigenvalue P_1 of the equation (4.6). Specifically, if I = constant, one wishes to find

$$P_1 = \min_{y \in Y} \left\{ \left(EI \int_0^l (y'')^2 \, dx \right) \bigg/ \left(\int_0^l (y')^2 \, dx \right) \right\}, \quad y \not\equiv 0 \tag{4.11}$$

where Y is the class of admissible deflections.

4.4 The Wirtinger-Poincaré-Almansi inequality

In 1905 Almansi proved the following result. Let f be a function which is continuously differentiable on [a,b], such that f(a) = f(b), and

$$\int_a^b f(x)\, dx = 0.$$

Then

$$\int_a^b \{f'(x)\}^2 \, dx \geqslant \left(\frac{2\pi}{b-a}\right)^2 \int_a^b \{f(x)\}^2 \, dx \tag{4.12}$$

This is a slight generalization of the better known Wirtinger's inequality. It can be regarded as a special (one-dimensional) case of Poincaré's inequality. In particular, Levi's form of

Poincare's inequality can be used to derive the Almansi or Wirtinger inequalities.

An extensive discussion of the various generalizations and of the history of inequalities of the type (4.12) is given in the monograph of Mitrinovič [12, pp. 141-151]. It also contains an excellent bibliography related to such inequalities. Here we shall need only the following special case of Wirtinger's inequality.

Let y(x) be a function defined on the interval $[0,\pi]$ such that $y(0) = y(\pi) = 0$ and $y'(x) \in L^2[0,\pi]$.

Then

$$\int_0^\pi (y')^2\,dx \geqslant \int_0^\pi y^2\,dx \quad . \tag{4.13}$$

The equality sign is true only if $y = A \sin x$, where A is a constant. A short proof of this inequality, which was given by Hans Lewy, appears as a footnote in the monograph of Hardy, Littlewood and Polya [13, p. 185].

We shall refer to this inequality as the Wirtinger-Poincaré-Almansi (WPA) inequality.

We observe that the inequality (4.5) can be restated as a modification of Almansi's inequality

$$\left(\frac{l}{\pi}\right)^2 \int_0^l (y')^2\,dx \geqslant \int_0^l y^2\,dx \tag{4.14}$$

if $y(0) = y(l) = 0$, $y' \in L^2[0,l]$, (for an arbitrary interval $[0,l]$) after a routine argument involving a change of variable. Here equality is attained only for the function $y = A\sin(\pi x/l)$. Similar inequalities can be defined on an interval $[0,l]$ for functions satisfying only a single condition $y(0) = 0$.

Hint: Symmetrize the problem by defining the function y(x) on the interval $[0,2l]$. For any $x \in [l,2l]$ define $y(x)=y(l+\xi) = -y(l-\xi)$, where $\xi=x-l$.

If end conditions are assigned to the derivative y'(x), rather than to the function y(x), it is not hard to derive an inequality similar to (4.13). For example, the following assertion is easily shown to be true.

Let y(x) be a function defined on [0,2Π],with $y'(x)\in L^2[0, 2\pi]$, y satisfying

$$\int_0^{2\pi} y(x)\, dx = 0 .$$

y' also satisfies boundary bonditions $y'(0) = y'(2\pi) = 0$.
Then,

$$\int_0^{2\pi} (y')^2\, dx \geqslant \int_0^{2\pi} y^2\, dx, \tag{4.14a}$$

and the equality sign is valid only if y =A cos x. The conclusions of this assertion are valid if we replace the interval [0,2π] by the interval [0, π]. In that case the requirement

$$\int_0^{\pi} y(x)\, dx = 0$$

may be replaced by the condition y(π/2) = 0.

4.5. Derivation of Euler's buckling load formula

We shall use Kirchhoff's energy argument. The stability (instability) of the compressed column depends on the non-existence (existence) of an admissible function $y(\tilde{x})$, $x\in[0,l]$ which satisfies the inequality (4.4). We are ready to state Euler's formula as a theorem.

Theorem 4.1

We consider a column having a constant cross-section which is freely supported at both ends and loaded as shown in figure 4.2 by a point load P applied at both ends. Then the trivial solution $y\equiv 0$ is stable if and only if $P<\pi^2 EI/l^2$.

Proof

Let us suppose that the trivial solution is stable. Then for every admissible function $\tilde{y}(x)\not\equiv 0$ we have the Kirchhoff energy inequality

$$V(\tilde{y}) \geq W(\tilde{y}), \tag{4.15}$$

that is

$$\frac{1}{2}\int_0^l EI(\tilde{y}'')^2\, dx \geq \frac{1}{2}\int_0^l P(y')^2\, dx . \tag{4.16}$$

Using the WPA inequality

$$\int_0^l (\tilde{y}'')^2\,dx \geqslant \left(\frac{\pi}{l}\right)^2 \int_0^l (y')^2\,dx$$

we obtain the desired inequality

$$P<\pi^2 EI/l^2 \qquad (4.17)$$

The converse argument is clear.

The equality V = W is attained when $P=P_{critical}=\pi^2 EI/l^2$ and when the WPA equality is true.

Two functions exist $y_1\equiv 0$ and $y_2=\sin(\pi x/l)$ such that $V(y_1)$ and $V(y_2) = W(y_2)$. The minimum of potential energy J is attained by $J(\tilde{y}), \tilde{y}\equiv 0$, when $P<P_{critical}$, but J is minimized by two functions $(y_1(x),\ y_2(x))$ when $P=P_{critical}$. In the terminology of Lagrange, the solutions of (4.3) bifurcate from the trivial solution as P increases past the critical value $P_{critical}=\pi^2 EI/l^2$. We comment that stability of a column which is subjected to different (possibly more complex) loads or boundary conditions may be examined by using almost identical arguments. For example, let us consider a column having one end rigidly held while the other end is free. A compressive load is applied to the free end, as shown in figure (4.3).

Let $\tilde{y}(x)$ denote an admissible deflection function for this column. $\tilde{y}(x)$ satisfies the following boundary conditions: $y(l)= y'(l) = 0$. We extend the function $\tilde{y}(x)$ to the interval $[0,2l]$ by defining $\tilde{y}(l+x)=\tilde{y}(l-x)$. We apply the WPA inequality to the extended function $\tilde{y}(x)$, $x\varepsilon[0,2l]$ and observe that this inequality must be valid on the interval $[0,\ell]$ as well as on the interval $[0,2\ell]$.

Thus we can derive the inequality

$$(4.18) \qquad \int_0^l (\tilde{y}')\,dx \geqslant \int_0^l \left(\frac{\pi}{2l}\right)^2 \tilde{y}^2\,dx \qquad \text{, i.e.}$$

Euler's buckling formula .

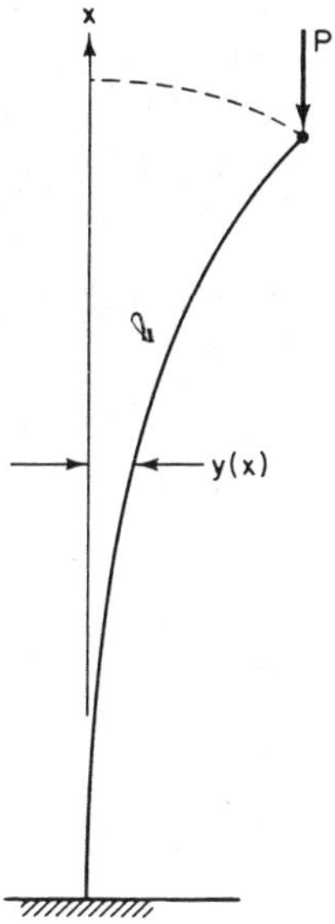

Figure 4.3

Substituting this inequality into Kirchhoff's law $V \geq W$, we obtain the stability condition $P \leq \pi^2 EI/(4l^2)$ which again agrees with Euler's estimate.

We make the following observation. One could use modern functional analytic arguments to arrive at the same conclusions. In the setting of Sobolëv space $H_0^2[0,l]$ the operator(functional) $J = V - W: H_0^2[0,l] \to \mathbb{R}$ is coercive if and only if P is smaller than the critical value. But if the operator J is coercive then the problem of minimizing $J(y)$, $y \in H_0^2[l]$ has a unique solution.

Consequently, the equation

$$(EIy'')'' + Py'' = 0 \qquad (4.19^a)$$

with appropriate boundary conditions

$$y(0) = y(l) = y''(0) = y''(l) = 0 \qquad (4.19^b)$$

has at most one solution. Since $y \equiv 0$ is a solution of (4.19a) with boundary conditions (4.19b) it is the only solution if $P < P_{\text{critical}}$.

P_{critical} is the first eigenvalue of the problem (4.19a) under (4.19b) and $y_1 = \sin(\pi x/l)$ is the corresponding eigenfunction.

Equation (4.19a) is the Euler-Lagrange equation for the problem of minimizing the functional $J(y) = V - W$.

4.6 A formulation of the eigenvalue problem for an elastic column.

A small deflection theory (almost quadratic energy form). We assume the validity of Hooke's law and of all assumption stated in the preceding part of this paper, with one exception. We are not going to ignore the effects of compression strain energy caused by the force P.

The compressive force acting on an element dx along with column is roughly $P \cdot \cos(y')$, causing a longitudinal deflection

$$(4.20)\quad \delta_c = \int_0^{\ell} \frac{P\cos(y')\,dx}{E\,A(x)} .$$

While we obtain nonlinear terms, we follow the basic linear theory. Replacing cos(y') by the series $1 - \frac{(y')^2}{2!} + \frac{(y')^4}{4!}$ we obtain an approximate expression

$$(4.21)\quad \delta_c \approx P\int_0^{\ell} \left(\frac{1-\frac{1}{2}(y')^2}{E\cdot A(x)}\right) dx,$$ that is almost a linear dependence. The axial deflection caused by combined geometric effects and compression effects is given by

$$(4.21^a)\quad \delta = \frac{1}{2}\int_0^{\ell} (y')^2 dx + P\int_0^{\ell} \left[\left(1-\frac{(y')^2}{2}\right)(E\,A(x))^{-1}\right]dx$$

The total strain energy of the column is

$$U_B + V_C = \frac{1}{2}\int_0^{\ell} (E\,I(x)\,(y'')^2)\,dx + \frac{1}{2}\int_0^{\ell} \frac{P^2\cos^2(y')}{E\,A(x)}\,dx$$

which is roughly equal to

$$U \approx \frac{1}{2}\int_0^{\ell} (E\,I(x)\,(y'')^2)\,dx + \frac{1}{2}\int_0^{\ell} \frac{P^2(1-(y')^2)}{E\,A(x)}\,dx \quad (4.22)$$

(Note: $\cos^2 t = 1 - t^2 + \frac{1}{3}t^4 \ldots$)

The work performed by the force P should be computed as a sum of two separate expressions.
The work of compressing an essentially straight shape is

$W_C = \frac{1}{2} \int_0^\ell \frac{P^2\, dx}{E\, A(x)} = \frac{1}{2} P \cdot \delta_C$, and the work of

deflecting the straight shape into the buckled shape is

$W_B = P \cdot \delta_B = \frac{1}{2} P \int_0^\ell (y')^2\, dx.$

In small deflection theory we can equate the work performed by the force P to the total strain energy of the column. Thus, we have(approximately)

$$\frac{1}{2} \int_0^\ell \frac{P^2\, dx}{E\, A(x)} + \frac{1}{2} \int_0^\ell P (y')^2\, dx =$$

$$\frac{1}{2} \int_0^\ell (E\, I\, (x)\ (y'')^2\, dx + \frac{1}{2} \int_0^\ell \frac{P^2 \cos^2(y')}{E\, A\, (x)}\, dx \ , \quad (4.22^a)$$

with the last term on the right hand side assumed to be roughly equal to

$$\frac{1}{2} \int_0^\ell \frac{P^2 (1 - (y')^2)}{E\, A\, (x)}\, dx. \qquad (4.22^b)$$

Hence, we have the following eigenvalue problem:

$$(4.23) \quad P^2 \int_0^\ell [(E\, A\, (x))^{-1}\ (y')^2]\, dx + P \int_0^\ell (y')^2\, dx$$
$$- \int_0^\ell E\, I\, (y'')^2\, dx = 0.$$

(Note that this value does not minimize V-W!)
The variational equation corresponding to (4.23) is

$$(4.24\ ) \quad P^2 [(E\, A\, (x))^{-1}\ y')'] + P\ y'' +$$
$$+ (E\, I\, (x)\ y'')'' = 0\ .$$

If $\tilde{y}(x)$ is a buckling deflection curve, the corresponding critical load is given by

$$(4.25)\ P_{cr} = \frac{1}{2} [\int_0^\ell (E\, A\, (x))^{-1} (\tilde{y}')^2 dx]^{-1} \cdot \{-\int_0^\ell (\tilde{y}')^2 dx +$$

$$[\int_0^{\ell} (\tilde{y}')^2 dx + 4(\int_0^{\ell} E\ I\ (x)(\tilde{y}'')^2\ dx)\cdot$$

$$(\int_0^{\ell} (E\ A\ (x))^{-1}(\tilde{y}')^2\ dx]^{\frac{1}{2}})\}$$

We comment that the other root of the quadratic equation (4.24) is ignored because only positive values of P_{cr} are of interest in the buckling problem.

In what follows, we assume a direct functional relation between $I(x)$ and $A(x)$, i.e. $I(x) = \phi\ (A(x))$ $\phi: \mathbb{R}_+ \to \mathbb{R}_+$, $\phi \in C^1\ (\mathbb{R}_+)$, in effect restricting the geometry of the design.

A formula which will prove useful in future considerations involves computation of the variation of P_{cr} with respect to changes in the design $A(x)$, subject to the "normality" constraint

$$\int_0^{\ell} (\tilde{y}')^2\ dx = 1.$$

$$\frac{\partial P_{cr}}{\partial A} = \tfrac{1}{2}[\int_0^{\ell} (E\ A)^{-1}(\tilde{y}')^2\ dx]^{-1}\ \{\tfrac{1}{2}\ [\int_0^{\ell} (\tilde{y}')^2 +$$

$$+4\int_0^{\ell}(E\phi(A)\ \cdot(\tilde{y}'')^2 dx \int_0^{\ell} (EA)^{-1}(\tilde{y}')^2 dx]^{\frac{1}{2}}[4E\phi'(A)(\tilde{y}'')^2$$

$$\cdot\int_0^{\ell} (EA)^{-1}\ (\tilde{y}')^2 dx - 4\int_0^{\ell} E\phi(A)(\tilde{y}'')^2 dx\ \cdot(EA)^{-2}(\tilde{y}')^2]$$

$$+ P_{cr}((EA)^{-2}\ (\tilde{y}')^2))\}.\quad (\ \frac{\partial\tilde{y}}{\partial A} \text{ is supposedly small}).$$

Denoting by $K = \int_0^{\ell} [EA)^{-1}\ (y')^2]\ dx$,

by $S = \int_0^{\ell} [E\ \phi(A)\ (y')^2]\ dx$,

and invoking the "normality" condition, we have

$$(4.26)\quad \Lambda_A = \frac{\partial P_{cr}}{\partial A} = \frac{1}{K}\ \{[1+4KS]^{-\frac{1}{2}}\ [E\phi'(A)(y'')^2\ \cdot\ K$$

$$-(EA)^{-2}(y')^2\ S] + \tfrac{1}{2}\ P_{cr}\ (EA)^{-2}\ (y')^2\}.$$

Λ_A is the sensitivity of the critical load to

the changes in the design. Before analyzing design problems, replacing the Rayleigh quotient by the non-linear form (4.25) , and deriving corresponding optimality criteria, we first point out some similarities and differences which are stated in the form of lemmas.

Lemma 4.1 If I(x) is a differentiable function of A(x), then P_{cr} is a differentiable function of A(x) and as $F(A)= \int_0^{\ell} (y')^2 \ (EA(x))^{-1} \, dx$

approaches zero, the value of P_{cr} approaches the value of the Rayleigh quotient, i.e.

$$\lim_{F(A)\to 0} (P_{cr}) = \frac{\int_0^{\ell} (E\, I\, (A(x))\ (\tilde{y}'')^2)\, dx}{\int_0^{\ell} (\hat{y}')^2 \, dx} \tag{4.27}$$

with the critical shape $\tilde{y}$ regarded as "fixed".

The statement of this lemma follows directly from an application of ℓ' Hospital's rule.

Lemma 4.2. P_{cr} is always smaller than the corresponding value of the Rayleigh quotient.

To prove this lemma we need to establish the following simple algebraic inequality. We consider the positive zero of the quadratic function $y = ax^2 + bx - c$, where a, b, c are positive real numbers, and the zero of the linear equation $z = bx - c$, and claim that the corresponding value of $x = \frac{c}{b}$ is a greater than the larger of

the two values of x corresponding to the zeros of y(x). (The case of equal roots is trivial.) Otherwise we would have the inequality

$$-\frac{b}{2a} + \sqrt{b^2 + 4ac} > \frac{c}{b} \quad , \text{ or}$$

$$\sqrt{b^2 + 4ac} > \frac{2ac}{b} + b > 0$$

and $b^2 + 4ac > b^2 + 4 \cdot ac + \frac{4a^2c^2}{b^2}$ which is impossible.

The proof of the lemma is complete if we identify

$$c = \int_0^{\ell} (EI(x)(y'')^2)dx$$

$$b = \int_0^{\ell} (y')^2 \, dx$$

$$a = \int_0^{\ell} (y')^2 (EA(x))^{-1} \, dx.$$

Note: The variation of $\tilde{y}$ was ignored in $\frac{\partial P_{cr}}{\partial A}$. Thus the limitations are quite severe.

So far, no conditions on smallness of y were explicitly stated, although, if they were specifically not regarded as small then the first term in the equation (4.23) should be replaced by the nonlinear energy form

$$(4.28) \quad \int_0^{\ell} [EI(x)\ (y'')^2 \cdot (1 + (y')^2)^{-3/2}]\,dx,$$

and other corrections may be necessary. For the time-being we shall retain only the first (constant) term in the series for $(1 + (y')^2)^{-3/2}$ remembering that this approximation should also be discarded in some discussions of optimal designs.
Theorem 4.2 The system (4.24) satisfies the Rayleigh quotient criterion and the form assumes a local minimum if among admissible

$H_0^4\ [o,\ell]$ functions y(x) is chosen so that it

satisfies the equation (4.25). Moreover the minimal value of the Rayleigh quotient for (4.25) approaches the classical value

$$\lambda_1 = \min \left\{ \frac{\int_0^{\ell} EI(x)\ (y'')^2\ dx}{\int_0^{\ell} (y')^2\ dx} \right\} \qquad (4.29)$$

as $\int_0^{\ell}\left((y')^2 \cdot (EA(x))^{-1}\right)dx \to 0.$

Proof. First we check on the necessary condition

$$\frac{\partial P_{cr}}{\partial \tilde{y}} = 0 \text{ if } y = \tilde{y}_{cr}.$$

Comments.

Let us stop, pause and consider the implications of our analysis.

The choice of non-linear terms was crude, the assumption that the term containing $\frac{\partial P_{cr}}{\partial \tilde{y}}$ can be ignored is easily knocked down by a suitable counter-example, the equality of work performed to strain energy increase is generally not valid, since the static equilibrium requires a minimum of potential energy that is of $U = V-W$, rather than the equality of V and W; that is of $U = 0$.

$U = 0$ corresponds to the minimum of $V - W$ when U is a quadratic form, but generally these two conditions are independent of each other. Thus, all kind of errors have been introduced into our analysis.

However, it is still an improvement on the linear eigenvalue problem in which we make assumptions in formulating a problem that is not obeyed by solutions to the same problem. The linear problem contains all of our errors and additional unjustified assumptions.

In shape optimization one routinely encounters such difficulties. Something is assumed to be "small" but the optimizing process produces non-negligible effects of this "small" quantity.

Among other implications, the statement $U = 0 \rightarrow \partial U/\partial y = 0$ is generally false if $y'(x)$ is not small. Some of these limitations will be discussed more carefully in volume 2 of this work in the contexts of shape optimization and sensitivity of the buckled shapes.

Differentiating (4.23) (using formal rules of Frechét differentiation as outlined in appendix A) we have the virtual work condition:

$$-2P \frac{\partial P}{\partial y} \left(\int_0^{\ell} [(EA)^{-1}(y')^2]\, dx \right) + P^2((EA)^{-1} y')'$$

$$- \frac{\partial P}{\partial y} \int_0^{\ell} (y')^2\, dx + P\, y'' + (EIy'')'' = 0 \qquad (4.30)$$

If $\frac{\partial P}{\partial y} = 0$, we recover the equation (4.24) as a necessary condition of the stationary behavior of P.

To show that this is a local minimum in $H_0^2\,[0,\ell]$ we check the behavior of the operator which represents the second derivative $\partial^2 P/\partial y^2$ at the point where the first derivative vanishes.

The operator $\frac{\partial^2 P}{\partial y^2} : H_0^{+2} \rightarrow H_0^{-2}$ is computed from equation (4.30), after we set $\frac{\partial P}{\partial y} = 0$.

The operator $\frac{\partial^2 P}{\partial y^2}$ satisfies the following equality at the critical point.

$$\frac{\partial^2 P}{\partial y^2} = -\left\{2\, P_{cr} \int_0^{\ell} [(EA)^{-1}(y')^2\, dx] + \int_0^{\ell} (y')^2 dx\right\}^{-1}$$

$$\cdot \left\{P^2 \frac{d}{dx} \left[(EA)^{-1} \frac{d}{dx}\ \right] + \frac{d^2}{dx^2} \left(P + EI \frac{d^2}{dx^2}\right)\right\}.$$

We observe that for a free beam subjected only to the end force P, the term

$\left(Py + EI \frac{d^2 y}{dx^2}\right)$ identically vanishes along the length of the beam in the equilibrium configuration. P_{cr} is positive, the operator

$P^2 \frac{d}{dx} [EA^{-1}) \frac{d}{dx} \cdot\ ]$ is a negative definite operator.

Hence, in some immediate neighborhood of an equilibrium configuration we have the inequality

$$\frac{\partial^2 P}{\partial y^2} > 0 .$$

This implies both local convexity of P as a functional of y, and local minimum property of P at the critical point where the Fréchet derivative of the nonlinear eigenvalue P, i.e. where $\frac{\partial P}{\partial y}$ vanishes.

The second part of our assertion follows from a direct application of ℓ-Hospital's rule to the formula (4.24), and from the well-known minimal property of Rayleigh's quotient.
The error $\tilde{e}$ committed by substituting ℓ as the upper limit in the integral, instead of $\ell - \delta_G$, is of the order

$$\tilde{e} \leq \max \{ \frac{1}{2} (y')^2 - \frac{1}{8} (y')^4 \} \cdot \delta_G ; \; x \in [0 \;\; , \delta_{G/2}] .$$

If the beam is clamped at both ends, the quantity in the curly brackets is very small and can be safely ignored.
The second effect that we do not ignore concerns the change in the total length of the column, caused by the change in the value of axial compressive force from P to P cos θ, where θ = arctan(y')). The actual change in the length of the column is

$$\delta_\ell' = \int_0^\ell \frac{P \cos \theta(s) - P}{EA(s)} dx.$$

The corresponding change in the deflection of the bent column along the x-axis is given by

$$\delta_c' = \int_0^{[\ell - \delta_G - \delta_c']} [\frac{P \cos^2 \theta(x) - P \cos \theta(x)}{EA(x)}] dx,$$

and we have

(4.31) $P^2 \int_0^{\ell} [\frac{1}{8}(y')^4 (E\,A(x))^{-1}]\,dx + P \int_0^{\ell} [\frac{(y')^2}{2} - \frac{(y')^4}{8}]\,dx - \frac{1}{2}\int_0^{\ell} [EI(x)(y'')^2 (1 + (y')^2)^{-3/2}]\,dx = 0$, or replacing

$(1 + (y')^2)^{-3/2}$ by

$1 - 3/2(y')^2 + \frac{15}{8}(y')^4 \cdot\cdot\ (y')^6$, we have

$$(4.32)\quad P^2 \int_0^{\ell} \frac{1}{8}[(y')^4 (EA)^{-1}]\,dx + P \int_0^{\ell} [\frac{(y')^2}{2} - \frac{(y')^4}{8}]\,dx - \frac{1}{2}\int_0^{\ell} (EI(y'')^2)[1 - 3/2(y')^2 + \frac{15}{8}(y')^4]\,dx = 0.$$

The positive root of the equation (4.32) is given by the usual quadratic formula which turns out to be unpleasant, but easily computed.

We comment that the negative root of the equation (4.32) is ignored for an obvious reason, since tensile forces are of no interest in the buckling process.

In what follows we shall assume a direct functional relation between I(x) and A(x), i.e. $I(x) = \phi(A(x))$.

$\Phi : \mathbb{R}_+ \to \mathbb{R}_+$, $\phi \in C^1(\mathbb{R}_+)$, in effect restricting the geometry of the design.

A basic formula which is necessary in considerations of optimality involves computation of sensitivity of the critical load P_{cr} with respect to the changes in the design function A(x) only.

While sensitivity with respect to design variables and optimality of design problems are not formulated properly in this volume, one has to

consider the sensitivity of eigenvalues with respect to design variable in designing of any computational algorithms, if we want to avoid some unpleasant surprises.
We denote by: $K = \int_0^{\ell} [(EA)^{-1} (y')^4]\, dx$

(4.35) $B = \int_0^{\ell} [\frac{(y')^2}{2} - \frac{(y')^4}{8}]\, dx$

(4.36) $S = \int_0^{\ell} [E\, I(x)\, (y'')^2\, (1 + (y')^2)^{-3/2}]\, dx.$

Then $P_{cr} = 4K^{-1} \{-B + \sqrt{B^2 + \frac{1}{4} KS}\}$ (4.37)

We compute the Fréchet derivatives

(4.38) $\frac{\partial K}{\partial A} = -(E^{-1}(y')^4\ A^{-2})$

(4.39) $\frac{\partial B}{\partial A} = 0$

(4.40) $\frac{\partial S}{\partial A} = E(y'')^2(1 + (y')^2)^{-3/2}\ \frac{dI(A)}{d\ A}$, and

(4.41) $\frac{\partial P_{cr}}{\partial A} = \Lambda_A = 4\ K^{-2}(E^{-1}(y')^4 A^{-2}\{ -B + (B^2 + \frac{1}{4}(KS))^{\frac{1}{2}}\} + 4K^{-1}\cdot(\frac{1}{2}B^2 + 1/4\ KS)^{-\frac{1}{2}}.$

$[(-\frac{1}{4}\ E^{-1}(y')^4\ A^{-2})\ \cdot\ \frac{1}{4}\ K\ (E(y'')^2(1 + (y')^2)^{-3/2}\ I'\ (A)]$

or

(4.42) $\Lambda_A = K^{-1}\ \cdot\ (E^{-1}(y')^4\ A^{-2})\ \cdot\ P_{cr} + K^{-1}(\frac{1}{2}\ B^2 + \frac{1}{4}\ KS)^{-\frac{1}{2}}\ [(E^{-1}\ A^{-2}(y')^4)\ S + (E(y'')^2(1 + (y')^2)^{-3/2}\ I'(A)\ K\ ]$

Λ_A is again the sensitivity function for the critical load with respect to the changes of the design vector A(x).

$\Lambda_A = 0$ is a necessary condition for the optimality of design, i.e. for extremal value of the critical load regarded as a functional of A(x).
Thus, we have

$$(4.43) \quad \Lambda_A K = (E^{-1}(y')^4 A^{-2})\, P_{cr} + (\tfrac{1}{2}B^2 + \tfrac{1}{4}KS)^{-\frac{1}{2}}.$$

$$[(E^{-1}A^{-2}(y')^4)S+E(y'')^2(1+(y')^2)^{-3/2}I'(A)K] = 0,$$

as a necessary condition for the extremum of P_{cr}. However, unless additional constraints are imposed on the design function A(x), it is easy to show that $\Lambda_A \neq 0$ for any continuous function A(x).

We note that $\Lambda_A K \to 0$ as $A \to \infty$, which makes good physical sense.
The formula 4.43 gives a clue to the difficulties associated with optimization procedures involving the quadratic eigenvalue problem (4.24) compared to the linear eigenvalue problem (4.11). The sensitivity functional of the Rayleigh quotient λ for the linear eigenvalue problem, subject to normality condition

$\int (y')^2 \, dx = 1$, is simply $\Lambda_L = \phi(A) \cdot (y'')^2$.

Hence if, for example, a constant weight constraint is imposed

$\int_0^{\ell} A(x)\, dx = C$, where C is a positive constant,

and we wish to optimize the Rayleigh quotient λ, the necessary condition for optimality of

$\lambda = \mu \left(\int_0^{\ell} A(x)dx - C\right)$ is given by the simple

formula $\Lambda_L = \mu$, or $\phi'(A) = \dfrac{\mu}{(y'')^2}$.

A much more complex analysis of the optimal buckling problem will result if the problem (4.37) is considered.
At this point we wish to point out some similar-

ities and some basic differences involved in replacing the Rayleigh quotient (4.11) by the highly nonlinear formula (4.25). First we point out some common properties stated in the form of two lemmas.

Lemma 4.4 If $I(A(x))$ is a differentiable function of $A(x)$ for all $x \in [0,\ell]$ then as

$F(A) = \int_0^\ell (y')^2 (EA(x))^{-1} dx$ approaches zero, the value

of P_{cr} approaches the value

$$P_{cr_0} = \frac{\int_0^\ell [EI(x)\,(y'')^2\,(1 + (y')^2)^{-3/2}]\,dx}{\int_0^\ell [(y')^2 - \frac{(y')^4}{4}]dx} . \qquad (4.44)$$

If (y') is "small" and only square terms are retained and powers higher than 4 ignored, then

$$P_{cr_0} \cong \frac{\int_0^\ell EI(x)\,(y'')^2\,dx}{\int_0^\ell (y')^2\,dx} ,$$

which is the classical form of the Rayleigh quotient. We observe that

$$\int_0^\ell EI(x)\,(y'')^2\,dx > \int_0^\ell [EI(x)(y'')^2\,(1 + (y')^2)^{-3/2}]\,dx, \text{ and}$$

$$\int_0^\ell ((y')^2 - \frac{(y')^4}{4})\,dx < \int_0^\ell (y')^2\,dx.$$

The resulting inequality

$$P_{cr} < \frac{\int_0^\ell [EI(y'')^2\,(1 + (y')^2)^{-3/2}]\,dx}{\int_0^\ell [(y')^2 - \frac{(y')^4}{4}]\,dx} < \frac{\int_0^\ell [EI(y'')^2]\,dx}{\int_0^\ell (y')^2\,dx}$$

completes the arguments.

Theorem 4.3 (an analogue of Rayleigh's theorem)

In the Sobolev space H_0^4 $[0,\ell]$ the function $\tilde{y}(x)$ minimizes the value of P_{cr} given by the equation (4.37) if and only if P_{cr} is the eigenvalue of and $\tilde{y}$ the corresponding eigenfunction of the non-linear eigenvalue problem.

(4.45) $$y'' - \frac{((y')^3)'}{2} + \frac{1}{2}(B^2 + \frac{1}{4}KS)^{-\frac{1}{2}} \cdot [2B(-y'' + \frac{((y')^3)'}{2}) +$$

$$\frac{1}{4}K\{[EI\, y''(1+(y')^2)^{3/2}]'' + 3\,E\,I\,(y'')^3 \cdot (1+(y')^2)^{-5/2} \cdot K$$

$$-\,4S\,(EA)^{-1}\,((y')^3)' + P_{cr}(EA)^{-1}\,((y')^3)' = 0.$$

Proof. A necessary condition for the minimum of (4.34) in H_0^4 $[0,\ell]$ is the vanishing of the Fréchet derivative.
Writing P_{cr} in the form $P_{cr} = \frac{N(y)}{M(y)}$ and applying the rules of Fréchet differentiation, as outlined in the appendix, we easily obtain the following formula for the vanishing of the Fréchet derivative $\frac{dP_{cr}}{dy} = 0$, that is

$$\frac{d}{dy}\{-B + (B^2 + \frac{1}{4}KS)^{\frac{1}{2}}\} - \frac{1}{4}P_{cr}\frac{dK}{dy} = 0, \text{ or} \qquad (4.47)$$

$$\frac{dB}{dy} + \frac{1}{2}(B^2 + \frac{1}{4}KS)^{-\frac{1}{2}} \cdot (2B\frac{dB}{dy} + \frac{1}{4}(S\frac{dK}{dy} + K\frac{dS}{dy})) - \frac{1}{4}P_{cr}\frac{dK}{dy} = 0$$

We compute

(4.48) $$\frac{dK}{dy} = -4\,((EA)^{-1}\,(y')^3)',$$

(4.49) $$\frac{dB}{dy} = -y'' + \frac{((y')^3)'}{2},$$

(4.50) $$\frac{dS}{dy} = (EI\,y''\,(1+(y')^2)^{-3/2})'' +$$

$$3(EI(y'')^3 \cdot (1 + (y')^2)^{-5/2}).$$

We substitute these expressions into (4.41) to obtain the required form (4.45). Hence, necessity of (4.45) is demonstrated. To prove sufficiency, we check on the sign of the second Fréchet derivative $\frac{d^2P_{cr}}{dy^2}$: = is represented by the following operator:

$$\frac{d^2P_{cr}}{dy^2} = -\frac{d^2}{dx^2} + \frac{1}{4} K\,(EI\,y'')\,(B^2 + \frac{1}{4}KS)^{-\frac{1}{2}}$$

$$+ \sigma((y')^2 : H_0^4 \to \mathcal{L}(H^4, L_2). \qquad (4.51)$$

This operator is positive for small values of y'. This sufficies to show that the equation (4.45) is both necessary and a sufficient condition for a local minimum of the functional P_{cr}.

(4.8) An Optimal design problem.

In a class of design problems which have been considered in structional design literature, the weight minimization problem and the constant weight problems are most frequently discussed. Let us consider the weight of the column to be the pay-off function. Our optimization criteria are:

Minimize $\rho \int_0^\ell A(x)\,dx$, where ρ is the density, subject to the constraint $P_{cr} \geq R$, where R is given.

We assign the normality condition $\int_0^\ell (y')^2\,dx = 1$.

The problem may be written formally as (4.42)

$$\text{Find } \min_{A(x)} \left\{ \rho \int_0^{\ell} A(x)dx \right\} \quad \begin{cases} \min\limits_{y(x)} P_{cr} = K \\ \int_0^1 (y')^2 dx = 1 \\ y(o) = y'(o) = 0 \\ y(\ell) = y'(\ell) = 0 \end{cases}$$

A necessary condition for solution of this problem is a stationary behavior of the functional

(4.43)

$$\Phi_W = \rho \int_0^{\ell} A(x)\, dx + \alpha\cdot(P_{cr} - K) + \beta \left(\int_0^{\ell} (y')^2 - 1\right),$$

where α , β are Lagrangian multipliers.
A dual problem of minimizing the buckling load P_{cr}, subject to a constant weight, results in the investigation of the stationary behavior of

$$\Phi_C = \alpha \left(\int_0^{\ell} A(x)\, dx - W + \beta \left(\int_0^{\ell} (y')^2 - 1\right) + P_{cr}\right., \tag{4.43}$$

where W is a given constant, α , β are Lagrangian multipliers. We tacitly assume the existence of only a single buckling mode corresponding to the critical value of the load P_{cr}. We shall offer later a brief discussion of the possibility of merging of the eigenvalues, corresponding to different modes of buckling, so that the lowest critical load does correspond to two different buckling modes. Some multiplications of this buckling condition were discussed in the bimodal optimality analysis part of the paper of N.Olhoff and S.H. Rasmussen [14]. However, the author of this article holds views which differ from those of Olhoff and Rasmussen on the significance of the miltimode case on the numerical analysis in optimization of design,

without disputing the correctness of their analysis.
The problem of minimization of the buckling load, subject to constant weight of the column involves the necessary condition

$$(4.52) \qquad \Lambda_A - \alpha = 0,$$

where Λ_A is given by the formula (4.42); writing it out in full, we have the necessary condition

(4.53)

$$K\alpha = [1 + 4KS]^{-\frac{1}{2}} [E\phi(A)(y'')^2 K - (EA)^{-2}(y')^2 S] + \frac{1}{2} P_{cr} (EA)^{-2} (y')^2, \text{ where}$$

$$K = \int_0^{\ell} [(EA)^{-1} (y')^2] dx,$$

$$S = \int_0^{\ell} [E\phi(A)(y'')^2] dx.$$

Minimization with respect to y(x) results in a restatement of the differential equation (4.11). Formula (4.52) has to be compared with the linear eigenvalue optimization formula

$$(4.54) \qquad \phi'(A) = \frac{\alpha}{(y'')^2}$$

before full assessment can be made of the significant differences regarding the geometric features of the optimal shape. For the sake of simplicity let us consider circular cross-section of the column, variations in design only affecting the radius of the circle as a function of the length.

Then $\phi(A) = \frac{A^2}{4\Pi}$, and $\phi'(A) = \frac{1}{2\Pi} A$.

Formula (4.52) implies that the optimal design satisfies the law $A(x) = \text{constant} \cdot (y'')^{-2}$.

There is no reason why y'(x) should not have locally large values. In fact a hinge at some point along the length of the column could exist with a discontinuous jump in the value of y'(x), implying a zero cross-sectional area at such point.
Numerical analysis fully confirms the occurence of such hinges in the absence of constraints on the maximum permissible value of stress. This is not the case in our analysis, as is shown by our next result.
Theorem 4.3 Suppose that the integral

$K = \int_0^{\ell} (EA)^{-1} (y')^2 \, dx$ exists. Then the cross-sectional area A(x) must be bounded away from zero at any interior point of the column in the optimal design.
Proof. First, let us multiply all terms of (4.46) by $(EA)^2$, obtaining

(4.48)

$$(EA)^2 K \alpha = [1 + 4\,KS\,]^{-\frac{1}{2}} \cdot [E\,\phi'(A)\,(y'')^2 K \cdot (EA)^2 - (y')^2 S\,] + \frac{1}{2} P_{cr} (y')^2.$$

Now, let us make the substitution

$P_{cr} = \frac{1}{2K} \{-1 + (1 + 4KS)^{\frac{1}{2}}\}$ and let $A(\tilde{x})$ approach zero at some point $\tilde{x}$ where $0 < \tilde{x} < \ell$.

We obtain

$$\frac{1}{4K} \{-1 + (1 + 4KS)^{\frac{1}{2}}\}\Big|_{\tilde{x}} = (1 + 4KS)^{-\frac{1}{2}} (y')^2 S\Big|_{\tilde{x}}, \text{ or}$$

$$(4.48)\quad (y')^2\Big|_{\tilde{x}} = \frac{(1 + 4KS)^{1/2} \cdot [-1 + (1 + 4KS)^{1/2}]}{4KS} =$$

$$= \frac{-(1 + 4KS)^{1/2} + (1 + 4KS)}{KS}\Big|_{\tilde{x}} \approx 1 \text{ for very large K.}$$

Hence, $|y'(\tilde{x})|$ is uniquely defined at a singular design point $\tilde{x}$. However, for extremely small values of $EA(\tilde{x})$, the solution of (4.47) approximates a trigonometric function behavior with a slope equal to approximately $y' \approx \cos(P\, E\, A^{-1/2}\, I^{-1/2})\,\tilde{x}$, which rapidly fluctuates, and which does not agree with the value given by (4.48). This completes the proof.

The contradiction arose from the fact that $y(x)$, hence $y'(x)$, $y''(x)$ must solve the equation (4.47) for any given design $A(x)$.

However, the solution which is unique for a simple eigenvalue case has not been informed about the optimality conditions imposed on $A(x)$, and predicts some value $y'(\tilde{x})$ at a singular point $\tilde{x}$ which does violate the optimality criterion (4.46) In fact, even a finite multiplicity of the eigenvalue λ_1 does not destroy this argument.

4.9 The dependence of eigenvalues on parameters.

Let us consider a simple eigenvalue problem

$$Ax = \lambda x$$

where A is an n×n matrix. Let us suppose that A is a function of a single vector $\underset{\sim}{u}$. Thus we wish to find eigenvalues and eigenfunctions of the eigenvalue problem

$$(4.49)\quad A(\underset{\sim}{u})\ \phi_i(\underset{\sim}{u}) = \lambda_i(\underset{\sim}{u})\ \phi_i(\underset{\sim}{u})$$

where $\underline{u}$ is a vector restricted to a subset U of the Banach space L_p, $1 \leq p \leq \infty$.

A is a self-adjoint real matrix (that is a symmetric matrix), and $\| \phi_i \| = 1$, i.e. $\langle \phi_i, \phi_i \rangle = 1$.

We compute the sensitivity of a simple eigenvalue λ_i to changes in the value of u noting that change of u will not affect orthonormality of $\{\phi_i(u)\}$. The following formal differentiation is easily justified.

$$\frac{\partial \lambda_i}{\partial \underset{\sim}{u}} \phi_i(\underset{\sim}{u}) + \lambda_i \frac{\partial \phi_i}{\partial \underset{\sim}{u}} = \frac{\partial A}{\partial \underset{\sim}{u}} \phi_i(\underset{\sim}{u}) + A(\underset{\sim}{u}) \frac{\partial \phi_i}{\partial \underset{\sim}{u}} .$$

The variational form of (4.49) is

$\langle A(\underset{\sim}{u}) \phi_i(\underset{\sim}{u}), \phi_i(\underset{\sim}{u}) \rangle = \lambda_i(u) \langle \phi_i, \phi_i \rangle = \lambda_i(\underset{\sim}{u})$ and

the sensitivity of λ_i is given by:

$$\frac{\partial \lambda_i}{\partial \underset{\sim}{u}} + \lambda_i \left\langle \frac{\partial \phi_i}{\partial \underset{\sim}{u}}, \phi_i \right\rangle = \left\langle \frac{\partial A}{\partial \underset{\sim}{u}} \phi_i, \phi_i \right\rangle +$$

$$\left\langle A(\underset{\sim}{u}) \frac{\partial \phi_i}{\partial \underset{\sim}{u}}, \phi_i \right\rangle \quad \text{or} \quad \frac{\partial \lambda_i}{\partial \underset{\sim}{u}} + \left\langle \frac{\partial \phi_i}{\partial u}, \lambda_i \phi_i \right\rangle =$$

$$\left\langle \frac{\partial A}{\partial \underset{\sim}{u}} \phi_i, \phi_i \right\rangle + \left\langle \frac{\partial \phi_i}{\partial u}, A \phi_i \right\rangle \qquad (4.50)$$

Since $A\phi_i = \lambda_i \phi_i$ and $\left\langle \frac{\partial \phi_i}{\partial \underset{\sim}{u}}, \lambda_i \phi_i \right\rangle = \left\langle \frac{\partial \phi_i}{\partial \underset{\sim}{u}}, A\phi_i \right\rangle$,

we derive a known result

$$\frac{\partial \lambda_i}{\partial u} = \left\langle \frac{\partial A}{\partial u} \phi_i, \phi_i \right\rangle \qquad (4,51)$$

We observe that nowhere in this development did we use the fact that A is a finite-dimensional operator.

We used the self-adjoint property of A, the fact that λ is a real number and that ϕ_i was a simple eigenvalue (that is of multiplicity one). This result is valid for the eignevalue problem when A is a self-adjoint operator in a Hilbert space, and λ_i is a simple eigenvalue of A.

A similar argument can be given in deriving the sensitivity of an eigenvalue λ_i with respect to the parameter $u \in \mathbb{R}$ for the eigenvalue equation

$$A\,\phi_i = \lambda_i\,B\,\phi_i \tag{4.52}$$

where A, B are self-adjoint operators on a Hilbert space H, λ_i is a simple eigenvalue, and ϕ_i the corresponding eigenvector. B is a positive definite operator and $\langle B\phi_i, \phi_i\rangle = 1$, $\langle B\phi_i, \phi_j\rangle = 0$ if $\phi_j \neq \phi_i$.
Let us assume that both A and B depend on a real parameter u.
An identical argument is used to derive the sensitivity formula

$$\frac{\partial \lambda_i}{\partial u} = \left\langle \frac{\partial A}{\partial u}\phi_i, \phi_i\right\rangle - \lambda_i\left\langle \frac{\partial B}{\partial u}\phi_i, \phi_i\right\rangle \tag{4.53}$$

This formula cannot be used if λ_i is of multiplicity greater than one.

For a simple eigenvalue λ_i of the equation $A(u)x = \lambda x$ the extremal property of λ_i regarded as a function of the parameter u can be attained at the value $\hat{u} \in U$ if either $\hat{u}$ lies on the boundary of the admissible region U of the "design space" or else if $\frac{\partial A(u)}{\partial u}\phi_i$ is orthogonal to ϕ_i.

If A is a self-adjoint operator in a Hilbert space H and the eigenfunctions ϕ_i, $i = 1,2,\ldots$ span H, then the completeness of the eigenfunction system $\{\phi_i\}$ implies that either

$\frac{\partial A}{\partial u}\phi_i = 0$, or else $\frac{\partial A}{\partial u}\phi_i = \phi_k$, $k \neq i$, that is

$\frac{\partial A}{\partial u}$ maps an eigenvalue vector ϕ_i into a (different) eigenvector ϕ_k.

Example.
The specific case which we shall examine is that of the vibrating beam. Let us formally vary the design. Under arbitrary changes in design we preserve the basic equation of motion.

This is done formally by writing

$$(4.54)\quad \frac{\partial}{\partial u}\{\langle L(u)\,W(u),\lambda\rangle - \langle \hat{q}(u),\lambda\rangle\} = 0.$$

Symbolically, this is expressed by the relation

$$(4.55)\quad \langle L(u)\,W_u,\lambda\rangle + \langle L_u W,\lambda\rangle - \langle \hat{q}_u,\lambda\rangle = 0\,,$$

where W is an element of $H_0^{2,1}$ in the case of a vibrating beam. Here the operator L_u is given by

$$(4.56^a)\quad L_u = \frac{\partial^2}{\partial x^2}\left(E\,\frac{\partial I(u)}{\partial u}\,\frac{\partial^2\cdot}{\partial x^2}\right) + \varrho\,\frac{\partial A}{\partial u}\,\frac{\partial^2\cdot}{\partial t^2}$$

for the vibrating beam, and

$$(4.56^b)\quad (L_u W) = \left[3h^2\,\frac{\partial h}{\partial u}\,(W_{xx} + \nu W_{yy})\right]_{xx} + \left[3h^2\,\frac{\partial h}{\partial u}\,(W_{yy} + \nu W_{xx})\right]_{yy} + 2(1-\nu)\left[3h^2\,\frac{\partial h}{\partial u}\,W_{xy}\right]_{xy} + \varrho\,\frac{\partial A}{\partial u}\,W_{tt}$$

for the vibrating thin plate.

In general, the operator L_u is constructed by formally differentiating the coefficients of the operator L with respect to the design function u. We discuss specifically the beam operator. However, nearly all results generalize to arbitrary hyperbolic operators.

We have not proved that our formula (4.51) is applicable to systems with distributed parameters. Subject to some minor restrictions it is, and the proof completely parallels the argument given in deriving (4.51).

We offer a non-trivial example of application of this formula to the problem of optimizing the natural frequency of a circular thin plate. The formula (4.51) is used, even though the derivative $\frac{\partial L}{\partial h}$ is an abstract (Fréchet) derivative of the operator L(h).

A free vibration of a thin circular plate is described by the approximate equation

$$(4.57)\quad \{r\,[D(w'' + w'/r)' + D'(w'' + \nu w'/r)]\}' = \omega^2 \rho\, h(r)\, w\, r$$

where (as before)

$D = Eh^3/(12(1-\nu^2))$ - the flexural rigidity
h - is the thickness
E - Young's modulus
W - lateral deflection
ω - natural angular frequency
ρ - mass density

The equation (4.57) can be restated as an eigenvalue problem

$$LW = \omega^2 W = \lambda W$$

$$(\lambda = \omega^2),$$

where

$$L = [\rho\, r\, h(r)]^{-1} \cdot \frac{\partial}{\partial x}\{r\,[D\,(\frac{\partial^2}{\partial x^2} \cdot + \frac{1}{r}\frac{\partial}{\partial x} \cdot) + \frac{\partial D}{\partial x}(\frac{\partial^2}{\partial x^2} + \frac{\nu}{r}\frac{\partial}{\partial x})]\}, \quad h_o \le h \le h_1 .$$

Then $L_h = \frac{\partial L}{\partial h}$ denotes the operator

$$L_h = [-\frac{1}{\rho r}(h(r))^{-2}] \cdot \frac{\partial}{\partial x}\{r\,[D\,\frac{\partial}{\partial x}(\frac{\partial^2}{\partial x^2} + \frac{1}{r}\frac{\partial \cdot}{\partial x}) + \frac{\partial D}{\partial x}(\frac{\partial^2}{\partial x^2} + \frac{\nu}{r}\frac{\partial}{\partial x})]\} + (\rho\, r\, h)^{-1} \cdot$$

$$\cdot \frac{\partial}{\partial x}\{r[D_h \frac{\partial}{\partial x}(\frac{\partial^2}{\partial x^2} + \frac{1}{r}\frac{\partial}{\partial x}) + \frac{\partial D_h}{\partial x}(\frac{\partial^2}{\partial x^2} + \frac{\nu}{r}\frac{\partial}{\partial x})]\}, \quad \text{with } D_h = \frac{Eh^2}{4(1-\nu^2)}$$

Let $w_1^2 = \lambda_1$ denote the lowest natural frequency of vibration.

The necessary condition for optimality of the first eigenvalue λ_1, whose corresponding eigenfunction is $\tilde{w}_1$ is either $L_k \tilde{w}_1 = \tilde{w}_k$ where w_k denotes an eigenfuction of the operator L satisfying

$$L\tilde{w}_k = \lambda_k \tilde{w}_k ,$$

$$<\tilde{w}_K, \tilde{w}_1 > = 0,$$

or else $h \equiv h_o$ or $h \equiv h_1$.

In [15] the author introduced a novel way of embedding the class of eigenvalue problems of continuum mechanics in a larger class of design problems. Only the buckling problem for the linear theory is treated in detail. However, the method is quite general and could be easily extended to stability considerations in a larger class of problems.

4.10 A simple example

An early forerunner of the energy approach to problems of elastic stability was the practical argument advanced by S. Timoshenko in 1910 paper [16]. His idea is best illustrated on the following simple example of a static equilibrium discussed in [11], pp. 83-84.

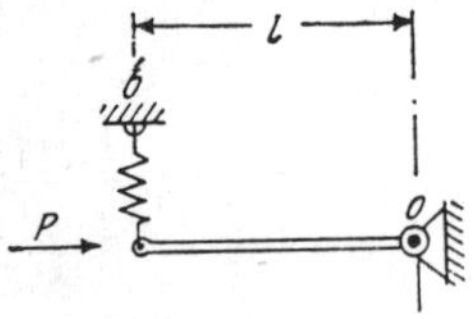

Fig. 1a

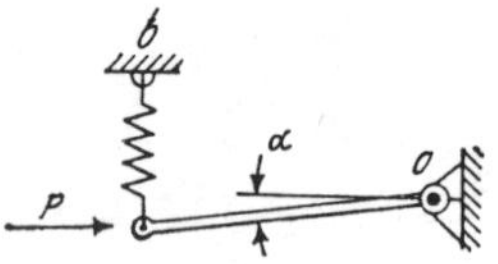

Fig. 1b

A weight P is acting axially on a rigid bar, which is restrained by a linear spring. The bar is free to rotate about the point σ in the plane of the bar and of the point q as shown. The force exerted by the spring is proportional to the increase in its length. No force is exerted by the spring in the equilibrium position shown on Figure 1a, the reaction of the bearing at the point σ being equal but oppositely directed to the applied force P.

The state of the system is completely described by specifying the generalized displacement (coordinate) α which is the angle of rotation of the bar. Choosing α to be an infinitesimal rotation angle, we compute the changes in two forms of energy.

The increase in the strain energy of the spring is given by $\delta V = k\dfrac{(\alpha l)^2}{2} + \textit{infinitesimal}$ of a higher order. The decrease in potential energy due to the lowering of the load P is given by

$$(4.58)\qquad \delta S = Pl(1-\cos\alpha) = Pl\frac{\alpha^2}{2} \quad + \textit{infinitesimal}$$

of a higher order. The equilibrium is stable if $\delta V > \delta S$, i.e. if $P < k\ell$, and is unstable if $P > k\ell$. $P = P_{cr} = k\ell$ is the critical load, corresponding to a zero of the difference of energy forms:

$$(4.59)\qquad (V-S) = Pl\frac{\alpha^2}{2} - k\frac{\alpha^2 l^2}{2} = 0\,.$$

Replacing P by a parameter λ, we could rewrite the equation (4.59) as an eigenvalue problem

$$(4.60)\qquad \langle A\alpha, \alpha\rangle = \langle \beta\alpha, \alpha\rangle\,\lambda\,.$$

This approach originated by Lord Rayleigh [17], is very well known, and will not be pursued immediately.

The change from equilibrium considerations with P regarded as a constant to the form (4.60), where P occurs as a parameter in an eigenvalue problem, usually slips by unobserved. Another basic parameter of this problem is completely overlooked in this formulation.

The stability of the trivial solution $\alpha \equiv 0$ depends on the parameter P and on the design variables k and ℓ

$$(4.61)\ \Phi(l,\lambda,\alpha) = \left[\frac{k\alpha^2\xi^3}{6} - \frac{\lambda\alpha^2\xi^2}{4}\right]_{\xi=l}, \qquad \left.\frac{\partial\Phi}{\partial\xi}\right|_{\xi=l} = \frac{k\alpha^2 l^2}{2} - \frac{\lambda\alpha^2 l}{2}\,.$$

Hence $\partial\Phi/\partial\ell>0$ corresponds to the subcritical condition of equilibrium, $\partial\Phi/\partial\ell<0$ to supercritical (unstable) equilibrium, and $\partial\Phi/\partial\ell = 0$ is the critical condition. The functional $\Phi(\ell,\alpha,\lambda)$ could have been constructed from the basic energy form by setting

$$\left.\frac{\partial\Phi}{\partial\xi}\right|_{\xi=l} = V(\alpha, l) - S(P, \alpha, l)\,, \quad \text{or} \quad \Phi = \int_0^l [V(\alpha, \xi) - S(P, \alpha, \xi)]\, \mathrm{d}\xi\,.$$

Checking on the sign of the second derivative of $\phi(\xi,\alpha,P)$ we find that in some neighborhood of the critical value of ξ, for a fixed P, $\partial^2\phi/\partial\xi^2>0$. Hence the critical point represents a local minimum of $\Phi(\xi,\alpha, P)$ with respect to ξ. Since it is linear in P and quadratic in α not much else need be said about its properties.

Of course, this example was extremely simple. However the basic considerations of this example carry over to much more complex situations.

4.11 The buckling of a beam

Again, we can approach the problem by considering two forms of energy. The strain energy of the beam trying to restore it to the initial configuration

(4.62) $$V = 1/2 \int_0^l EI(x)\, (w'')^2\, \mathrm{d}x$$

while the work done by the action of the axial load is

(4.63) $$1/2 \int_0^l P(w')^2\, \mathrm{d}x\,.$$

The work of transverse load q(x) is

(4.64) $$\int_0^l q(x)\, w(x)\, \mathrm{d}x\,.$$

Suppose that $\psi(x)\in C^2[0,\ell]$ is an arbitrary (trial) function obeying the assigned boundary conditions as $x = 0$ and $x = \ell$. Suppose that the following inequality is obeyed for a given value of $P = \lambda_\psi$:

$$1/2 \int_0^l S(x)\,(\psi'')^2\,dx - \int_0^l q(x)\,\psi(x)\,dx > 1/2 \int_0^l \lambda_\psi(\psi)^2\,dx\,, \qquad (S(x) = EI(x))\,. \tag{4.67}$$

Then λ_ψ will be called a subcritical value for the trial function $\psi(x)$. This is equivalent to an inequality

$$\lambda_\varphi < \frac{\int_0^l S(x)\,(\varphi'')^2\,dx - 2\int_0^l q(x)\,\varphi(x)\,dx}{\int_0^l (\varphi')^2\,dx}\,. \tag{4.68}$$

The right hand side of the inequality (4.68) is the Rayleigh quotient which is an upper bound on the critical value of the axial force. (See the classical paper of W. Ritz [18], and [19], chapter 10 or references [22], [24], [11] .

This classical problem is usually formulated as an eigenvalue problem

$$\langle A\varphi, \varphi\rangle - 2\,\langle q, \varphi\rangle = \lambda\langle\beta\varphi, \varphi\rangle\,, \qquad A = \frac{d^2}{dx^2}\left(S(x)\,\frac{d^2}{dx^2}\right), \qquad \beta = -\frac{d^2}{dx^2} \tag{4.69}$$

and is generally attacked by methods related to [19] with $S(x)$ and ℓ regard as fixed.

We shall attempt to embed this classical problem in a more general class of design problems. $S(x)$, $q(x)$ and ℓ will be permitted some variation. We shall assume

$$S(x) \in L_2[0,l]\,, \qquad L > l > 0\,, \qquad q(x) \in L_2[0,l]\,, \qquad w(x) \in H_0^2[0,l]\,. \tag{4.70}$$

Note: $H_0^2[0,\ell]$ is the Sobolev space obtained by closure of $C_0^\infty[0,\ell]$ functions in the topology generated by the Sobolëv norm:

$$||y||^2_{H^2} = \int_0^l [(y)^2 + (y')^2 + (y'')^2]\,dx. \tag{4.71}$$

It follows from the Sobolëv imbedding lemma that $w \in H_0^2[0,\ell]$ is at least once continuously differentiable, and $w''(x) \in L_2[0,\ell]$.

We assign either simple support, or built-in boundary conditions at x = 0 and x = ℓ. For an arbitrary trial function $\psi(x)$ we construct the following functional:

$$\Phi(S, \lambda, l, \psi(x)) = \tfrac{1}{2}\int_0^l S(x)\,(\psi''(x))^2\,dx - \int_0^l q(x)\,\psi(x)\,dx -$$

$$\tfrac{1}{2}\lambda\int_0^l (\psi')^2\,dx = V(S, l, \psi) - W(\lambda, \psi)\,. \qquad (4.73)$$

Again, the Kirchhoff's energy inequality can be asserted.

Here, let us repeat that we consider the one dimensional case for the sake of simplicity and that higher dimensional cases (plates and shells) and more complex boundary conditions can be handled by raising almost identical arguments involving a change in the sign of a specific energy form.

For the sake of simplicity let us assign simple support boundary conditions

$$w(0) = w''(0) = w(l) = w''(l) = 0\,.$$

Since $\Psi(x) \in H_0^2[0,\ell]$ and $w \in C^1[0,\ell]$, $\Psi(x)$ can be expanded in Fourier series converging pointwise to the values of $\Psi(x)$ for all $x \in [0,\ell]$. We choose $\phi_n(x,\ell) = 1/(\pi)^{\frac{1}{2}}\sin(n\pi x/\ell)$ and write the Fourier expansion

$$\psi(x, l) = \sum_{n=1}^{\infty} a_n\varphi_n(x, l)\,. \qquad (4.73)$$

Because of the uniform convergence, we can replace infinite sums by their finite approximations

$$\psi(x, l) = \sum_{n=1}^{N} a_n\varphi_n(x, l). \qquad (4.74)$$

To obtain simple estimates, let us first assume that $S(x) = EI(x)$ = constant = S. Then the strain energy of the beam is given by

$$V(S,l)=\frac{\pi^3}{4}\left(\frac{S}{l^3}\right)\sum_{n=1}^{N}n^4a_n^2-\int_0^l\left(q(x)\sum_{n=1}^{N}a_n\varphi_n(x)\right)dx=\frac{\pi^3}{4}\left(\frac{S}{l^3}\right)\sum_{n=1}^{N}n^4a_n^2-\sum_{n=1}^{N}a_nk_n(l)\,,$$

where $k_n(\ell) = \int_0^\ell q(x)\phi_n(x)\, dx$, i.e. $k_n(\ell)$ are the Fourier coefficients of the $L_2[0, \ell]$ function $q(x)$,
and $\phi_n = 1/(\pi^{\frac{1}{2}})\sin(n\pi x/\ell)$ in this computation.

The work done by the axial force P is

$$W(l)=\frac{P\pi}{4l}\sum_{n=1}^{N}n^2a_n^2\,.$$

Condition of elastic stability is

$$V(S,l)-W(l,\lambda)>0\,,$$

or in terms of the Fourier coefficients

$$\frac{\pi^3}{4}\frac{S}{\xi^3}\sum_{n=1}^{N}\left(n^4a_n^2-a_nk_n(\xi)-\lambda\frac{\pi}{4\xi}n^2a_n^2\right)>0|_{\xi=l}\,. \tag{4.75}$$

We now compute the functional

$$\Phi(S,\lambda,l,a_n)=\int_0^l\left(\frac{\pi^3}{4}\frac{S}{\xi^3}\sum_{n=1}^{N}n^4a_n^2-\sum_{n=1}^{N}a_nk_n(\xi)-\lambda\frac{\pi}{4\xi}\sum_{n=1}^{N}n^2a_n^2\right)d\xi\,,\qquad 0\leq\xi\leq l\,.$$

Then $\partial\Phi/\partial\ell = V(S, \ell, a_n) - W(\ell, a_n, \lambda)$. $\partial\Phi/\partial\ell > 0$ indicates elastic instability, $\partial\Phi/\partial\ell = 0$ indicates the critical condition

$$\frac{\partial^2\Phi}{\partial l^2}=\frac{-3\pi^3S}{4l^4}\sum^{N}n^4a_n^2-\sum^{N}a_nk_n'(l)+\frac{\pi}{4l^2}\sum^{N}n^2a_n^2\,,\qquad\left(k_n'\equiv\frac{dk_n}{dl}\right).$$

Suppose $k_n(\ell)$ are constants independent of ℓ so that $k_n(\ell) \equiv 0$. It is easily checked that in the neighborhood of the critical point we have

$$\frac{\partial^2\Phi}{\partial l^2}=\frac{1}{l}(-3V+W) \tag{4.76}$$

Hence, it is negative, and the critical point corresponds to a local maximum of $\Phi(\ell)$. Moreover, Φ is a linear fuction of S and λ, and in the absence of the applied load $q(x)$, it is a quadratic function of $a_n, n = 1, 2, \ldots, N$.

4.12 A detailed analysis of the critical load, and some two-sided estimates

Using the functional $\Phi(S, \lambda, \ell, a_n)$ and equating to zero the derivative $\partial\Phi/\partial\ell = 0$, we have obtained a necessary condition for a critical state with the trial function $\Psi(x)$ representing displacement. Since the subcritical condition is given by the inequality

$$\frac{\partial \Phi}{\partial l} > 0 ,$$

$$\lambda < \frac{(\pi^3/4)\,(s/l^3)\,\sum^N n^4 a_n^2 - \sum^N a_n k_n(l)}{(\pi/(4l))\,\sum^N n^2 a_n^2} = R(a_n), \qquad (4.77)$$

the first critical load corresponds to $P_{cr} = \inf_{a_n}\{R(a_n)\}$ resulting in a necessary condition

$$\frac{\partial R(a_n)}{\partial a_n} = 0 , \qquad n = 1, 2, \dots , N , \qquad (4.78)$$

in the finite-dimensional case.

In fact we have re-derived another version of Rayleigh quotient. Since the load $q(x)$, and therefore the coefficients $k_n(\ell)$ are known, an upper bound on λ follows from the particular application of RAYLEIGH and RITZ techniques. See [18] [19]

Hence, the usual Rayleigh quotient technique could be restated in the formulation of this chapter as follows

$$\left.\begin{array}{l} \text{Find } \inf_{a_n} R(a_n) \text{ subject to the constraint } \partial\Phi/\partial l\ (l, \lambda, a_n) \geqq 0 \\ \\ \text{or } \sup_{a_n} R(a_n) \text{ subject to the constraint } \partial\tilde{\Phi}/\partial\tilde{l}\ (l, \lambda, a_n) \leqq 0. \end{array}\right\} \quad 4.79$$

In each case by trying some values of a_n, we obtain an upper bound on λ_1. Of course, since a_n were coefficients of an arbitrary trial function $\Psi(x)$, only upper bounds on λ can be derived from the behaviour of the function $\Phi(S, \ell, \lambda, a_n)$. The computational advantages of dealing directly with the function Φ are considerable. All values S, λ, ℓ, a_n can be adjusted simultaneously knowing that $\partial\Phi/\partial\ell \geqq 0$, $\partial R/\partial a_n = 0$, $n = 1, 2, \ldots, N$, constitute design criteria for subcritical value of λ corresponding to the best approximation to an actual deflection by a trial function represented by the first N terms of the Fourier series.

4.13 Lower bounds on the critical loads

Suppose we solve the well-known eigenvalue problem on a fixed interval $[0, \ell]$: (4.80)

$$\frac{d^2\psi}{dx^2} = \mu\psi\,, \quad \psi \in H_0^1[0, l]\,, \quad \psi'(l) = 0\,, \quad \text{or} \quad \psi(0) = \psi(l) = 0\,,$$

or equivalently find a sequence fo eigenvalues and eigenfunctions μ_i, Ψ_i satisfying

$$\mu_k = \inf_{\psi \in H_0^1} \left\{ \int_0^l (\psi')^2 \Big/ \int_0^l \psi^2\, dx \right\} \quad (4.81)$$

subject to orthogonality conditions

$$\int_0^l (\psi, \psi_i)\, dx = 0\,, \qquad \text{(for all } i < k)\,,$$

where Ψ_i are the eigenfunctions of (4.80)

This problem is easily identified with the beam problem, if S = constant and ℓ is fixed, if we put $\Psi(x) = w'(x)$, $\Psi_i = \eta_i'(x)$, and restate the eigenvalue problem in the variational form

$$(4.82) \quad \mu_k = \inf_{w \in H_0^2} \left\{ \int_0^l (w'')^2\, dx \Big/ \int_0^l (w')^2\, dx \right\},$$

subject to orthogonality conditions

$$(4.83) \quad \int_0^l w'(x)\, \eta_i'(x)\, dx = 0\,, \qquad (i = 1, 2, \ldots, k-1)\,.$$

The solutions of (4.80) are sines and cosines of the form $\sin n\pi x/\ell$, $\cos n\pi k/\ell$, $n = 0, 1, \ldots$ which form an orthogonal, complete set of base vectors of $L_2[0, \ell]$. Obviously, their derivatives and second derivatives form another orthogonal subset of $L_2[0, \ell]$. Hence, for each value of ℓ

we have a complete orthonormal set $\Phi_i(x,\ell)$ which solves the eigenvalue problem (4.69)

Regarding ℓ as a variable we formulate our functional

$$\Phi\left(S,\lambda,l,\psi(x)\right)=\int_0^l\left(V(S,\xi,\lambda,\psi(x)\right)\mathrm{d}\xi\,, \tag{4.84}$$

where as before

$$V\left(S,l,\psi(x)\right)=1/2\int_0^l\left(S(x)\,(\psi'')^2\right)\mathrm{d}x-\int_0^l\left(q(x)\,\psi(x)\right)\mathrm{d}x\,,$$

$$W\left(l,\lambda,\psi(x)\right)=1/2\lambda\int_0^l(\psi')^2\,\mathrm{d}x\,.$$

The value $\lambda_\Psi(\ell)$ denotes that the equality

$$V\left(S,l,\psi(x)\right)-\lambda_\psi W\left(l,\lambda,\psi(x)\right)=0 \tag{4.85}$$

is true. We choose $w_k(x,\ell)$ such that

$$\int_0^l(w_k'\eta_i')\,\mathrm{d}x=0\,,\qquad\int_0^l S(x,w_k''\eta_i'')\,\mathrm{d}x=0\quad\text{if}\quad i\le k\,. \tag{4.86}$$

We define

$$u(x,l)=w_k(x,l)+\sum_{i=k+1}^{\infty}C_i\eta_i(x,l)\,,\qquad\psi(x,l)=w_k(x,l)+\sum_{l=1}^{\infty}C_i\eta_i(x,l)\,.$$

Then

$$\frac{\partial\Phi(\psi,l,\lambda,u)}{\partial l}=\int_0^l-q(x)\,w_k+\sum_{i=1}^{\infty}C_i\eta_i\,\mathrm{d}x+\frac{1}{2}\int_0^l S(x)\sum_{i=1}^{k}(C_i\eta_i'')^2\,\mathrm{d}x+$$

$$+\frac{1}{2}\int_0^l\left[S(x)\left(w_k''+\sum_{i=k+1}^{\infty}C_i\eta_i''\right)^2\right]\mathrm{d}x-$$

$$\frac{1}{2}\int_0^l\lambda_l\left(w_k'+\sum_{i=k+1}^{\infty}C_i\eta_i'\right)^2\mathrm{d}x-\frac{1}{2}\int_0^l\lambda_u\sum_{c=1}^{k}(C_i\eta_i')^2=$$

$$=\int_0^l\left(-q\cdot\sum_{i=1}^{k}C_i\eta_i\right)\mathrm{d}x+V(l,u)-W(u,\lambda_u)+\frac{1}{2}\int_0^l\left(S(x)\sum_{i=1}^{k}(C_i\eta_i'')^2\right)\mathrm{d}x$$

$$= \big(V(l, u) + V(l, \varrho)\big) - \big(W(u, \lambda_u) - W(\varrho, \lambda_u)\big)\,, \qquad (4.87)$$

with $\varrho = \sum_{i=1}^{k} C_i \eta_i.$

Suppose that (for some L) $V(L,\rho) > 0$, which is certainly the case if for any $L > 0$ $q(x)$ is sufficiently small . Then

$$(4.88) \quad \frac{\partial \Phi(\psi, l, \lambda_u)}{\partial l} > 0\,, \quad \forall l < L\,,$$

and $\lambda_u < \lambda_\Psi$ where λ_Ψ is the critical value corresponding to the choice of H_0^2 trial function $\Psi(x)$. Obviously, any trial function can be written in the form (5.7b) ($\Psi(x) \in H_0^2[0, \ell]$). Let $k = 1$, and let $\tilde{\lambda} = \inf(\lambda_u)$ be chosen so that : $\partial\Phi(u, l, \lambda_u)/\partial l = 0$, and $u(x, l) = w(x, l) + C_1\eta_1(x, l)$. Then $\tilde{\lambda}$ is a lower bound on the first critical value of the beam problem

$$(4.89) \quad \frac{d^2}{dx^2}\left(S(x)\frac{d^2 w}{dx^2}\right) - q(x)\, w(x) = -\frac{\lambda d^2 w}{dx^2}.$$

4.14 Remarks on variational techniques used for approximating the eigenvalues of operators in elastodynamics.

An extensive literature exists on computation of upper and lower bounds for eigenvalues of self-adjoint operators in a Hilbert space.

For an outline of the Ritz-Rayleigh and Galerkin variational methods we refer the reader to Bickley and Temple [22], Kato [19], Dunford and Schwartz [23] , Weinberger [24.]

The approximations to eigenvalues and eigenvectors of an operator on a Hilbert space H are given in the Ritz and Galerkin schemes by restricting the range of the operator to a finite dimensional subspace $S \subset H$.

The main idea of all these approximations is quite simple. To estimate the first eigenvalue of a linear operator $L: H \to H$ we can use the Rayleigh quotient. For example, let us estimate the first eigenvalue of the eigenvalue problem for a vibrating membrane.

$$(4.90) \quad \Delta u = -\lambda \rho u \quad \text{in } \Omega \subset \mathbb{R}^2$$

$$(\Delta = \frac{\partial^2}{\partial x^2} + \frac{\partial}{\partial y^2}) \quad ,$$

$$u = 0 \quad \text{on } \partial\Omega,$$

where $u \in H^1_0 (\Omega)$,

$$\rho(\underset{\sim}{x}) \in C(\Omega), \ \rho(\underset{\sim}{x}) > 0 \text{ in } \Omega.$$

$$\underset{\sim}{x} \in \mathbb{R}^2, \ (d\underset{\sim}{x} = dx \cdot dy),$$

and Ω is a closed bounded region in the Euclidean plane. As is well known (see Temple and Bickley [22]), the eigenvalues λ_i, $0 < \lambda_1 < \lambda_2 \ldots < \lambda < \ldots \to \infty$ are the stationary points of the Rayleigh quotient

$$J(u) = \iint_\Omega \| \text{grad } u\|^2 dx \Big/ \iint_\Omega (\rho(x)u^2) dx \ .$$

Let H denote the Hilbert space of functions which is the closure of $C^\infty(\Omega)$ functions in the Sobolev norm $\| f\| = \iint_\Omega \{(\rho(x) \ f^2) + \| \text{grad } f\|^2\} \, d\underset{\sim}{x}$ that vanish identically on $\partial\Omega$. Then λ_1 is the fundamental eigenvalue and $u_1 \in H, \|u_1\| \neq 0$ a corresponding eigenfunction if $J(u_1) \leq J(v)$ for any $v \in H$.

Since u_1 provides a minimum of J(in H), it is clear that choosing any $v \in H$ we obtain an upper bound on λ_1 by computing $J(v)$.

<u>An example</u>.

A more systematic approach due to Ritz consists of restricting H to a finite dimensional subspace S_n and minimizing $J(u)$ $(u \in H)$ in S_n. Consider a vibrating circular disc of radius R, density ρ (of constant cross-section), clamped at the edge, and vibrating freely. Find the lowest natural frequency $\omega_1 > 0$, or a bound on ω_1.

<u>Solution</u>: The strain energy is given by

$$\int_{\Omega} D[(\nabla^2 Z_0)^2 - 2(1-\nu) \frac{\partial^2 Z}{\partial x^2} \frac{\partial^2 Z_0}{\partial y^2} - (\frac{\partial^2 Z_0}{\partial x \partial y})]$$

Since the vibration occurs at constant frequency, the kinetic energy is

$$T = \frac{1}{4} \omega^2 \rho \int_{\Omega} \left(Z_0^2 \right) \cdot dA$$

(confirm this by separating variables, or by assuming $Z = e^{i\omega t} Z_0(x,y)$!)

Changing to polar coordinates, we have

$$V = \int_0^R D\{[\frac{\partial^2 Z_0}{\partial r^2} + \frac{1}{r} \frac{\partial Z_0}{\partial r} + \frac{1}{r^2} \frac{\partial^2 Z_0}{\partial \theta^2}]^2 - 2(1 - \nu)$$

$$[\frac{\partial^2 Z_0}{\partial r^2} (\frac{1}{r} \frac{\partial Z_0}{\partial r} + \frac{1}{r^2} \frac{\partial^2 Z_0}{\partial \theta^2}) - (\frac{\partial}{\partial r} (\frac{1}{r} \frac{\partial Z_0}{\partial \theta}))^2] \} \, dr$$

$= \langle AZ_0, Z_0 \rangle$, where the operator A is positive definite, provided ν satisfies the physically necessary condition $0 < \nu < \frac{1}{2}$.

Let us neglect the terms in V containing $\frac{\partial Z_0}{\partial \theta}$, or $\frac{\partial^2 Z}{\partial \theta^2}$. Then the kinetic energy is:

$$T = \frac{\Pi}{2} \rho \omega^2 \int_0^R r Z_0^2 (r) \, dr.$$

Now we substitute any $\tilde{Z}_0(r)$ obeying the boundary conditions.

We refer to $\tilde{Z}_0$ as a "trial function".

For example, choose

$$\tilde{Z}_0 = C_1 r(R-r) \cdot \sin (\frac{\Pi r}{R}) + C_2(R^3 - 3r^2 R + 2r^3).$$

Let us choose $C_1 = C_2$ for simplicity of computation.

Computing: $\frac{\| Z \|_A^2}{\| Z \|^2}$, we

obtain approximately $\omega^2 = 104.6\ D/\rho R^4$ which is extremely close to the known value of ~ $104.4 \cdot D/\rho R^4$.

Again, $\min_{u \in H} J(u) \leq \min_{u \in S_n} J(u)$

A minimum over some set is clearly "better" than a minimum over a smaller subset.

Specifically, let Ω be a rectangle $0 < x < a, 0 < y < b$.

We choose a function

$$v = \sum_{m,n=1}^{M,N} C_{mn} \sin \frac{m\pi x}{a} \sin \frac{n\pi y}{b}$$ and we

compute $J(v) = \tilde{J}(c_{mn})$

A necessary condition for the minimum of $\tilde{J}(c_{mn})$ is:

$$\frac{\partial \tilde{J}}{\partial c_{mn}} = 0, \; m = 1,2,\ldots M; \; n = 1,2,\ldots N.$$

Solving this system of algebraic equation we determine a $\min_{u \in S_n} \{J(u)\}$ attained in the finite dimensional space $S_n \subset H$.

This is the essence of the Rayleigh-Ritz approach. We minimize the value of the Rayleigh functional in some suitably chosen subset of the Hilbert space H.

The Galerkin approach consists of selecting a specific subspace of H and minimizing the appropriate distance to that subspace.

The details can be found in most texts on numerical analysis.

If the space in which the approximate solution is "located" and the space chosen for minimizing the approximation are the same, the Galerkin's approach is identical with the Ritz approximation. Other refinements of such techniques include the Kantorovich method (see [27]) and its several variants ([28], [29]). The bibliography concerning more recent developments that are making use of finite elements and spline approximations may be found in the references of [2] and [25].

As we indicated, all of these techniques when applied to the Rayleigh quotient produce the upper bounds on the eigenvalues.

Lower bounds are generally harder to find and produce less accurate approximations when used. An earlier idea of estimating from below the eigenvalues for the Laplace operator is due to Trefftz.

The idea is fairly simple.

If we can find a minimum of a Rayleigh quotient in a class of functions containing H as a subset, such minimum is a lower bound on the first eigenvalue. We could choose as test functions the solutions of the initial equation

$$Au = 0 \text{ in } \Omega$$

but <u>not</u> satisfying boundary conditions

$$u = f \text{ in } \partial\Omega.$$

Instead they may be required to satisfy "the relaxed conditions" such as

$$u(\underset{\sim}{x}_i) = f(\underset{\sim}{x}_i)$$

at some points $\underset{\sim}{x}_i \in \partial\Omega$.

Then the minimum of Rayleigh quotient over such test functions gives us a lower bound on the first eigenvalue of A.

Instead of discussing several versions of this theory and of the theory of Weinstein, Fichera, Bazley, Fox, and Kato introducing most ingenious variations of this simple concept, we shall offer a specific example based on author's ideas originally published in [30].

4.15 Lower bounds of the natural frequencies of inhomogeneous plates

4.15.1 Introduction. This part is a review of author's paper [30]. It deals with an application of the intermediate problem technique to the computation of the natural frequencies of a vibrating inhomogeneous plate. The intermediate problem technique was first introduced in 1935 by Alexander Weinstein [43]. An abstract formulation of this technique was developed by N. Aronszajn [31], [32], and is related to the theories of N. Bazley and D.W. Fox [34], [35], [36], N. Aronszajn [33] and to further developments of A. Weinstein in [44], [45], [46], G. Fichera [38], [39], J.B. Diaz[37], T. Kato[41], and S.T. Kuroda [42]. An expository article of A. Weinstein [47] and a monograph of S.H. Gould [40] provide a review of this theory.

We formulate the Weinstein determinant for the plate problem in terms of the eigenfunctions and eigenvalues of the corresponding membrane problem.

4.15.2. Physical assumptions and notation. We assume the correctness of Hooke's law and the usual linear hypothesis of thin plate theory. The plate occupies a finitely connected compact subset $\bar{\Omega}$ of the Euclidean plane E^2. The boundary $\partial\Omega$ of $\bar{\Omega}$ consists of a finite number of straight lines, whose union will be denoted by Γ_1, and of finite number of smooth arcs, whose union will be denoted by Γ_2. Also $\partial\bar{\Omega} = \overline{\Gamma_1 \cup \Gamma_2}$ (where the bar denotes the closure operation). The Interior of $\bar{\Omega}$ will be denoted by Ω. n will denote the unit vector in the direction of the exterior normal to the boundary $\partial\Omega$, whenever it is defined. x, y are the Cartesian coordinates, t is the time variable, ∇^2 denotes the Laplacian:

$$\nabla^2 = \frac{\partial^2}{\partial x^2} + \frac{\partial^2}{\partial y^2}.$$

$\Diamond^4$ denotes the operator defined by:

$$\Diamond^4(A, B) = \frac{\partial^2 A}{\partial x^2}\frac{\partial^2 B}{\partial y^2} - 2\frac{\partial^2 A}{\partial x\,\partial y}\frac{\partial^2 B}{\partial x\,\partial y} + \frac{\partial^2 A}{\partial y^2}\frac{\partial^2 B}{\partial x^2}. \tag{4.91}$$

The physical meanings of symbols used here are: $h(x, y)$ is thickness of the plate ($h > 0$ in Ω), ν is Poisson's ration ($0 < \nu < \frac{1}{2}$), ρ is mass density of the plate, $\rho(x, y) > 0$ in $\bar{\Omega}$, E is Young's modulus ($E > 0$), D is flexural rigidity: $D = Eh^3/12(1 - \nu^2)$, and $w(x, y, t)$ is the transverse displacement. The displacement function $w(x, y, t)$ is assumed to be twice continuously differentiable on $\bar{\Omega}$. (Note: We do *not* assume $w \in C^4(\Omega)$.)

4.15.3. Definitions of Hilbert spaces $L_2(\Omega)$, $L_2(\partial\Omega)$, $L_{2,4}(\partial\Omega)$. We consider a class of real-valued functions of two real variables which are square integrable in the region $\Omega \subset \mathbb{R}^2$, and form the space $L_2(\Omega)$ in the usual manner by identifying all functions which offer only on a set of measure zero in Ω. For any functions $f(x,y)$, $g(x,y)$ $L_2(\Omega)$ in inner product

$$\iint_\Omega f(x, y)\cdot g(x, y)\, dx\, dy = (f, g)_\Omega$$

is defined. $((f, f)_\Omega)^{1/2} = ||f||_\Omega$ is the usual $L_2(\Omega)$ norm of $f(x, y)$.

In a similar way we define the space $L_2(\partial\Omega)$ of functions $f(x, y)$ defined on $\partial\Omega$ for which the integral $\oint_{\partial\Omega} f^2(x, y)\, ds$ exists. The inner product of $f, g \in L_2(\partial\Omega)$ is defined by the contour integral:

$$(f, g)_{\partial\Omega} = \oint_{\partial\Omega} f(x, y)\cdot g(x, y)\, ds,$$

and $((f, f)_{\partial\Omega})^{1/2} = ||f||_{\partial\Omega}$ defines the $L_2(\partial\Omega)$ norm of $f(x, y)$.

We could also consider vector-valued functions $A(x, y) = [A_1(x, y), A_2(x, y), A_3(x, y), A_4(x, y)]$, that is vectors with four components, such that each component $A_i(x, y)$, $i = 1, 2, 3, 4$, belongs to $L_2(\Omega)$. We define $L_{2,4}(\Omega)$ as the Hilbert space of all such vectors with the inner product:

$$\langle \mathbf{A}, \mathbf{B}\rangle_\Omega = \iint_\Omega [A_1(x, y)B_1(x, y) + A_2(x, y)B_2(x, y) + A_3(x, y)B_3(x, y) + A_4(x, y)B_4(x, y)]\, dxdy.$$

The $L_{2,4}$ norm is: $||\mathbf{A}|| = (\langle \mathbf{A}, \mathbf{A}\rangle_\Omega)^{1/2}$. The definition of $L_{2,4}(\partial\Omega)$ is analogous.

4.15.4. The basic equation and the boundary conditions. The displacement function $w(x,y)$ obeys in Ω the differential equation:

$$\nabla^2[D(x, y)\nabla^2 w(x, y, t)] - (1 - \nu)\Diamond^4(D(x, y), w(x, y, t)) + \rho(x, y)\,\frac{\partial^2 w(x, y, t)}{\partial t^2} = 0.$$

A separation of variables reduces Eq. (1) to the form of two simultaneous equations:

$$\nabla^2[D(x, y)\nabla^2 w_0(x, y)] - (1 - \nu)\Diamond^4(D(x, y), w_0(x, y)) - \omega^2\rho(x, y)w_0(x, y) = 0 \quad (92a)$$

$$w(x, y, t) = w_0(x, y)\exp(j\omega t). \quad (92b)$$

Equation (2a) can be rewritten as:

$$\nabla^2[D(x, y)\nabla^2(w_0(x, y)] - (1 - \nu)\Diamond^4(D, w_0) = \lambda\rho w_0(x, y).$$

This equation may not be satisfied in the classical sense, because the physically motivated differentiability assumptions do not specify the existence of the first term of this equation everywhere in Ω, and we may replace (92a) by its weak form:

$$\iint_\Omega \{\nabla^2[D(x, y)\nabla^2 w_0(x, y)]\cdot\psi(x, y) - (1 - \nu)\Diamond^4(D(x, y), w_0(x, y))\cdot\psi(x, y)\}\, dx\, dy$$

$$= \lambda \iint_\Omega \rho(x, y) w_0(x, y)\psi(x, y)\, dx\, dy$$

for any $\psi(x, y) \in L_2(\Omega)$. The use of the fundamental theorem of calculus of variations gives us:

$$([\nabla^2(D\nabla^2 w_0) - (1 - \nu)\Diamond^4(D, w_0)], \psi)_\Omega = \lambda(\rho w_0, \psi)_\Omega . \qquad (93)$$

On the Γ_1 part of the boundary $\partial\Omega$ the function $w_0(x, y)$ obeys the simple support conditions:

$$w_0(x, y) = 0, \qquad (93a)$$

$$\nu\nabla^2 w_0(x, y) + (1 - \nu)\frac{\partial^2 w_0(x, y)}{\partial n^2} = 0. \qquad (93b)$$

On the Γ_2 part of the boundary the plate is clamped, i.e.:

$$w_0(x, y) = 0, \quad \partial w_0(x, y)/\partial n = 0 \qquad (94a,b)$$

We make the following assumptions on the nature of solutions $\omega_0(x,y)$ in $\bar{\Omega}$:

i) $\omega_0(x,y) \in C^2(\bar{\Omega})$

ii) $\nabla^2(D\nabla^2\omega_0)$ is regarded in Ω as a generalized derivative in the sense of Sobolëv.

4.15.5 Solution of the intermediate problem for the case when $\Diamond^4(D,\omega) = 0$ and Γ_1 consists of straight lines.

The basic equation (92a) is reduced to the form

$$\frac{1}{\rho(x, y)}\nabla^2[D(x, y)\nabla^2 w_0(x, y)] - \lambda w_0(x, y) = 0. \qquad (95)$$

We now replace the boundary condition (4b) by the requirement:

$$\int_{\Gamma_2}\left[D(x, y)p_j(x, y)\frac{\partial w_0(x, y)}{\partial n}\right] ds = 0, \qquad j = 1, 2, 3, \cdots, m, \qquad (96)$$

where the functions $p_j(x, y)$ are selected to obey the following conditions:

(6a) The collection of functions $(D(x, y)p_j(x, y))$, $j = 1, 2, 3, \cdots \infty$, forms a dense subset of $L_2[(\partial\Omega) \cap \Gamma_2]$.

(6b) Each function $(D(x, y)p_j(x, y))$ is continuously differentiable on Γ_2.

(6c) Each function $(D(x, y)p_j(x, y))$ is harmonic in Ω.

(6d) Each function $p_j(x, y)$ is identically equal to zero on Γ_1.

Using (6b), (6c), (6d), (4a) and Green's theorem, we obtain:

$$\int_{\Gamma_2}\left(Dp_j\frac{\partial w_0}{\partial n}\right) ds = \int_{\partial\Omega}\left(Dp_j\frac{\partial w_0}{\partial n}\right) ds$$

$$= \int_{\partial\Omega}\left[Dp_j\frac{\partial w_0}{\partial n} - w_0\frac{\partial}{\partial n}(Dp_j)\right] ds$$

$$= (Dp_j, \nabla^2 w_0)_\Omega - (\nabla^2(Dp_j), w_0)_\Omega$$

$$= (p_j, D\nabla^2 w_0)_\Omega = 0 \qquad (97)$$

which is now valid for $p_1, p_2, \cdots, p_n$. We observe that the equality (7) and Eq. (5) can be derived as a necessary condition for the weak extremum (see for example [18], page 21 for the definition) of the functional

$$I(w_0) = (D\nabla^2 w_0, \nabla^2 w_0)_\Omega - \lambda(\rho w_0, w_0)_\Omega - 2\sum_{i=1}^{m} \mu i(p_i, D\nabla^2 w_0)_\Omega, \quad (98)$$

where λ, μ_i are Lagrangian multipliers. By an elementary variational argument we can derive the necessary conditions for the extremum of $I(w_0)$ to be:

$$\frac{1}{\rho}\nabla^2(D\nabla^2 w_0) - \lambda w_0 = 0 \quad \text{in} \quad \Omega, \quad (99a)$$

$$w_0 \equiv \text{constant on } \partial\Omega, \quad (99b)$$

$$\nabla^2 w_0 = \sum_{i=1}^{m} \mu_i p_i \quad \text{on} \quad \partial\Omega. \quad (99c)$$

It follows from (99c) that $\nabla^2 \omega_0 \equiv 0$ on Γ_1, since all $p_i(x,y)$ vanish on Γ_1. Then our assumption that Γ_1 consists of straight lines implies that $\partial^2\omega_0/\partial n^2 \equiv 0$ on Γ_1, since $\omega_0 \equiv$ on Γ_1, $\partial^2\omega/\partial s^2 \equiv 0$ on Γ_1, and $\nabla^2\omega_0 = (\partial^2\omega/\partial s^2) + (\partial^2\omega/\partial n^2)$. (This would be false if Γ_1 had non-vanishing curvature.) Hence condition (3b) is automatically satisfied, and becomes a natural condition of the corresponding variational problem.

We introduce the "base problem," namely the problem of vibrating inhomogeneous membrane

$$\nabla^2\left(\frac{u(x, y)}{\rho(x, y)}\right) = -\sqrt{\lambda}\, u(x, y) \quad \text{in} \quad \Omega, \quad (4.100a)$$

$$u \equiv 0 \qquad \text{on} \quad \partial\Omega. \quad (4.100b)$$

We assume that we know all we need to know about the "base problem." That is, we know the eigenvalues $\lambda_1, \lambda_2, \cdots, \lambda_m$ and the corresponding eigenfunctions $\zeta_1, \zeta_2, \cdots, \zeta_n$, which satisfy

$$\nabla^2(\zeta_i/\rho) = -\sqrt{\lambda_i}\zeta_i \quad \text{in} \quad \Omega, \quad (4.101a)$$

$$\zeta_i \equiv 0 \qquad \text{on} \quad \partial\Omega. \quad (4.101b)$$

Using the relationships (9) and Green's formula we have:

$$(D\nabla^2 w_0, \zeta_i)_\Omega = -\frac{1}{\sqrt{\lambda_i}}(D\nabla^2 w_0, \nabla^2(\zeta_i/\rho))_\Omega$$

$$= -\frac{1}{\sqrt{\lambda_i}}\left\{\left(\nabla^2(D\nabla^2 w_0), \frac{\zeta_i}{\rho}\right)_\Omega\right.$$

$$+ \int_{\partial\Omega} \left[D\nabla^2 w_0 \cdot \frac{\partial}{\partial n}\left(\frac{\zeta_i}{\rho}\right)\right] ds - \int_{\partial\Omega} \left[\frac{\zeta_i}{\rho} \cdot \frac{\partial}{\partial n}(D\nabla^2 w_0)\right] ds \Big\}$$

$$= -\frac{1}{\sqrt{\lambda_i}} \left(\lambda \cdot \rho \cdot w_0 , \frac{\zeta_i}{\rho}\right)_\Omega - \frac{1}{\sqrt{\lambda_i}} \int_{\partial\Omega} \left[D\nabla^2 w_0 \cdot \frac{\partial}{\partial n}\left(\frac{\zeta_i}{\rho}\right)\right] ds$$

$$= \frac{1}{\lambda_i} \left(\lambda w_0 , \nabla^2\left(\frac{\zeta_i}{\rho}\right)\right)_\Omega - \frac{1}{\sqrt{\lambda_i}} \int_{\partial\Omega} \left[D\nabla^2 w_0 \cdot \frac{\partial}{\partial n}\left(\frac{\zeta_i}{\rho}\right)\right] ds$$

$$= \frac{\lambda}{\lambda_i} \left\{ \left(\nabla^2 w_0 , \frac{\zeta_i}{\rho}\right)_\Omega + \int_{\partial\Omega} \left(\frac{\zeta}{\rho} \cdot \frac{\partial w_0}{\partial n}\right) ds \right\}$$

$$- \frac{1}{\sqrt{\lambda_i}} \int_{\partial\Omega} \left[\sum_{j=1}^{m} \mu_j \cdot D \cdot p_j \cdot \frac{\partial}{\partial n}\left(\frac{\zeta_i}{\rho}\right)\right] ds$$

$$= \frac{\lambda}{\lambda_i} \left(\nabla^2 w_0 , \frac{\zeta_i}{\rho}\right)_\Omega - \frac{1}{\sqrt{\lambda_i}} \sum_{i=1}^{m} \left(D \cdot p_i , \nabla^2\left(\frac{\zeta_i}{\rho}\right)\right)_\Omega$$

$$= \frac{\lambda}{\lambda_i} \left(\frac{\nabla^2 w_0}{\rho} , \zeta_i\right)_\Omega + \sum_{j=1}^{m} (Dp_j , \zeta_i)_\Omega .$$

Comparing the first and last terms in our chain of equalities, we have

$$\lambda_i (D\nabla^2 w_0 , \zeta_i)_\Omega = \lambda \left(\frac{D\nabla^2 w_0}{D \cdot \rho} , \zeta_i\right)_\Omega + \lambda_i \sum_{j=1}^{m} \mu_j (Dp_j , \zeta_i)_\Omega . \qquad (4.102)$$

We now observe that the eigenfunctions ζ_i form orthogonal bases of the space $L_2(\Omega)$. The completeness follows from the completeness of the eigenfunctions of the Laplace's operator, and from the positive definite property of $\rho(x, y)$. The orthogonality is checked by considering for $\lambda_i \neq \lambda_j$:

$$\sqrt{\lambda_i}\, (\zeta_i , \zeta_j)_\Omega - \sqrt{\lambda_j}\, (\zeta_i , \zeta_j)_\Omega = (-\nabla^2(\zeta_i/\rho), \zeta_j)_\Omega$$

$$+ (\zeta_i , \nabla^2(\zeta_j/\rho))_\Omega = \int_{\partial\Omega} \left[\zeta_i \cdot \frac{\partial}{\partial n} (\zeta_j/\rho) - \zeta_j \cdot \frac{\partial}{\partial n} (\zeta_i/\rho)\right] ds = 0.$$

Hence:

$$\left(\left(\lambda_i - \frac{\lambda}{\rho D}\right) D\nabla^2 w_0 , \zeta_i\right)_\Omega = \lambda_i \sum_{j=1}^{m} \mu_j (D \cdot p_j , \zeta_i)_\Omega .$$

We observe that for an arbitrary N

$$\sum_{i=1}^{N} \left(\left(\lambda_i - \frac{\lambda}{\rho D}\right) D\nabla^2 w_0 , \zeta_i\right) \zeta_i = \sum_{i=1}^{N} \left(\lambda_i - \frac{\lambda}{\rho D}\right) D\nabla^2 w_0 = \sum_{i=1}^{N} \lambda_i \sum_{j=1}^{m} \mu_j (Dp_j , \zeta_i)_\Omega \cdot \zeta_i$$

(assuming $||\zeta_i|| = 1$ without any loss of generality), and therefore

$$D\nabla^2 w_0 = \sum_{i=1}^{\infty} \left\{ \frac{\lambda_i \sum_{j=1}^{m} \mu_j (Dp_j , \zeta_i)\zeta_i}{\lambda_i - \dfrac{\lambda}{\rho \cdot D}} \right\} \qquad (103)$$

with convergence of the right-hand side assured. But by Eq. (97) we have

$$(D\nabla^2 w_0 , p_k)_\Omega = 0, \quad k = 1, 2, \cdots m.$$

Hence:

$$\sum_{j=1}^{m} \mu_j \sum_{i=1}^{\infty} \lambda_i (Dp_j , \zeta_i)_\Omega \left(\frac{\zeta_i}{\lambda_i - (\lambda/\rho D)} , p_k\right)_\Omega = 0, \qquad k = 1, 2, \cdots , m. \quad (104)$$

This can be regarded as a system of m equations in m unknowns, which can have a nontrivial solution only if the determinant of the coefficients is equal to zero:

$$\det\left[\sum_{i=1}^{\infty} \lambda_i (D\cdot\zeta_i\,,p_j)_\Omega \left(\frac{D\zeta_i}{D\lambda_i - (\lambda/\rho)}\,,p_k\right)_\Omega\right] = 0, \qquad j,k = 1,2,\cdots,m. \tag{105}$$

This is the Weinstein determinant for our problem. In the case when ρ and D are constant it reduces to the result of Weinstein. See [47] and [32].

$$\det \sum_{i=1}^{\infty}\left\{(\rho\lambda_i)\cdot\frac{(\zeta_i\,,p_j)_\Omega(\zeta_i\,,p_k)_\Omega}{(\rho\lambda_i)-(\lambda/D)}\right\} = 0, \qquad j,k = 1,2,\cdots,m. \tag{105a}$$

Since our $(\rho\lambda_i)$ corresponds to λ_i of the Weinstein article, and our λ/D to his λ, the formulas are clearly identical.

We note that changing to r, θ coordinates (cylindrical polars), a possible choice of the functions $p_i(\theta)$ for a starlike region Ω could be

$$\sin(n\theta),\ \cos(k\theta),\quad n = 1,2,\cdots,\ k = 0,1,2,\cdots,\ \text{on } \Gamma_1,$$

$$p_i(\theta) \equiv 0 \text{ on } \Gamma_2,\quad p_i(\theta) \text{ harmonic in } \Omega.$$

This implies that we have to solve first the Dirichlet's problem for the region Ω. That is, however, easily done by well-known numerical procedures (for example, the relaxation technique). In a private communication Professor Weinstein has suggested to me a variant of a numerical computation for a region Ω, such that a conformal map is known mapping Ω into the unit circle. Unfortunately, I did not attempt to follow his recipe.

In this paragraph I would like to offer a comment on the unusual nature of the result obtained. The base problem considered in the usual Weinstein technique as originally given in [43] retains the differential operator and only changes the boundary conditions to construct a simpler boundary-value problem. In the papers of Bazley and Fox (see, for example, [34], [35] the differential operator is changed effecting the splitting of the operator into a sum of two operators, so that one of them turns out to be manageable, but then the boundary conditions remain unchanged.

Here we have replaced in the "base problem", the operator $(1/\rho)\nabla^2(D\nabla^2\cdot)$, by the operator $((\nabla^2/\rho)\cdot)$, but also the very complex boundary conditions assigned to the plate (specificed separately on the subarcs of Γ_1 and Γ_2) were replaced by the very simple condition $u \equiv 0$ on $\partial\Omega$. This is a unique simplification. Unfortunately, more general results did not follow the original paper of 1974, and no software exists for implementing this technique.

4.16 Outline of Ramm's theory.

In 1981 Alexander Ramm published a very short note on max min properties of eigenvalues of a linear compact operator T acting on a Hilbert space H. He proved the following variational generalization of Rayleigh's formula.

Let λ_j be the eigenvalues of T, $|\lambda_1|\geq|\lambda_2|\geq$, ... let r_j be the moduli of the real parts of the eigenvalues order so that $r_1\geq r_2$... Let L_j be the eigensubspace of T corresponding to λ_j, M_j be the eigensubspace of T corresponding to r_j, $\tilde{L}_j = \sum_{k=1}^{j} \dot{+} L_k$, $\tilde{M}_j = \sum_{k=1}^{j} \dot{+} M_k$. Let t_j be the moduli of the imaginary parts of the eigenvalues, $t_1\geq t_2\geq,\ldots,\tilde{N}_j = \sum_{k=1}^{j} \dot{+} N_k$, N_j be the eigensubspace of T corresponding to t_j, $L_j \dot{+} \tilde{L}_j^{\amalg} = H$, where $\dot{+}$ denotes the direct sum, $\amalg$ denotes the direct complement in H.

Then the following formulas hold:

$$|\lambda_j| = \max_{x\in\tilde{L}_{j-1}^{\amalg}} \min_{\substack{y\in H\\(x,y)=1}} |(Tx, y)|, \tag{1}$$

$$r_j = \max_{x\in\tilde{M}_{j-1}^{\amalg}} \min_{\substack{y\in H\\(x,y)=1}} |\mathrm{Re}(Tx, y)|, \tag{2}$$

$$t_j = \max_{x \in \tilde{N}_{j-1}^{\perp}} \min_{\substack{y \in H \\ (x,y)=1}} |\mathrm{Im}(Tx, y)|. \qquad (2')$$

Remark 1. There is a one-to-one correspondence between $\{M_j\}$ and $\{L_j\}$. Namely, take $M_i = L_{j(i)}$, where $j(i)$ is so chosen that the eigenvalue $\lambda_{j(i)}$ has $|\mathrm{Re}\lambda_{j(i)}| = r_i$.

Remark 2. If $\| T_n - T\| \to 0$ and T_n are compact, then $\lambda_j(T_n) \to \lambda_j(T)$, j. Thus $|\lambda_j(T_n)| \to |\lambda_j(T)|$, $\mathrm{Re}\lambda_j(T_n) \to \mathrm{Re}\lambda_j(T)$, This fact permits an approximate calculation of the spectrum of T using in (1), (2) the operator T_n instead of T. One can take, e.g., n-dimensional operator T_n, i.e., dim range $(T_n) = n$.

Remark 3. For the moduli of the imaginary parts of the eigenvalues $t_1 \le t_2 \le \ldots$, the following formula holds:

$$t_j = \min_{x \in \tilde{N}_{j-1}^{\perp}} \min_{\substack{y \in H \\ (x,y)=1}} |\mathrm{Im}(Tx, y)|.$$

Similar arguments can be applied to derive variational principles for the spectrum of any unbounded linear operator with a discrete spectrum. Let operator A be a closed linear operator densely defined on H, $\lambda_j = \lambda_j(A)$ its eigenvalues, $\sigma(A) = \{\lambda_j\}$, $\sigma(A)$ denotes the spectrum of A. Each λ_j is an isolated eigenvalue of finite algebraic multiplicity. The eigenvalues are ordered so that $|\lambda_1| \le |\lambda_2| \le \ldots$. Let $r_1 \le r_2 \le \ldots$ be the moduli of real parts of the eigenvalues of A. Again we emphasize the fact that r_j is not necessarily equal to $|\mathrm{Re}\lambda_j|$, but it is possible to establish a one-to-one correspondence between $\{\lambda_j\}$ and $\{r_j\}$ by setting $|\mathrm{Re}\lambda_{j(i)}| = r_i$, as above. The variational principles read

$$|\lambda_j| = \min_{x \in \tilde{L}_{j-1}^{\perp}} \min_{\substack{y \in H \\ (x,y)=1}} |(Tx, y)|,$$

$$r_j = \min_{x \in \tilde{M}_{j-1}^{\perp}} \min_{\substack{y \in H \\ (x,y)=1}} |\mathrm{Re}(Tx, y)|.$$

Here, $\tilde{L}_j = \sum_{k=1}^{j} \dot{+} L_k$ and K_k is the eigensubspace corresponding to λ_k. A similar meaning is ascribed to M_j. For original articles see [55], [56].

4.17 Buckling with unilateral constraints.

Let us consider a plate subjected to end loads applied to the edges.

The plane of the plate shall be called "horizontal" the transverse displacement will be called "vertical." Effects of inequalities imposed on the vertical displacements were discussed in Miersemann papers [50], [51], [52]. Buckling with horizontal unilateral constraints can be found in the papers of Do [53], [54]. He considers problems of two-dimensional elasticity in which horizontal displacements are constrained by a variational inequality. The state equation is of the form $(L-\lambda I)\, W + C(\lambda, w) = 0$ where I is the identity operator, L a linear operator, C a non-linear operator as introduced by von Karman, and w denotes the vertical displacement. The parameter λ denotes magnitude of the load.

In this type of problems non-uniqueness of solutions can be demonstrated. Only under some restrictive assumptions can a critical load be evaluated and some admissible bifurcated shapes can be exhibited. At the time of writing no theory exists that would even estimate the number of admissible solutions. Even the result that is obvious in the nonlinear buckling of columns, namely that the bifurcation loads can be computed from the linearalized theory is in doubt unless some assumptions are introduced. The paper of Do [54] partially answers this problem, but, generally this area of research is open.

References

[1] J.P. Aubin, Approximation of elliptic boundary value problems, J. Wiley and Sons, New York, 1972.

[2] P.G. Ciarlet, Numerical analysis of the finite element methods, Université de Montreal Press, 1976.

[3] R.A. Adams, Sobolëv Spaces, Academic Press, New York, 1975.

[4] Leonhardi Euleri Opera Omnia, Seriei Secundae. Introduction to Vol. X and XI by C. Truesdell, 1960.

[5] J.L. Lagrange, Sur le figure de la colonnes, Miscelenea Taurinensia, 1970 reprint, Vol. V.

[6] E.L. Nikolaǐ, Collected works, Moscow, Gostekhizdat, 1955.

[7] J.P. Rebiére and S. Sahraoui, Parametric resonance by combined frequencies of cantilever bars under a periodic load, Mech. Research Comm. Vol. 15, #1(1978) p. 39-44.

[8] J.P. den Hartog, Advanced strength of materials, McGraw-Hill, New York, 1952.

[9] Ya. G. Panovko and I.I. Gubanova, Stability and vibration of elastic systems, 3rd edition, Nauka, Moscow, 1974.

[10] A.E. Taylor, Introduction to Functional analysis, J. Wiley and Sons, New York, 1958.

[11] S. Timoshenko and Gere, The theory of elastic stability, McGraw-Hill, New York, 1961, 1st edition.

[12] D.S. Mitrinović, Analytic inequalities, Grundlagen der Math. Wissenschaften series, Vol. 165, Springer Verlag, 1970.

[13] G.H. Hardy, J.E. Littlewood and G. Polya, Inequalities, 2nd edition, Cambridge University Press, Cambridge, 1952.

[14] N. Olhoff and S.H. Rassmussen, On single and bimodal optimum buckling loads of clamped columns , Int. J. Solids Structures (1977), Vol. 13, pp. 605-614, Pergamon Press.

[15] V. Komkov, An embedding technique in elastic stability, ZAMM, 60, (1980), p. 503-507.

[16] S. Timoshenko, Sur stabilité des systemes elastiques, Bulletin Polytechn. Instit. Kiev, 1910.

[17] Lord Rayleigh, The theory of sound, London 1894, reprinted by Dover Press, N.Y., 1961.

[18] W. Ritz, Gesammelte Werke, Paris 1911.

[19] T. Kato, Perturbation theory for linear operators, Springer Verlag, Berlin, 1966.

[20] K. Yosida, Functional analysis, 2nd edition, Springer, Berlin, 1967.

[21] E.L. Reiss, Column buckling an elementary example of bifurcation, in: Bifurcation theory and nonlinear eigenvalue problems, J.B. Keller and S. Antman, editors, Benjamin Publishers, New York and Amsterdam, 1969.

[22] W. Bickley and G. Temple, Rayleigh principle, Dover Press, N.Y., 1956.

[23] N. Dunford and J.T. Schwartz, Linear operators, Vol. II Spectral theory, self-adjoint operators in Hilbert space, Interscience, New York, 1963.

[24] H-Weinberger, Variational methods for eigenvalue approximation, S.I.A.M., Philadelphia, 1974.

[25] P.G. Ciarlet, Topics in mathematical elasticity, Studies in Mathematics and its applications, North Holland, Amsterdam, 1985.

[26] M. Kitahara, Applications of boundary integral methods to eigenvalue problems, North Holland, Amsterdam, 1985.

[27] L.V. Kantorovich and V.L. Krylov, Approximate methods of higher analysis, Noordhoff, Groningen, Holland 1958.

[28] V. Komkov, An iterative method for obtaining decreasing sequences of upper bounds on functionals, Math. Research Center, Univ. of Wisconsin.

[29] V. Komkov, Review of the Kantorovich variational method, Ann. Soc. Math. Pol. Series I, Comm. Math. XVII, (1973),105-117.

[30] V. Komkov, On lower bounds to the natural frequencies of inhomogeneous plates, Quarterly of Applied Mathematics, 1974, #1, p. 395-401.

[31] N. Aronszajn, The Raleigh-Ritz method and A. Weinstein method of approximation of eigenvalues, I, Proc. Nat. Acad. Sci. U.S.A. 34(1948), 474-480.

[32] N. Aronszajn, II, Proc. Nat. Acad. Sci U.S.A. 34(1948), 594-601.

[33] N. Aronszajn, Approximation methods for eigenvalues of completely continuous symmetric operators, in Proc. Symp. Spectral Theory and Differential Problems, Oklahoma State Univ., Stillwater, Okla,1951.

[34] N. Bazley and D.W. Fox, Truncation in the method of intermediate problems for lower bounds to eigenvalues, J. Res. Nat. Bur. Stds. 65B, (1961).

[35] N. Bazley and D.W. Fox, Methods for lower bounds to frequencies of continuous elastic systems, John Hopkins Univ. Applied Physics Lab. Report TG 609, 1964.

[36] N. Bazley, Lower bounds for eigenvalues, J. Math. Mech. 10 (1961), 289-308.

[37] J.B. Diaz, Upper and lower bounds on eigenvalues, in 8th Symposium in Applied Mathematics, A.M.S., pp. 53-58, New York, 1958.

[38] G. Fichera, Linear elliptic differential systems and eigenvalue problems, Lecture Notes in Mathematics, Springer-Verlag,1964.

[39] G. Fichera, Lezioni sulle transformazioni lineari, Inst. Math. Univ. Trieste, 1954.

[40] S.H. Gould, Variational methods in eigenvalue problems, Univ. of Toronto Press, Toronto 1957.

[41] T. Kato, Quadratic forms in Hilbert space and asymptotic perturbation series, Univ. of California, lecture notes, Berkeley, Calif., 1955.

[42] L.E. Payne, Inequalities for eigenvalues of membranes and plates, J. Rat. Mech. Anal. 4 (1955), 517-529.

[43] A. Weinstein, Sur la stabilité des plaques encastrées, Compt. Rend. 200 (1935), 107-109.

[44] A. Weinstein, Intermediate problem and the maximum-minimum theory of eigenvalues, J. Math. Mech. 12, 235-246, (1963).

[45] A. Weinstein, Some applications of the new maximum-minimum theory of eigenvalues, J. Math. Anal. Applic. 12 (1965), 58-64.

[46] A. Weinstein, A necessary and sufficient condition in the maximum-minimum theory of eigenvalues, studies in mathematical analysis and related topics. Stanford, Cal., Stanford Univ. Press, 1962.

[47] A. Weinstein, Bounds for eignevalues and the method of intermediate problems, in proceedings of the international conference on partial differential equations and continuum mechanics, Madison, University of Wisconsin Press, 1961, pp. 39-53.

[48] L.E. Elsgolc, Calculus of variations, Addison-Wesley, Reading Mass, 1962.

[49] S.T. Kuroda, Finite dimensional perturbation and representation of the scattering operator, Pacific J. of Math. 13 (1963), 1305-1318.

[50] E. Miersemann, Eigenvalue problems for variational inequalities, in Contemporary Mathematics Series, Vol. 4, Problems of elastic stability and vibrations, American Mathematics Society, Providence, R.I., 1980, V. Komkov, editor.

[51] E. Miersemann, Verzwigungprobleme für Variationsungleichungen, Math. Nachr., 65 (1975), 187-209.

[52] E. Miersemann, Über nichtlineare Eigenwertaufgaben in konvexen Mengen, Math. Nachr., 88(1978), 195-213.

[53] Claude Do, Nonlinear bifurcation of an elastic plate subjected to unilateral conditions in its plane, in Contemporary Developments in Continuum Mechanics, and Partial Differential Equations, G.M. de la Penba and L.A. Madeiros editors, North Holland Publ. Co., Groningen, Holland, 1978.

[54] Claude Do, Bifurcation theory for elastic plates subjected to unilateral conditions, J. Math. Anal. Appl., 60(1977), 435-448.

[55] A.G. Ramm, Variational principles for compact, non-selfadjoint operators, J. Math. Anal. Applic. 80 (1981), p. 291-293.

[56] I.P. Popov, Variational principles for the spectrum of non-selfadjoint operators, Math. Doklady USSR, 208 (1973), p. 290-292.

Chapter 5

AN EXAMPLE OF A VARIATIONAL APPROACH TO MECHANICS USING "DIFFERENT" RULES OF ALGEBRA

5.0 Preliminary remarks

In many instances duality does not help in deriving complementary variational principles due to "wrong" signs of abstract second derivatives. The Lagrangian formulation "works", the Hamilton's canonical equations are displayed, everything seems to be in the right form. Upon a closer examination only one variational principle can be formulated at best. Even if the signs of the second Fréchet derivatives can be determined in the vicinity of a critical point of the cost functional, both are of the same sign. Suppose that the Lagrangian for the problem $\mathcal{L}(x,Ax)$ vanishes at $x = \bar{x}$.

The Hamiltonian system (for $H = H(\underset{\sim}{x},\underset{\sim}{p})$

$\dfrac{\partial H}{\partial p} = Ax$, $\quad \dfrac{\partial H}{\partial x} = A^*p$ is satisfied at the point

$x = \bar{x}$, $p = \bar{p}$, but H fails to be convex at some

neighborhood of $x = \bar{x}$ for fixed $\bar{p}$, and concave at some neighborhood of $p = \bar{p}$ for fixed $\bar{x}$, (or vice versa). Then we do not have an obvious way of formulating dual variational principles, or of deriving two-sided bounds on the value of the Lagrangian. Let us first consider a system where the dual variational principles can be applied and both upper and lower bounds on the Lagrangian can be found from such principles.

<u>Example</u>

We shall consider the one-dimensional non-linear static system whose equation of deflection is:

$$\frac{d^2y(x)}{dx^2} - kf(y(x)) = 0.\ 0 \leq x \leq l, \tag{5.1}$$

with the boundary conditions:

$$y(0) = y(l) = 0.$$

We assume that the force kf(y(x)) becomes greater with increasing displacement, that is

$$\frac{d}{d\xi} f(\xi) > 0.$$

We rewrite the equation(5.1) as a pair of equations:

$$\frac{dy}{dx} = p \qquad \text{(A)}$$

$$-\frac{dp}{dx} = -kf(y). \qquad \text{(B)}$$

The Hamiltonian W(y,p) is given by:

$$W(x, p) = \tfrac{1}{2}p^2 - k \int_0^y f(\xi)\, d\xi = \tfrac{1}{2}p^2 - kF(y).$$

and the equations (A). (B) may be rewritten in the form

$$\frac{dy(x)}{dx} = W_p$$

$$-\frac{dp(x)}{dx} = W_y.$$

To demonstrate the techniques of finding bounds for the Lagrangian, and then effecting "gradual improvement" in an approximate solution of the problem we select a trial function:

$$\tilde{y}(x) = 2\alpha x(l-x) \qquad \text{(C)}$$

where α is a constant which is to be determined later.

We see that the choice of displacement function (C) satisfies the boundary conditions.

We now require that the equation (A) should also be satisfied. (Clearly we cannot expect that by a coincidence the equation (B) would also be satisfied, since in that case we have discovered a solution of our problem and numerical approximations are no longer necessary.) Equation (A) gives us the expression for $\tilde{p}(x)$:

$$\tilde{p}(x) = 2\alpha l - 4\alpha x.$$

The corresonding value of the Lagrangian integral is:

$$\mathscr{L}_1(\tilde{y},\tilde{p}) = \int_0^l \mathscr{L}(\tilde{y},\tilde{p},x)\,\mathrm{d}x = \int_0^l [\tfrac{1}{2}\tilde{p}^2 - KF(2\alpha x(l-x))]\,\mathrm{d}x +$$

$$\int_0^l \left[2\alpha x(l-x)\frac{\mathrm{d}}{\mathrm{d}x}\tilde{p}(x)\right]\mathrm{d}x + \{\tilde{y}(x)\tilde{p}(x)\}_0$$

$$= \int_0^l [\tfrac{1}{2}(2\alpha l - 4\alpha x)^2 - KF(2\alpha x(l-x))]\,\mathrm{d}x - 2\alpha^2 \int_0^l x(l-x)\,\mathrm{d}x.$$

This value of $\mathfrak{L}_1$, which depends on the parameter α is now chosen to satisfy the additional relationship

$$\frac{\mathrm{d}\mathfrak{L}_1(\alpha)}{\mathrm{d}\alpha} = 0.$$

We note that the equation (A) can be rewritten in the form:

$$L_p = \left(\frac{\mathrm{d}}{\mathrm{d}x}y, p\right)_p - \tfrac{1}{2}(p,p)_p = 0.$$

where the subscript p denotes Fréchet differentiation of the functional $\mathfrak{L}(p,y)$ with respect to p. As before:

$$\mathfrak{L}(p,y) \overset{\Delta}{=} \int_0^l \mathscr{L}(p,y)\,\mathrm{d}x.$$

We note that

$$\mathfrak{L}_{pp} = -1 < 0.$$

Hence, we can make a conclusion concerning

$\mathfrak{L}(\hat{y},\hat{p})$, where $\hat{y}(x)\cdot\hat{p}(x)$ are the solutions of the system (A-B) obeying the boundary conditions.

The value of the Lagrangian $\mathfrak{L}_1(\tilde{y}(x))$ obtained by substituting the trial function $\tilde{y}(x)$ and the corresponding $\tilde{p}(x)$ computed from equation (A) is an <u>upper</u> <u>bound</u> on the critical value $\mathcal{L}(\hat{y},\hat{p})$ where $\{\hat{y},\hat{p}\}$ is the critical pair that solves the equations (A) and (B), and $\hat{y}$ is the solution of the original system (*).

Since $\mathcal{L}_{yy}$ is positive, we conclude that choosing a trial function $\tilde{\tilde{p}}(x)$, computing $\tilde{\tilde{y}}(x)$ from equation (B), and computing the value of the Lagrangian $\mathcal{L}_2(\tilde{\tilde{p}},\tilde{\tilde{y}})$ we would obtain <u>a</u> <u>lower</u> <u>bound</u> on the value of $\mathcal{L}(\hat{y},\hat{p})$.

A remedy for "wrong" signs of derivatives occurred to the author in 1979 when he suggested in a talk to the A.M.S. that dual variational principles could be produced by altering the rules of multiplication for some vector valued functions. The "right" algebra for some problems turned out to be the quaternionic algebra of Hamilton. The contents of the next section were published in [13].

5.1 Quaternions and Fréchet Differentiation. An Introductory Discussion

By a quaternion we shall understand a form

$$q = a_0(\mathbf{x})\mathbf{1} + a_1(\mathbf{x})\mathbf{i} + a_2(\mathbf{x})\mathbf{j} + a_3(\mathbf{x})\mathbf{K},\ \mathbf{x} \in \mathbf{R}^n,$$

where the a_i, $i = 0, 1, 2, 3$, are functions from $\mathbf{R}^n$ into $\mathbf{C}$, satisfying the usual algebraic rules of a quaternion ring. See [13]. In what follows we shall consider only the cases where $n = 0, 1, 2,$ or 3, i.e., $\mathbf{x} = \{x_0, x_1, x_2, x_3\}$ at most. Suppose that $f: Q \to Q$, and the $a_i(\mathbf{x})$, $i = 0, 1, 2, 3$, are functions which belong to the Hilbert space $L_2(\Omega)$, $\Omega \subset \mathbf{R}^n$.

5.2 <u>The Quaternion Algebra</u>

The quaternion units $\{\mathbf{1}, \mathbf{i}, \mathbf{j}, \mathbf{K}\}$ obey the multiplication table:

	1	i	j	K
1	1	i	j	K
i	i	−1	K	−j
j	j	−K	−1	i
K	K	j	−i	−1

The usual (componentwise) addition rules are postulated for quaternions. Ring structure is assigned with respect to the operations of addition and multiplication.

Some algebraic properties of quaternions are listed below.

(a) Multiplication is associative; i.e.,

$$\mathbf{q}_1 \times (\mathbf{q}_2 \times \mathbf{q}_3) = (\mathbf{q}_1 \times \mathbf{q}_2) \times \mathbf{q}_3$$

(obviously it is not commutative).

(b) Division is defined, except division by the zero quaternion

$$\varnothing = \mathbf{1} \cdot 0 + \mathbf{i} \cdot 0 + \mathbf{j} \cdot 0 + \mathbf{K} \cdot 0.$$

In fact, the division formula is easily derived. Given

$$\mathbf{A} = \mathbf{1}a_0 + \mathbf{i}a_1 + \mathbf{j}a_2 + \mathbf{K}a_3 \qquad (a_i \in \mathbf{R})$$

and

$$\varnothing \neq \mathbf{C} = \mathbf{1}c_0 + \mathbf{i}c_1 + \mathbf{j}c_2 + \mathbf{K}c_3; \qquad (c_i \in \mathbf{R})$$

there exist quaternions β_l and β_r such that

$$\mathbf{A} = \boldsymbol{\beta}_l \mathbf{C} = \mathbf{C} \boldsymbol{\beta}_r .$$

The proof involves the solution of four equations in four unknowns with the determinant of the coefficients being equal to

$$D = \sum_{i=0}^{3} c_i^4 + \sum_{i=0}^{3} c_i^2 \left(\sum_{j \neq i} c_j^2 \right),$$

which is nonvanishing, unless $c_1 = c_2 = c_3 = c_4 = 0$, i.e., unless $\mathbf{C} = \varnothing$ (the zero quaternion).

The vectors of the coefficients of β_l and β_r are given by

$$\boldsymbol{\beta}_l = \gamma^{-1}\alpha,$$

$$\boldsymbol{\beta}_r = (\gamma^T)^{-1}\alpha,$$

respectively, where

$$\alpha = \begin{pmatrix} a_0 \\ a_1 \\ a_2 \\ a_3 \end{pmatrix},$$

$$\gamma = \begin{bmatrix} c_0 & -c_1 & -c_2 & -c_3 \\ c_1 & c_0 & c_3 & -c_2 \\ c_2 & -c_3 & c_0 & c_1 \\ c_3 & c_2 & -c_1 & c_0 \end{bmatrix},$$

and γ^T is the transpose of γ.

5.3 Fréchet Derivatives of quaternionic functions

We illustrate the definition of a left(right) Fréchet derivative by offering easy examples.

(a) Compute the derivative f(q) in the direction of q_2, where

$$q = a_0(x)\mathbf{1} + a_1(x)\mathbf{i} + a_2(x)\mathbf{j} + a_3(x)\mathbf{K}, \qquad x \in \mathbf{R}^n,$$

and

$$f(q) = c_1(q \times q) + c_2 q,$$

and q_2 is an arbitrary nonzero quaternion.

We compute

$$f(q + \epsilon q_2) - f(q) = \epsilon(2a_0c_1 + c_2)q_2 .$$

Hence, the Fréchet left derivative exists and, as asserted in Lemma 2, it is a scalar $2a_0c_1 + c_2$.

(b) It takes an elementary computation to show that, for a product

$$Q = \{q_1, q_2\} = \int_\Omega (q_1 \times q_2)\, dx,$$

the Fréchet left derivative with respect to q_1 is defined and is equal to q_2. Observe that

$$Q(q_1 + \epsilon\bar{q}) - Q(q_1) = \{\epsilon\bar{q}, q_2\}.$$

Example 1. Consider the case of the beam equation

$$Lw = AA^*w = s(x, t),$$

where s(x,t) is the applied load.

The Legendre transformation takes the form:

$$A^*W = \left[\mathbf{i}\frac{\partial}{\partial\tau} + \mathbf{j}\left(\frac{\partial^2}{\partial x^2}(D(x, y))^{1/2}\frac{\partial^2}{\partial x^2}\right)\right] W = p,$$

$$Ap = \left[\mathbf{i}\frac{\partial}{\partial t} + \mathbf{j}\frac{\partial^2}{\partial x^2}(D(x, y))^{1/2}\frac{\partial^2}{\partial x^2}\right](\mathbf{i}p_0 + \mathbf{j}p_1)$$

$$= \mathbf{1}(s(x, \tau)),$$

$$A^*w = \partial H/\partial p, \qquad Ap = \partial H/\partial w,$$

where

$$H = \tfrac{1}{2}\{w, AA^*w\} = \tfrac{1}{2}\{A^*w, A^*w\}$$

$$\underset{\text{def}}{\equiv} \tfrac{1}{2}\int_0^T\int_0^l (w(x, t) \cdot s(x, t))\, dx\, dt = \tfrac{1}{2}\{p, p\}.$$

As before, {,} denotes the quaternionic product, while · is the ordinary pointwise multiplication of scalars. The apparent negative sign of the energy product {,} is retained in the pointwise product

$$\tfrac{1}{2}\int_0^T\int_0^l w(x, t) \cdot s(x, t)\, dx\, dt \quad ,$$

since s(x,t) is the negative of the force applied to the beam.

The Lagrangian is given by

$$\mathscr{L} = \{A^*w, p\} - \mathscr{H} = \{w, Ap\} - \mathscr{H}.$$

A solution of these "canonical" equations coincides with the critical point $\{\hat{p},\hat{w}\}$, exactly as in the classical theory.

5.4 Bilinear Maps

Properties of bilinear maps

$$\int_\Omega q_1(\mathbf{x}) \ast q_2(\mathbf{x})\, dx = \mathbf{J}(q_1(x), q_2(x))$$

(where J belongs to the ring of constant quaternions).

Basic Lemma. A necessary and sufficient condition for the stationary behavior of J is $q_1 = q_2 = \emptyset$. There are no other critical points.

Proof. Regard $\hat{\psi}$ as fixed. Suppose that $\mathbf{J}(\hat{\phi})$ is a critical point of $\mathbf{J}(\phi, \hat{\psi})$. We perform Fréchet differentiation, i.e., we vary ϕ by substituting $\phi = \hat{\phi} + \epsilon\xi$, $\epsilon \in \mathbf{R}$, ϵ controlling the magnitude of ξ,

$$\mathbf{J}(\hat{\phi} + \epsilon\xi, \hat{\psi}) = \int_\Omega (\hat{\phi} + \epsilon\xi) \times \hat{\psi})\, dx,$$

$$\Delta\mathbf{J} = \mathbf{J}(\hat{\phi} + \epsilon\xi) - \mathbf{J}(\hat{\phi}) = \epsilon \int_\Omega (\xi \times \hat{\psi})\, dx$$

$$= \epsilon \left[\mathbf{1} \left(\int_\Omega (\xi_0 \hat{\psi}_0)\, dx \sum_{i=1}^{3} \int \xi_i \hat{\psi}_i\, dx \right) + \mathbf{i} \int_\Omega [(\xi_2 \hat{\psi}_3) - (\xi_3 \hat{\psi}_2)]\, dx + \cdots \right];$$

labeling

$$\epsilon \int_\Omega (\xi_0 \hat{\psi}_0)\, dx = \Delta \mathbf{J}_0 ,$$

etc., we obtain

$$\epsilon \left[\int_\Omega (\xi \times \hat{\psi}) \right] dx = \mathbf{1}\Delta J_{\phi_0} + \mathbf{i}\Delta J_{\phi_1} + \mathbf{j}\Delta J_{\phi_2} + \mathbf{K}\Delta J_{\phi_3} .$$

Stationary behavior of the quaternionic integral $J(\phi, \hat{\psi})$ implies stationary behavior of each component. Hence, dividing by ϵ and letting ϵ approach zero, we obtain the required result, namely $\Delta\mathbf{J}\phi = 0$ implies that for every ξ, $\int_\Omega (\xi \times \hat{\psi})\, dx = 0$, which is possible only if $\hat{\psi} = 0$.

A similar result is obtained by fixing q_1 and varying q_2.

The converse is trivial.

COROLLARY. *$Aq_1 = 0$ is a critical point of*

$$J(Aq_1, q_2) = \int_\Omega (Aq_1 \times q_2)\, dx$$

and

$$Aq_1 = \partial\mathscr{H}/\partial q_2, \qquad A^*q_2 = \partial\mathscr{H}/\partial q_1$$

are critical points of

$$\{Aq_1, q_2\} - \mathscr{H}(q_1, q_2),$$

where $\mathscr{H}(q_1, q_2)$ is an arbitrary map of quaternion-valued functions $q_1(x)$, $q_2(x)$ into the complex numbers under the symmetric differentiation definition.

Note. $\partial\mathscr{H}/\partial q$ is uniquely defined if the map $\mathscr{H}$: $(Q(x) \oplus Q(x)) \to Q$ is one dimensional, i.e., it is either real or complex valued, since then the "right" and "left" derivatives coincide.

THE RIESZ REPRESENTATION THEOREM, LAX-MILGRAM THEOREM: *Let $\mathbf{Q}$ $(=a_0\mathbf{1} + a_1\mathbf{i} + a_2\mathbf{j} + a_3\mathbf{K})$ be a quaternion-valued functional $\mathbf{Q} = \mathbf{Q}(q(\mathbf{x}))$, $\mathbf{x} \in \Omega$, which is linear in $q(\mathbf{x})$ and continuous in the $L_2^{(4)}(\Omega)$ topology (i.e., $q(\mathbf{x}) \to_{L_2^{(4)}} \varnothing$ implies $\mathbf{Q} \to \varnothing$), then there exists quaternions $\phi_1(\mathbf{x})$, $\phi_2(\mathbf{x})$, such that*

$$Q(q) = \int_\Omega \phi_1(\mathbf{x}) \times q(\mathbf{x})\, d\mathbf{x} = \int_\Omega q(\mathbf{x}) \times \phi_2(\mathbf{x})\, d\mathbf{x}$$

for every $q(\mathbf{x}) \in L_2^{(4)}(\Omega)$. Moreover, ϕ_1, ϕ_2 are unique in the $L_2^{(4)}(\Omega)$ sense.

Proof. We shall prove only the existence of $\phi_1(\mathbf{x})$ since the argument concerning the existence of $\phi_2(\mathbf{x})$ is identical. Since $L_2^{(4)}(\Omega)$ is a Hilbert space (observe that algebraic properties were not used in the definition), the continuous linear map $q(\mathbf{x}) \to a_i(q(\mathbf{x}))$ $(i = 0, 1, 2, 3)$ satisfies the requirements of the Riesz representation theorem.

We shall simply construct a quaternion $\phi(x)$ with the required properties. If $\phi(\mathbf{x}) = \mathscr{C}_0(x)\mathbf{1} + \mathscr{C}_1(\mathbf{x})\mathbf{i} + \mathscr{C}_2(\mathbf{x})\mathbf{j} + \mathscr{C}_3(\mathbf{x})\mathbf{K}$

$$\int_\Omega \phi(\mathbf{x}) \times q(\mathbf{x})\, d\mathbf{x} = (\mathscr{C}_0 c_3 - \mathscr{C}_3 c_2)]\,\mathbf{i} + [(\mathscr{C}_0 c_2 + c_0\mathscr{C}_2) + (\mathscr{C}_3 c_1 - \mathscr{C}_1 c_3)]\,\mathbf{j}$$
$$+ [(\mathscr{C}_0 c_3 + c_0\mathscr{C}_3) + (\mathscr{C}_1 c_2 - \mathscr{C}_2 c_1)]\,\mathbf{K} = \mathbf{Q}.$$

Obviously, $\mathbf{Q}$ linearly depends on $q(\mathbf{x})$. Let us select one of the components of $\mathbf{Q}$, say $a_2 = (\mathscr{C}_0 c_2 + c_0\mathscr{C}_2) + (\mathscr{C}_3 c_1 - \mathscr{C}_1 c_3)$. a_2 does continuously depend on $q(\mathbf{x})$ regarded as a four-component vector

$$\begin{bmatrix} c_0 \\ c_1 \\ c_2 \\ c_3 \end{bmatrix}$$

Hence, by the classical version of the Riesz representation theorem there exists a vector B such that $\mathbf{Q} = \int_\Omega \sum_{i=0}^{3} (c_i B_i)\, d\mathbf{x}$. B is identified with

$$\begin{bmatrix} \mathscr{C}_0 \\ -\mathscr{C}_1 \\ \mathscr{C}_2 \\ \mathscr{C}_3 \end{bmatrix}.$$

Repeàting this argument for each component of Q, we conclude that there exists $\phi(x)$ satisfying the left multiplication property. Q.E.D.

An almost identical argument works in the proof of the Lax-Milgram property, which is stated in Lemma 5. We shall not repeat the arguments.

5.5 Gâteaux Differentiability

We define the Gâteaux derivative of f(q) (with respect to q) at the point $q = \tilde{q}$ in the direction of q_1 to be the quantity $f'_{q_1}(q)|_{q=\tilde{q}} = q_2$ such that, for $\varepsilon > 0$,

$f(q \pm \epsilon q_1) - f(q) = \epsilon(q_1 \times q_2)$, i.e., a "left" linear form in q_1. (See Lemma 1 for the reason why terms in higher power of ε are not even mentioned.) If $q = q(x)$, we would define

$$\{q_0 \times q_1\}_\Omega = \int_\Omega (q_0 \times q_1)\, d\mathbf{x}$$

and find $q_2(\mathbf{x})$ by the rule

$$\epsilon \int (q_1(\mathbf{x}) \times q_2(\mathbf{x}))\, d\mathbf{x} \approx f(q(\mathbf{x}) + \epsilon q_1(\mathbf{x})) - f(q(\mathbf{x})).$$

The symbol $\approx$ implying that the remainder is of the order $\mathcal{O}(\varepsilon^2 q_1(x))$ for small ε. $\times$ denotes quaternion multiplication.

It will become apparent from physical examples that the choice of the quaternionic product has to be unique if physical laws are to be represented by Fréchet differentiation.

Let us define the Gâteaux derivative of $f(q), f: Q \to Q$, in the direction of a quaternion $q_1 \neq \varnothing$, to be q_2 such that $\epsilon(q_2 \times q_3) = f(q + \epsilon q_1) - f(q)$ for any $\epsilon > 0$.

LEMMA 1. *q_2 is uniquely defined.*

Proof. This follows immediately from the divisibility property of the ring of quaternions. For any fixed $\mathbf{x} = \bar{\mathbf{x}}$ we have a unique quaternion

$$q_2 = \mathbf{1}c_0 + \mathbf{i}c_1 + \mathbf{j}c_2 + \mathbf{K}c_3$$

such that

$$q_2 \times \epsilon q_1(\bar{\mathbf{x}}) = f(q(\bar{\mathbf{x}} + \epsilon q_1(\bar{\mathbf{x}})) = -f(q(\bar{\mathbf{x}})).$$

Defining $q_2(\bar{\mathbf{x}})$ to be that quaternion for each $\bar{\mathbf{x}}$ completes the proof of the uniqueness of the definition.

Comment. The choice of the "left" definition $\partial f(q)/\partial q = q_2$ such that $\epsilon(q_2 \times q_1) = f(q + \epsilon q_1) - f(q)$ is arbitrary. A different value would have resulted if we had defined $\partial f/\partial q = q_2$ such that $\epsilon(q_1 \times q_2) = f(q + \epsilon q_1) - f(q)$. As long as the definition remains consistent, and "right" or "left" differentiation definitions are used consistently, either one can be applied. For bilinear products, such as $Q = \{q_0, q_1\} = \int_\Omega (q_0(\mathbf{x}) \times q_1(\mathbf{x}))\, d\mathbf{x}$, the most convenient definition mixing the "left" and the "right" definitions results in $\partial Q/\partial q_0 = q_1$, $\partial Q/\partial q_1 = q_0$.

This turns out to be the definition which confirms the formal manipulation rules of mathematical physics. Of course, we can define the Fréchet derivative of any quaternion $q_0(\mathbf{x})$ with respect to any other quaternion $q_1(\mathbf{x}) \neq \varnothing$, in an identical manner.

Say $q_1 \neq \varnothing$. Thee there exists q_3 such that $q_0(q_1) = q_3 \times q_1$. We form the difference in the direction of an arbitrary quaternion q_2

$$q_3 \times (q_1 + \epsilon q_2) - q_3 \times q_1 = \epsilon q_3 \times q_2$$

Hence $\partial q_0/\partial q_1 = q_3$ by definition. Using the "right" instead of the "left" representation $q_0(q_1) = q_1 \times q_3$, we obtain an identical result

$$(q_1 + \epsilon q_2) \times q_3 - q_1 \times q_3 = \epsilon q_2 \times q_3.$$

Hence $\partial q_0/\partial q_1 = q_3$. Similarly, $\partial(Aq_0)/\partial q_0 = A$ for any linear operator A, since $A(q_0 + \epsilon q_1) - Aq_1 = \epsilon A q_1$ and $\partial(A \times q_0)/\partial q_0 = A$ if the "right" definition is adopted, while $\partial(A \times q_0)/\partial q_0 = A^*$ if the "left" definition is used.

From the above discussion the definition of second derivatives of a bilinear functional of the form $F(q_0, q_1) \to \mathbf{R}$ follows logically.

By the quaternionic version of the Lax–Milgram theorem, $F(q_0, q_1)$ must be of the form $F(q_0, q_1) = \int (Aq_0 \times q)\, d\mathbf{x} = \{Aq_0, q_1\}$, where A is a linear operator; i.e., the mapping of $q_0(\mathbf{x})$ into $Aq_0(\mathbf{x})$ is linear in Ω.

It follows from the definition that

$$\partial F(q_0, q_1)/\partial q_0 = A^* q_1,$$

$$\partial F(q_0, q_1)/\partial q_1 = Aq_0.$$

We observe that second derivatives do not obey Tonelli's law, i.e.,

$$\partial^2 F/\partial q_0\, \partial q_1 \neq \partial^2 F/\partial q_1\, \partial q_0.$$

LEMMA. *Every binary function $f\colon Q \times Q \to Q$ which linearly depends on a given quaternion $\hat{q}$ can be written in the form $f(q, \hat{q}) = \hat{q} \times \phi(q)$.*

Proof. The hypothesis implies that $f(q, c\hat{q}) = cf(q, \hat{q})$ for any constant c. Clearly if $\hat{q} = \varnothing$ then $f = \varnothing$. If $\hat{q} \neq \varnothing$, define $\phi(q) = \hat{q}^{-1} \times f(q, \hat{q})$ and define $\phi(q)$ arbitrarily if $\hat{q} = \varnothing$. Q.E.D.

LEMMA 2. *If f is a polynomial, then the Gâteaux derivative of f is independent of the direction, i.e., f is Fréchet differentiable, and the derivative is a scalar.*

Proof. A straightforward computation using only the definition.

Comment. This statement does not have to be true in general. Consider the following counterexample:

$$f(q) = q_+(\mathbf{x}) = a_{0_+}(\mathbf{x})\, 1 + a_{1_+}(\mathbf{x})\, \mathbf{i} + a_{2_+}(\mathbf{x})\, \mathbf{j} + a_{3_+}(\mathbf{x})\, \mathbf{K},$$

where

$$a_{i_+}(\mathbf{x}) = 0 \quad \text{if} \quad a_i(\mathbf{x}) < 0$$
$$= a_i(\mathbf{x}) \quad \text{if} \quad a_i(\mathbf{x}) \geqslant 0, \quad i = 0, 1, 2, 3.$$

Trying to solve for q_2 the equation

$$f(q + \epsilon q_1) - f(q) = (q + \epsilon q_1)_+ - q_+ = q_1 \times q_2$$

may result in different values of q_2, depending on the choice of q_1.

Defining the product $\{q_0, q_1\}$ to be $\int_\Omega q_0(\mathbf{x}) \times q_1(\mathbf{x})\, d\mathbf{x}$ we arrive at conclusions similar to Lemmas 1 and 2.

LEMMA 3. *The Fréchet derivative of $Q(q) = \int_\Omega (q \times \mathrm{e})\, d\mathbf{x}$ is defined and is equal to* $\mathrm{e}(\mathbf{x})$.

The proof is elementary (See Appendix 2 for an outline).

LEMMA 4 (The Lax–Milgram theorem). *If $Q(q_0, q_1)$ is a bilinear quaternion form linearly depending on $q_0(\mathbf{x})$, then Q is of the form $\{Lq_0, q_1\}$.*

For the proof, see Appendix 4.

We are now ready to recognize some well-known equations of physics in the framework of differentiation of quaternion products.

5.2 Quaternion Representation of Some Operator Equations of Physics in the Form $T = A^*A$

a. *The 3-Dimensional Laplace Operator*

We introduce the operator

$$A = \mathbf{1} \cdot 0 + \mathbf{i}\frac{\partial}{\partial x} + \mathbf{j}\frac{\partial}{\partial y} + \mathbf{K}\frac{\partial}{\partial z}$$

and its formal adjoint

$$A^* = \mathbf{1} \cdot 0 - \mathbf{i}\frac{\partial}{\partial x} - \mathbf{j}\frac{\partial}{\partial y} - \mathbf{K}\frac{\partial}{\partial z}.$$

Then AA^* is negative definite and is given by

$$AA^* = \Delta_3 = \mathbf{1}\left(\frac{\partial^2}{\partial x^2} + \frac{\partial^2}{\partial y^2} + \frac{\partial^2}{\partial z^2}\right).$$

If we feel that AA^* should be positive definite, rather than negative definite, we define accordingly

$$A = \mathbf{1} \cdot 0 + \mathbf{i}\left(\frac{i\partial}{\partial x}\right) + \mathbf{j}\left(\frac{i\partial}{\partial y}\right) + \mathbf{K}\left(\frac{i\partial}{\partial z}\right).$$

Then $A^* = A$ and $AA^* = -\Delta_3$. Here, as usual, $i^2 = -1$. Fir purposes of complex analysis it is more convenient to rewrite the two-dimensional Laplace equation in the formalism:

$$h = \mathbf{1}\frac{\partial}{\partial t} + \mathbf{i}\frac{\partial}{\partial x} + \mathbf{j}\frac{\partial}{\partial y} + \mathbf{K} \cdot 0,$$

$$\bar{h} = -\mathbf{1}\frac{\partial}{\partial t} - \mathbf{i}\frac{\partial}{\partial x} - \mathbf{j}\frac{\partial}{\partial y} + \mathbf{K} \cdot 0,$$

where the operator

$$h\bar{h} = \bar{h}h = \mathbf{1}\left(\frac{\partial^2}{\partial t^2} + \frac{\partial^2}{\partial x^2} + \frac{\partial^2}{\partial y^2}\right)$$

is applied to functions of the form $u(x, y) + iv(x, y)$, the first term being utterly superfluous. On the other hand, applying h to a quaternion of the form

$$q = \mathbf{1}c_0 + \mathbf{i}f_1(x, y) + \mathbf{j}f_2(x, y) + \mathbf{K} \cdot 0,$$

we obtain

$$hq = \mathbf{1}\left(\frac{\partial}{\partial x} f_1(x, y) + \frac{\partial}{\partial y} f_2(x, y)\right) + \mathbf{K}\left(\frac{\partial}{\partial x} f_2(x, y) - \frac{\partial}{\partial y} f_1(x, y)\right).$$

If $\overline{f_1(x, y) + if_2(x, y)}$ is analytic, we have $hq \equiv 0$, i.e., h acts like the $\bar{\partial}$ operator.

b. *The Beam Equation*

We define the operator

$$A = \mathbf{1}\left(i\frac{\partial}{\partial t}\right) + \mathbf{i}\left((D(x, y))^{1/2}\frac{\partial^2}{\partial x^2}\right), \qquad D(x, y) > 0;$$

then

$$A^* = \mathbf{1}\left(i\frac{\partial}{\partial t}\right) + \mathbf{i}\left(\frac{\partial^2}{\partial x^2} D(x, y)^{1/2}\right) \tag{b1}$$

and

$$A^*A = \mathbf{1}\left[\left(-\frac{\partial^2}{\partial t^2}\right) - \frac{\partial^2}{\partial x^2}\left(D(x, y)\frac{\partial^2}{\partial x^2}\right)\right], \tag{b2}$$

which is the classical Lagrange operator. It is an easy exercise to check that if the boundary conditions are natural then A^* is the true adjoint of A, and $A^*A = AA^*$, even though the operator AA^* looks strange at a first glance.

Note. We have omitted the **j**, **K** terms, but clearly we could write

$$A = \mathbf{1}(i\partial/\partial t) + \mathbf{i}((D(x, y)^{1/2})\partial^2/\partial x^2) + \mathbf{j}\cdot 0 + \mathbf{K}\cdot 0.$$

c. *The Klein–Gordon Operator*

We introduce the operator

$$A = \mathbf{1}\left(i\frac{\partial}{\partial(ct)}\right) + \mathbf{i}\frac{\partial}{\partial x} + \mathbf{j}\frac{\partial}{\partial y} + \mathbf{K}\frac{\partial}{\partial z} \tag{c1a}$$

and its adjoint

$$A^* = \mathbf{1}\left(i\frac{\partial}{\partial(ct)}\right) - \mathbf{i}\frac{\partial}{\partial x} - \mathbf{j}\frac{\partial}{\partial y} - \mathbf{K}\frac{\partial}{\partial z}; \tag{c1b}$$

clearly

$$AA^* = A^*A = -\partial^2/\partial(ct)^2 + \Delta = \Box. \tag{c2}$$

This decomposition also leads to a novel representation of Maxwell's equations.

5.3 Some General Remarks on Applications to Quantum Mechanics

It is interesting to observe that the quaternion notation has several advantages. From the Hamiltonian decomposition of $AA^*\mathbf{u} = \mathbf{f}$, we derive separate equations of a generalized system

$$A^*\psi = \phi, \qquad A\phi = \mathbf{f},$$

where, in the quantum mechanical formalism,

$$\mathbf{f} = m^2c^2\Psi.$$

A slight modification gives us

$$A = i\hbar\left[1\left(i\frac{\partial}{\partial(ct)}\right) + \mathbf{i}\frac{\partial}{\partial x} + \mathbf{j}\frac{\partial}{\partial y} + \mathbf{K}\frac{\partial}{\partial z}\right].$$
$$A^* = i\hbar\left[1\left(i\frac{\partial}{\partial(ct)}\right) - \mathbf{i}\frac{\partial}{\partial x} - \mathbf{j}\frac{\partial}{\partial y} - \mathbf{K}\frac{\partial}{\partial z}\right], \tag{c3}$$

and the canonical equations of Hamilton in the form

$$A^*\psi = \partial W/\partial\phi,$$
$$A\phi = \partial W/\partial\psi, \tag{c4}$$

with

$$W = \{\phi, \phi\} + \{f, \psi\} \tag{c5}$$

defining the Hamiltonian.

The corresponding Lagrangian is given by

$$\mathscr{L} = W - \{A^*\psi, \phi\}. \tag{c6}$$

So far ϕ, ψ are undefined quaternions, which can be easily identified with the solutions of the Dirac equation, if another "factor" mc is introduced into the operators A and A^*. If mc is regarded as a scalar (i.e., $mc = 1(mc) + \mathbf{i}\cdot 0 + \mathbf{j}\cdot 0 + \mathbf{K}\cdot 0$), no complications arise with respect to formal adjoints of mc ($\overline{mc} \equiv mc$). At this point it natural to regard mass as a general quaternion arriving in a natural way at mass plus "spin up" or "spin down" states as conjectured by Edmonds [1], i.e., we replace m by $\mathbf{m}$.

One could form several conjectures regarding quantum mechanical interpretation of Eqs. (c4)–(c6) with mc defined in the general quaternionic form. System (c4) can be rewritten as

$$i\hbar\bar{\partial}\psi = \mathbf{m}c \times \phi,$$
$$i\hbar\partial\phi = \mathbf{m}c \times \psi, \tag{c7}$$
$$\bar{\partial} \equiv \bar{A}, \qquad \partial \equiv A,$$

to bring them into a more familiar appearance of Dirac equations.

We postpone this discussion, preferring to treat the Dirac equation separately (Section f). It follows easily from Vaînberg's theorem that systems (c4) or (c7) represent a critical point of the Lagrangian, corresponding to dual variational principles, allowing one to introduce completely different computational techniques, based on Noble's two sided variational inequalities (See [4]).

d. *Maxwell's Equations*

Let the electric current and charge density be represented by the quaternion

$$Q_e = 4\pi i[\mathbf{1}(i\rho c) + \mathbf{i}j_x + \mathbf{j}j_y + \mathbf{K}j_z] \tag{d1}$$

and the magnetic (monopole) current and magnetic charge density be represented by the quaternion

$$Q_m = 4\pi[\mathbf{1}(i\bar{\rho}c) + \mathbf{i}\bar{j}_x + \mathbf{j}\bar{j}_x + \mathbf{K}\bar{j}_z]. \tag{d2}$$

Now using the properties of the operator A, and applying it to the complex vector $\mathbf{E} + i\mathbf{H}$, we have

$$\begin{aligned}
&\left[\mathbf{1}\left(i\frac{\partial}{\partial(ct)}\right) + \mathbf{i}\frac{\partial}{\partial x} + \mathbf{j}\frac{\partial}{\partial y} + \mathbf{K}\frac{\partial}{\partial z}\right](\mathbf{E} + i\mathbf{H}) \\
&= \left(-\frac{\partial E_0}{\partial(ct)} - \frac{\partial E_x}{\partial x} - \frac{\partial E_y}{\partial y} - \frac{\partial E_z}{\partial z}\right)\mathbf{1} + \left(\frac{\partial E_0}{\partial x} - \frac{\partial H_x}{\partial(ct)} - \frac{\partial E_y}{\partial z} + \frac{\partial E}{\partial y}\right)\mathbf{i} \\
&+ \left(\frac{\partial E_0}{\partial y} + \frac{\partial E_x}{\partial z} - \frac{\partial E_z}{\partial x} - \frac{\partial H_y}{\partial(ct)}\right)\mathbf{j} + \left(\frac{\partial H_z}{\partial(ct)} + \frac{\partial E_y}{\partial x} + \frac{\partial E_x}{\partial y} + \frac{\partial E_0}{\partial z}\right)\mathbf{K} \\
&+ i\left\{\left(\frac{\partial H_0}{\partial(ct)} - \frac{\partial H_x}{\partial x} - \frac{\partial H_y}{\partial y} - \frac{\partial H_z}{\partial z}\right)\mathbf{1}\right. \\
&+ \left(\frac{\partial E_x}{\partial(ct)} + \frac{\partial H_0}{\partial x} + \frac{\partial H_z}{\partial y} - \frac{\partial H_y}{\partial z}\right)\mathbf{i} + \left(\frac{\partial E_y}{\partial(ct)} - \frac{\partial H_z}{\partial x} + \frac{\partial H_0}{\partial y} + \frac{\partial H_x}{\partial z}\right)\mathbf{j} \\
&+ \left.\left(\frac{\partial E_z}{\partial(ct)} + \frac{\partial H_y}{\partial x} - \frac{\partial H_x}{\partial y} + \frac{\partial H_0}{\partial z}\right)\mathbf{K}\right\}.
\end{aligned} \tag{d3}$$

Equating $A(\mathbf{E} + i\mathbf{H})$ with $(1/c)(Q_e + iQ_m)$ we obtain the usual set of Maxwell's equations, after setting all space derivatives of E_0 and H_0 equal to zero.

However, if we assume that the quadruples (ρ, j_x, j_y, j_z), $(\bar{\rho}, \bar{j}_x, \bar{j}_y, \bar{j}_z)$ can be derived from a potential, the following argument gives another version of Maxwell's equations. We introduce the usual electric and magnetic vectors

$$\mathbf{E} = \begin{pmatrix} E_x \\ E_y \\ E_z \end{pmatrix} \quad \text{and} \quad \mathbf{H} = \begin{pmatrix} H_x \\ H_y \\ H_z \end{pmatrix}$$

as well as a fourth component E_0, H_0, such that

$$
\begin{aligned}
\frac{\partial E_0}{\partial(ct)} &= -4\pi\rho, \\
\frac{\partial E_0}{\partial x} &= -\frac{4\pi}{c}\dot{j}_x, \\
\frac{\partial E_0}{\partial y} &= -\frac{4\pi}{c}\dot{j}_y, \\
\frac{\partial E_0}{\partial z} &= -\frac{4\pi}{c}\dot{j}_z,
\end{aligned} \tag{d4}
$$

where, as before, ρ is the charge density, while

$$
J = \begin{pmatrix} j_x \\ j_y \\ j_z \end{pmatrix}
$$

is the electric current. Similarly, we conjecture the possibility that there exists an H_0 such that

$$
\begin{aligned}
\frac{\partial H_0}{\partial(ct)} &= -4\pi\bar{\rho}, \\
\frac{\partial H_0}{\partial x} &= -\frac{4\pi}{c}\bar{j}_x, \\
\frac{\partial H_0}{\partial y} &= -\frac{4\pi}{c}\bar{j}_y, \\
\frac{\partial H_0}{\partial z} &= -\frac{4\pi}{c}\bar{j}_z,
\end{aligned} \tag{d5}
$$

then $A(\mathbf{E} + i\mathbf{H}) = 0$ is the set of Maxwell's equations.

It is straightforward computation that

$$
A^*A(\mathbf{E} + i\mathbf{H}) \equiv 0
$$

or

$$
\Box(\mathbf{E} + i\mathbf{H}) \equiv 0,
$$

whether the assumptions concerning the existence of E_0, H_0 satisfying (d4), (d5) are true or false.

e. *Rigid Body Mechanics*

If we adopt the Euler angles θ, ϕ, ψ as the generalized coordinates describing a motion of a rigid body with a fixed point, the equations of motion are derived from Hamilton's principle of least action

$$
\delta \int_{t_0}^{t_1} (T - V)^{1/2}\, ds = 0. \tag{e1}
$$

The kinematic metric g^{ij} is defined by the quadratic from of the kinetic energy

$$
T = \tfrac{1}{2}(g^{ij}\dot{q}_i\dot{q}_j). \tag{e2}
$$

For a spinning top this means, unfortunately, that the metric g^{ij} is of the form

$$g^{ij} = \begin{pmatrix} 1 & 0 & 0 \\ 0 & 1 & \cos\theta \\ 0 & \cos\theta & 1 \end{pmatrix}, \tag{e3}$$

i.e., the path described by a geodesic (e1) has a locally nonorthogonal coordinate system. This is hardly surprising, since the local map $(\theta, \phi, \psi) \to (x, y, z)$,

$$\begin{aligned} z &= \cos\theta, \\ y &= -\sin\theta\cos\phi, \\ x &= \sin\theta\sin\phi, \end{aligned} \tag{e4}$$

satisfies the constraint

$$x^2 + y^2 + z^2 = 1, \tag{e5}$$

i.e. the motion is restricted to the surface of a unit sphere, and we can not produce a global orthogonal system restricted to the surface of a unit sphere in $\mathbf{R}^3$.

Alternate sets of variables have been proposed by various authors, for example, Sansò [5] or Lattman, the general idea being to embed the problem in a four-dimensional space, so that the motion of a rigid body takes place in a three-dimensional subspace equipped with a global orthogonal system of coordinates.

Following Sansò we introduce coordinates q_0, q_1, q_2, q_3 (which may be complex valued) such that

$$\sum_{i=0}^{3} q_i^2 = 1. \tag{e6}$$

We simply postulate that the kinematic metric is Euclidean (and now we get away with it), i.e.,

$$ds^2 = C^2\left(\sum_{i=0}^{3} (dq_i)^2\right), \tag{e7}$$

where C is a real number which may be chosen later for the sake of physical convenience. Sansò and Evans have discovered that writing the equations of the rotation vector in terms of the four quantities q_i, we have the following representation:

$$\begin{pmatrix} \omega_x \\ \omega_y \\ \omega_z \\ 0 \end{pmatrix} = A \begin{pmatrix} \dot{q}_3 \\ \dot{q}_1 \\ \dot{q}_2 \\ \dot{q}_0 \end{pmatrix} \tag{e8}$$

with

$$A = 2\begin{bmatrix} -q_3 & q_0 & q_2 & q_1 \\ -q_0 & -q_3 & -q_1 & q_2 \\ -q_1 & -q_2 & q_0 & -q_3 \\ q_2 & -q_1 & q_3 & q_0 \end{bmatrix}. \tag{e9}$$

The constraint (e6) becomes:

$$\sum_{i=0}^{3} q_i \dot{q}_i = 0. \tag{e10}$$

The quantities q_i can be regarded as components of the quaternion $\underset{\sim}{Q} = \mathbf{1}q_0 - \mathbf{i}q_1 - \mathbf{j}q_2 - \mathbf{K}q_3$.

We notice the following facts. The matrix A is orthogonal. However, in the corresponding three-dimensional representation, the rotation matrix assumes the following form:

$$B = \begin{bmatrix} q_0^2 + q_1^2 - q_2^2 + q_3^2 & -2(q_0q_3 - q_2q_1) & 2(q_0q_2 - q_3q_1) \\ 2(q_0q_3 + q_2q_1) & q_0^2 - q_1^2 + q_2^2 - q_3^2 & -2(q_0q_1 - q_2q_3) \\ -2(q_0q_2 - q_1q_3) & 2(q_0q_1 + q_2q_3) & q_0^2 - q_1^2 - q_2^2 + q_3^2 \end{bmatrix},$$

which is in general nonorthogonal.

The advantages of quaternionic representation become apparent.

The required variational representation is an easy consequence of the orthogonality of A. Since A^*A is positive if A is of the form $a_0 + i(a_1\mathbf{i} + a_2\mathbf{j} + a_3\mathbf{K})$ where the a_i are real, we can adopt a convention that rotations will be represented by quaternions of that form. This does not violate any previously made assumptions, and is exactly the representation of Sansò [5]. Denoting $(A^*A)^{1/2}$ by $\mathcal{A}$, we can rewrite the equations of motion for the rigid body in the form

$$\mathcal{A}y = \omega,$$

where

$$\mathcal{A} = T^*T,$$

or, factoring $\mathcal{A}$ out,

$$Ty = p,$$
$$T^*p = w,$$

which is a standard canonical form (See Noble [4] or Arthurs [7]).

In classical mechanics p does not have an obvious physical interpretation. We could refer to it as the $\frac{1}{2}$ spin, or simply as the generalized momentum conjugate to the generalized coordinate $y(t)$, following the Legendre transformation

$$p = \partial\mathscr{L}/\partial(Ty).$$

Comment. Evans and Sansò defined their quaternionic coordinates in terms of the Euler angles (within a sign change) as follows:

$$q_0 = \cos(\theta/2)\cos(\phi + \psi)/2,$$
$$q_1 = \sin(\theta/2)\cos(\phi - \psi)/2,$$
$$q_2 = \sin(\theta/2)\cos(\phi - \psi)/2,$$
$$q_3 = \cos(\theta/2)\sin(\phi + \psi)/2.$$

f. *Dirac Equation, van der Waerden Equations*

Let α be an operator represented by components

$$\alpha_x = \begin{bmatrix} 0 & 0 & 0 & 1 \\ 0 & 0 & 1 & 0 \\ 0 & 1 & 0 & 0 \\ 1 & 0 & 0 & 0 \end{bmatrix},$$

$$\alpha_y = \begin{bmatrix} 0 & 0 & 0 & i \\ 0 & 0 & -i & 0 \\ 0 & i & 0 & 0 \\ -i & 0 & 0 & 0 \end{bmatrix}, \tag{f1}$$

$$\alpha_z = \begin{bmatrix} 0 & 0 & 1 & 0 \\ 0 & 0 & 0 & -1 \\ 1 & 0 & 0 & 0 \\ 0 & -1 & 0 & 0 \end{bmatrix};$$

let β be the matrix

$$\beta = \begin{bmatrix} 1 & 0 & 0 & 0 \\ 0 & 1 & 0 & 0 \\ 0 & 0 & -1 & 0 \\ 0 & 0 & 0 & -1 \end{bmatrix}. \tag{f2}$$

Then the equation

$$i\hbar\partial\psi/\partial t = (c\boldsymbol{\alpha} \cdot \boldsymbol{\beta} + mc^2\beta)\,\psi, \tag{f3}$$

$$\psi = \begin{pmatrix} \psi_1 \\ \psi_2 \\ \psi_3 \\ \psi_4 \end{pmatrix}$$

is the original form of Dirac's equation. It was at first regarded as a relativistic version of the Schrödinger equation.

The original form of Dirac's equation (f3) can be manipulated into various alternative forms. For example, defining matrices

$$\gamma^0 = \begin{bmatrix} 1 & 0 \\ 0 & 1 \end{bmatrix}, \qquad \gamma = \begin{bmatrix} 0 & \sigma_i \\ \sigma_i & 0 \end{bmatrix}, \tag{f4}$$

where the σ_i are the Pauli matrices ($i = 1, 2, 3$), the Dirac equation may be written as a pair of equations

$$\hbar(i/m)\,(i\partial/\partial(ct) - \nabla\cdot\sigma)\,\chi^L = \chi^R,$$
$$\hbar(i/m)\,(i\partial/\partial(ct) + \nabla\cdot\sigma)\,\chi^R = \chi^L. \quad \text{(f5)}$$

By eliminating either χ^L or χ^R, one obtains the van der Waerden equations

$$-(1/m^2)\,\Box\,\chi^L(x) = \chi^L,$$
$$-(1/m^2)\,\Box\,\chi^R(x) = \chi^R. \quad \text{(f6)}$$

We define the operators

$$\Delta = \frac{i\hbar}{m}\left[1\left(i\frac{\partial}{\partial(ct)}\right) - \mathbf{i}\left(\frac{\partial}{\partial x}\cdot\sigma_1\right) - \mathbf{j}\left(\frac{\partial}{\partial y}\cdot\sigma_2\right) - \mathbf{K}\left(\frac{\partial}{\partial z}\cdot\sigma_3\right)\right],$$

$$\Delta^* = \frac{i\hbar}{m}\left[1\left(i\frac{\partial}{\partial(ct)}\right) + \mathbf{i}\left(\frac{\partial}{\partial x}\cdot\sigma_1\right) + \mathbf{j}\left(\frac{\partial}{\partial y}\cdot\sigma_2\right) + \mathbf{K}\left(\frac{\partial}{\partial z}\cdot\sigma_3\right)\right].$$

This enables us to rewrite the Dirac equations (f5) in the generalized canonical form

$$A\chi^L = \chi^R = \partial H/\partial\chi^R,$$
$$A^*\chi^R = \chi^L = \partial H/\partial\chi^L \quad \text{(f7)}$$

with the Hamiltonian given by

$$H = \tfrac{1}{2}[\{A\chi^L, \chi^R\} + \{\chi^L, A^*\chi^R\}]. \quad \text{(f8)}$$

5.4 A Theory of Critical Points of Quaternionic Functionals

The theory derived here parallels the known results summarized in Vaînberg's monograph [6]. The difficulties which arise are of a purely algebraic nature. The arguments of purely analytic nature can be reproduced without many changes from the well known texts. For an excellent review of some topics in applications of Fréchet differentiation we recommend the article by Nashed [9], the original reports of Noble [4], and the monograph by Arthurs [7].

Notation. $\mathbf{q}(x) \in L_2^{(4)}(\Omega)$ denotes that each component of $\mathbf{q}(x) = c_0(x)\mathbf{1} + c_1(x)\mathbf{i} + c_2(x)\mathbf{j} + c_3(x)\mathbf{K}$ is an $L_2(\Omega)$ function. The norm of $\mathbf{q}(x)$ is $\|\mathbf{q}(x)\| = \{\sum_{i=0}^{b} \|c_i(x)\|^2\}^{1/2}$. With this norm topology we define continuity of a map from $L_2^{(4)}(\Omega)$ to the scalar quaternion field by requiring that each component of $\mathbf{Q}(\mathbf{q}(x))$ is continuous, that is, $\mathbf{Q}(q(x)) = a_0\mathbf{1} + a_1\mathbf{i} + a_2\mathbf{j} + a_3\mathbf{K}$ is a continuous functional of $\mathbf{q}(x)$ if each element of $\mathbf{Q}(\mathbf{q}(x))$ continuously depends on $\mathbf{q}(x)$. This implies that whenever $\{q_i(x)\} \to^{L_2^{(4)}(\Omega)} \bar{\mathbf{q}}(x)$ then the corresponding sequences of numbers converge: $a_i(q_n(x)) \to^{L_2(\Omega)} a_i(\bar{\mathbf{q}}(x))$, $i = 0, 1, 2, 3$. We shall consider quite arbitrary functionals generally mapping quaternions $q_1(x), q_2(x), \ldots, q_n(x)$ into a constant quaternion $\mathbf{Q}$.

The simplest case involves a bilinear, or a sequilinear form

$$\mathbf{Q} = \{q_1, q_2\} \overset{\text{def}}{=} \int_\Omega (q_1(x) \times \overline{q_2(x)})\, dx,$$

i.e.,

$$\begin{aligned}\mathbf{Q} &= \tfrac{1}{2}Q_0 + \mathbf{i}Q_1 + \mathbf{j}Q_2 + \mathbf{K}Q_3 \\ &= \tfrac{1}{2}\int_\Omega a_0(x)\, dx + \mathbf{i}\int_\Omega a_1(x)\, dx + \mathbf{j}\int_\Omega a_2(x)\, dx + \mathbf{K}\int_\Omega a_2(x)\, dx,\end{aligned}$$

where, for almost each $x \in \Omega$, $\mathbf{1}a_0(x) + \mathbf{i}a_1(x) + \mathbf{j}a_2(x) + \mathbf{K}a_3(x) = q_1(x) \times q_2(x)$. In general the product $\{\ ,\ \}_\Omega$ will be defined by the formula $\{q_1, q_2\} = \int_\Omega (q_1(x) \times \overline{q_2(x)})\, dx$—where as before $\times$ denotes quaternionic multiplication and the bar denotes conjugation. The sequilinear property of $\{q_1, q_2\}_\Omega$ arises naturally. We postulate $c\mathbf{Q}(q_1, q_2) = \mathbf{Q}(cq_1, q_2) = \mathbf{Q}(q_1, \bar{c}q_2)$ to ensure the usual inner product property of $\{q_1, q_2\}$ if the $\mathbf{i}$, $\mathbf{j}$, and $\mathbf{K}$ components of q_1 and q_2 are equal to zero.

Unfortunately we have to distinguish between right and left Fréchet derivatives. The rules of differentiation are introduced as follows.

The identities

$$\mathbf{Q}(q_1 - \epsilon\eta) - \mathbf{Q}(q_1) = \epsilon\{\eta, \phi\} + o(\epsilon^2)$$
$$\mathbf{Q}(q_1 + \epsilon\eta) - \mathbf{Q}(q_1) = \epsilon\overline{\{\psi, \eta\}} + o(\epsilon^2)$$

define, respectively, the left and right Fréchet derivatives ϕ, ψ, whenever they are satisfied.

If Q has only one nonzero component (assume without any loss of generality that $\mathbf{i}$, $\mathbf{j}$, $\mathbf{K}$ components are equal to zero), then the sequilinear property of the product $\{\ ,\ \}$ implies $\psi = \phi$. Hence Fréchet differentiation of sequilinear complex- or real-valued quaternionic functionals is uniquely defined. So far in mathematical physics we have only encountered the case when the functionals were either real or complex valued. However, the general theory requires considerable discussion, carefully separating the right and left differentiation.

For this reason we shall denote the right and left factors of sequilinear quaternionic products by $|\, q_2\}$ and $\{q_1\,|$, respectively, in a manner borrowed from Dirac's bra-Ket notation. Then

$$\frac{\partial \mathbf{Q}}{\partial\{q_1|} = |q_2\}, \qquad \frac{\partial \mathbf{Q}}{\partial|q_2\}} = \{q_1|$$

implies the correct left or right multiplication in the reconstruction of the original functional Q. Moreover we can borrow the physics notation and denote

$$\{Aq_1, q_2\} \quad \text{by} \quad \{q_1 \mid A \mid q_2\}.$$

This is equal by the definition of A^* to $\{q_1, A^*q_2\}$. The "right" or "left" definitions are promptly ignoted if the functional is complex or real valued. For the scalar case, when the quaternion-valued functional possesses only one nonzero component, the entire Fréchet differentiation theory as outlined by Vaĭnberg [6] can be reproduced with only minor modifications. For the general case this theory needs to be rederived. We shall start this project with a technical lemma.

LEMMA 1. *Let $Q(q(x))$ be a functional possessing the left Fréchet derivative. Then it possesses the right Fréchet derivative and the critical points of Q with respect to left differentiation are the critical points with respect to the right differentiation. (A critical point* e *is the point at which the derivative vanishes.)*

Proof (a direct computation). In what follows we may consider quaternion valued functionals of n-copies of $L_2^{(4)}(\Omega)$, which may be considered a product space $L_{2,n}^{(4)}(\Omega) = L_2^{(4)}(\Omega) \times L_2^{(4)}(\Omega) \cdots L_2^{(4)}(\Omega)$ with an obvious definition of a quaternionic n-product $\mathbf{q}^{(1)} \times \mathbf{q}^{(2)} = \{q_1^{(1)} \times q_1^{(2)}, q_2^{(1)} \times q_2^{(2)} \cdots q_n^{(1)} \times q_n^{(2)}\}$, where $\mathbf{q}$ is the n-tuple $\mathbf{q} = \{q_1, q_2, \dots, q_n\}$, $\mathbf{q}_1 \in L_2^{(4)}(\Omega)$, $\mathbf{q}_2 \in L_2^{(4)}(\Omega), \dots$, etc. Left and right Fréchet differentiability is defined in an analogous manner.

Before proving the basic Theorem 1, we need to introduce some definitions.

DEFINITION. A curve in $L_{2,n}^{(4)}(\Omega)$ is a continuous mapping from $\mathbf{R}$ into $L_2^{(4)}(\Omega)$. (Continuity is defined with respect to $L_2^{(4)}(\Omega)$ norm.) A line passing through a point $\tilde{\mathbf{q}}$ in the direction of $\eta \in L_{2,n}^{(4)}(\Omega)$ is the collection of all points of the form $q + t\eta$, $-\infty < t < +\infty$, $q = q_1, q_2, \dots, q_n$.

LEMMA 2. *Suppose that $\Phi(\mathbf{q}(x))$ is a real-valued functional defined for all $\mathbf{q} \in L_{2,n}^{(4)}(\Omega)$. Then a necessary and sufficient condition for $\Phi(\mathbf{q}(x))$ to attain a local minimum (maximum) at $\mathbf{q} = \tilde{\mathbf{q}}$ is that $\Phi(\mathbf{q})$ attains a local minimum (maximum) at $\tilde{\mathbf{q}}$ on any line passing through $\dot{\mathbf{q}}$ in $L_{2,n}^{(4)}(\Omega)$.*

Proof. This follows almost directly from the Bing–Anderson theorem on the structure of a separable Hilbert space.

COROLLARY. *$\Phi(\mathbf{q}(x))$ attains a local minimum at $\mathbf{q} = \tilde{\mathbf{q}}$ if it attains a local minimum on any curve passing through $\tilde{\mathbf{q}}$.*

THEOREM 1. *Let us consider a real-valued functional $\Phi(q)$, $q \in L_{2,n}^{(4)}(\Omega)$, possessing Fréchet derivatives in some neighborhood of $\tilde{\mathbf{q}} \in L_{2,n}^{(4)}(\Omega)$. A necessary condition for a local extremum of $\Phi(\mathbf{q})$ at $\tilde{\mathbf{q}}$ is the vanishing at $\tilde{\mathbf{q}}$ of the left (or right) gradient of Φ.*

Proof (almost trivial). Suppose that $\mathbf{0}$ is a local extremum of Φ; then $\Phi(\tilde{\mathbf{q}} + + t\eta) - \Phi(\mathbf{q}) = t\{\eta, \partial\Phi/\partial\mathbf{q}\} + o(t^2)$ and for sufficiently small $|t|$ if $\partial\Phi/\partial\mathbf{q} \neq \varnothing$, we can choose η such that $\{\eta, \partial\Phi/\partial\mathbf{q}\}_{(n)}$ is positive (or is negative) denying the fact that $\tilde{\mathbf{q}}$ was an extremal point of Φ.

Note 1. Clearly this is not a sufficient condition for an extremum of Φ.

Note 2. The right or left gradient of Φ is not generally real- or complex-valued in any region of $L^{(4)}_{2,n}(\Omega)$ even if Φ is.

AN EXAMPLE. The Hamiltonian (f7) generates the corresponding Lagrangian

$$\mathscr{L} = H - \{A\chi^L, \chi^R\}.$$

If $\{A\chi^L, \chi^R\}$ is a scalar we need not distinguish between right and left derivatives, and the vanishing of the derivative of the Lagrangian is equivalent to the existence of a solution of the system of equations (f7). If $\{A\chi^L, \chi^R\}$ has more than one nonvanishing quaternionic component, clearly this simple analysis is no longer applicable.

5.5. A Class of Boundary Value Problems

Let T be a quaternion operator of the form

$$T = \tau_0\mathbf{1} + \tau_1\mathbf{i} + \tau_2\mathbf{j} + \tau_3\mathbf{K},$$

where $\tau_0, \tau_1, \tau_2, \tau_3$ are operators mapping a Hilbert space H_ϕ into a Hilbert space H_ϕ. (Note: H_ϕ may be actually a direct product of Hilbert spaces: $H_\phi = \sum_{i=1}^{4} H_{\phi_i}$, with τ_i acting only on H_{ϕ_i}, $i = 1, 2, 3, 4$, but this is an unimportant technical detail, which will be ignored here.) T^* denotes the quaternionic adjoint of T in the same sense as in Sections 3 and 4, i.e., for any quaternions $\phi \in H_\phi{}^4$, $\psi \in H_\psi{}^4$ ($H^4 = H \oplus H \oplus H \oplus H$), the following equality holds

$$\{T\phi, \psi\}_{H_\psi{}^4} = \{\phi, T^*\psi\}_{H_\phi{}^4},$$

for all $\phi \in H_\phi{}^4$, $\psi \in H_\psi{}^4$.

Suppose that Hilbert spaces H_{ϕ_i}, H_{ψ_i} are spaces of functions confined to regions Ω_1, Ω_2 of $\mathbf{R}^n$, $\mathbf{R}^m$, respectively. For the sake of simplicity we shall take $n = m$ and $\Omega_1 = \Omega_2 = \Omega$, with $H_{\phi_i} = H_{\psi_i} = \phi_{\phi_j} = H_{\psi_j}$, $i, j = 1, 2, 3, 4$. The boundary of Ω will be denoted by $\partial\Omega$, and $\Omega \cup \partial\Omega$ by $\bar{\Omega}$.

Again T and T^* are quaternionic operators such that the products $\{T\phi, \psi)_{H_\psi{}^4}$, $\{\phi, T^*\psi\}_{H_\phi{}^4}$ are real valued and are related by the following equality

$$\{T\phi, \phi\}_{H_\psi{}^4(\Omega)} = \{\phi, T^*\psi\}_{H_\phi{}^4(\Omega)} + B(\phi, \psi)_{\partial\Omega}, \tag{4.1}$$

where $B(\phi, \psi)_{\partial\Omega}$ is a number which only depends on the behavior of the components of ϕ and ψ on $\partial\Omega$.

Moreover we shall postulate that there exists a linear map σ such that all components of quaternions $\sigma\phi$, ψ restricted to $\partial\Omega$ are square integrable, and the product

$$\{\sigma\phi, \psi\}_{\partial\Omega} = \int_{\partial\Omega} (\sigma\phi \times \bar{\psi})\, ds$$

exists and is real valued for all $\psi \in H_\phi^4$, all $\psi \in H_\psi^4$. Then we can define the adjoint operator σ^* by the equality

$$\int_{\partial\Omega} (\sigma\phi \times \bar{\psi})\, dx = \int_{\partial\Omega} (\psi \times \overline{\sigma^*\psi})\, dx.$$

We shall consider a class of boundary value problems.

$$T\phi = \psi \in H_\psi^4(\Omega), \tag{4.2a}$$

$$T^*\psi = g \in H_\phi^4(\Omega), \tag{4.2b}$$

$$\begin{aligned} \sigma\phi\,|_{\partial\Omega} &= \mu(x)|_{x\in\partial\Omega}\,, \\ \phi^*\phi\,|_{\partial\Omega} &= 0. \end{aligned} \tag{4.3}$$

Assuming that all products are real valued, the existing critical point theory can be applied immediately to this class of problems. See [6, 7, 9]. The corresponding Hamiltonian and Langrangian are given, respectively, by

$$H = \tfrac{1}{2}\{\psi, \psi\}_{H_\psi^4(\Omega)} + \{\phi, g\}_{H_\phi^4(\Omega)}\,. \tag{4.4}$$

The Lagrangian is given by

$$\begin{aligned} &\{T\phi, \psi\}_{H_\psi^4(\Omega)} - \tfrac{1}{2}\{\psi, \psi\}_{H_\phi^4(\Omega)} - \{\phi, g\} + \{(\sigma\phi - \mu), \psi\}_{\partial\Omega} \\ &= \{T\phi, \psi\}_{H_\psi^4(\Omega)} - H + \{(\sigma\psi - \mu), \psi\}_{\partial\Omega} \\ &= \{T\phi, \psi\}_{H_\psi^4(\Omega)} - H - \bar{H}. \end{aligned}$$

These equations can be written in the following form:

$$\begin{aligned} \{T\phi, \psi\}_{H_\psi^4} &= \left\{\frac{\partial H}{\partial\psi}, \psi\right\}_{H_\psi^4} \quad \text{in} \quad \Omega \\ \{\psi, T^*\psi\}_{H_\phi^4} &= \left\{\frac{\partial H}{\partial\phi}, \phi\right\}_{H_\phi^4} \quad \text{in} \quad \Omega \end{aligned} \tag{4.5}$$

$$\begin{aligned} \{(\sigma\phi - \mu), \psi\}_{\partial\Omega} &= \left\{\frac{\partial H}{\partial\psi}, \psi\right\}_{\partial\Omega} \quad \text{in} \quad \partial\Omega \\ \{\phi, \sigma^*\psi\}_{\partial\Omega} &= \left\{\phi, \frac{\partial H}{\partial\phi}\right\}_{\partial\Omega} \quad \text{in} \quad \partial\Omega. \end{aligned} \tag{4.6}$$

The following variational problems can be formulated.

If L is convex in ϕ and concave in ψ in a neighborhood of a point $z_0 = (\phi_0\,, \psi_0)$ such that

$$\frac{\partial L}{\partial\phi\,|_{z_0}} = 0, \qquad \frac{\partial L}{\partial\psi\,|_{z_0}} = 0,$$

then the point z_0 is a min-max point of L. We need to comment that aside from conditions $\phi \in H^4$, $\psi \in H^4$, the functions ϕ, ψ need to be continuous in some neighborhood of the boundary $\partial\Omega$, in some sense, otherwise the boundary value problems lose all physical meaning.

In the case of real-valued functionals major parts of known boundary value theories may be routinely rederived with changes effected only in the algebraic rules. See, for example, [14], or [12]. Moreover, as was shown by M. Gurtin, [14] the inner products for quaternionic components in the formulation of the $L_2^{(4)}(\Omega)$ theory can be replaced by convolution products or other bilinear forms whenever the operator occurring in the physical problem acquires symmetry with respect to the new bilinear form.

The numerical bounds and estimates similar to [15] are deliberately omitted at this stage of research. However, arguments parallel to Arthurs [15][11] with computation of upper and lower bounds of Lagrangian functionals should be derived for specific quaternionic formulations of physical problems to establish the usefulness of the general theory.

REFERENCES

[1] J.D. Edmonds, Complex energies in relativistic quantum theory, Found. Physics 4 No. 4(1973), 473-479.

[2] D.J. Evans, On the representation of orientation space, Molecular Phys. 34 No. 2(1977), 317-325.

[3] L. Euler, "Novi Comment.," Sect. 6, p.208. St. Petersburg, Russia, 1776.

[4] B. Noble, "Complementary Variational Principles for Boundary Value Problems," Technical Report No. 558, M.R.C. University of Wisconsin, Madison, 1965.

[5] F. Sansó, A further account of roto-translations and the use of the method of conditioned observations.

[6] M.M. Vainberg, "Variational Methods for Investigation of Non-linear Operators," (transl. from Russian), Holden-Day, San Francisco, 1963.

[7] A.M. Arthurs, "Complementary variational principles," Oxford Mathematical Monographs, Oxford Univ. Press(Clarendon) London, 1970.

[8] J. Dieudonné, "Linear algebra and geometry," Hermann, Paris and Houghton- Mifflin Boston, 1969.

[9] M.Z. Nashed, Differentiability and related properties of nonlinear operators: some aspects of the role differentials in nonlinear functional analysis, in "Nonlinear Functional Analysis and Applications" (L.B. Rall, Ed.), pp. 103-309, Academic Press, New York, 1971.

[10] V. Komkov, Application of Rall's theorem to classical elastodynamics, J. Math. Anal. Appl. 14(1966), 511-521.

[11] A.M. Arthurs, A note on Komkov's class of boundary value problems and associated variational principles, J. Math. Anal. Appl. 33(1971), 402-407.

[12] M.E. Gurtin, Variational principles for linear initial value problems. Quart. Appl. Math. 22(1964), 252-256.

[13] V. Komkov, Quaternions, Fréchet differentation and some equations of mathematical physics. Critical Point Theory. J. Math. Anal. Applic., Vol. 71, #1, 1979, p. 187-209.

[14] M. Gurtin, Variational principles for linear initial value problems, Quart. Appl. Math. 22. (1964) p. 252-256.

[15] A.M. Arthurs, Dual extremum principles and error bounds for a class of boundary value problems, J. Math. Anal. and Applic. 41, (1973), p.781-795.

VARIATIONAL PRINCIPLES OF CONTINUUM MECHANICS

Appendix A

Abstract Differentiation and Integration.

A. 1.0 Introductory comments.

We shall consider primarily problems posed in a Hilbert space setting. Most statements offered here are easily translated into Banach space terminology, where we have to keep careful count of which elements belong to the space $\mathcal{B}$ and which belong to its dual $\mathcal{B}^*$, or even to $\mathcal{B}^{**}$. In the Hilbert space setting we can identify H with its dual H^*, and all arguments are generally simplified.

1.1 The concept of a derivative.

To offer generalizations of the concept of a derivative and of the usual necessary condition for the extremum of a function, we need to have a look at the concept of a derivative for a function of a single (real) variable and for a function of two variables. Differentiability of a function f whose domain is $\mathcal{D}(f)$, $f(x)$: $\mathcal{D}(f) \subset \mathbb{R} \to \mathbb{R}$, can be stated as follows: given $x_0 \in \mathcal{D}(f)$ and given h, we can express the difference $f(x_0 + h) - f(x_0) = \Delta_h f_{x_0}$ in the form

(A.1) $\Delta_h f_{x_0} = K(x_0;h) \cdot h + r(x_0;h)$,

where K is some constant (depending on x_0 and on h) and $r(x_0,h)$ has the property

$$\lim_{h\to 0} \left|\frac{1}{h}\right| \, |r(x_0;h)| = 0.$$

Moreover (regarding K as a function of h) $\lim_{h\to 0} K(x_0;h)$ exists. We could, of course, generalize this by considering $\lim_{h\to 0_+}$, $\lim_{h\to 0_-}$ or $\limsup_{h\to 0} \{K(x_0;h)\}$ etc., obtaining right, left, upper and lower derivatives, corresponding to the original definitions of Dini in the one-dimensional case.

We shall deliberately discuss only the simplest concepts.

The equation (A1) can be rewritten as :

$$\text{(A.2)} \quad \Delta_h f_{x_0} = \langle K(x_0;h), h\rangle + r(x_0,h).$$

where in one dimension the inner product $\langle , \rangle$ is interpreted as a simple multiplication of real numbers.

At this point we have deliberately avoided the form $\frac{\Delta f}{h}$, $\lim_{h\to 0} (\frac{\Delta f}{h})$, because generalizations of this form are possible only in spaces where quotients of this type are defined. In general h will be a vector in an infinite dimensional vector space, f may be some mapping, and the quotient $\frac{\Delta f}{h}$ makes no sense at all. However, the analog of formula (A1) has an easy interpretation. It is clear that for the infinite dimensional case one dimensional concepts are inadequate, and we shall consider the concept of a derivative in $\mathbb{R}^2$, and attempt to generalize the basic notions from $\mathbb{R}^2$ to an arbitrary Hilbert space.

We shall consider two basic pointwise maps:

$$f: \ \mathbb{R}^2 \to \mathbb{R}^2$$

$$\phi: \ \mathbb{R}^2 \to \mathbb{R}^1 .$$

f will be called an operator, a map, or a transformation, ϕ will be called a functional. Clearly, a functional is a map $\mathbb{R}^2 \to \mathbb{R}^2$ ($\mathbb{R}^1 \subset \mathbb{R}^2$). However, the one dimensional range of this map simplifies

many arguments, and such maps should be considered separately.
The map f may have the following types of derivatives in $\mathbb{R}^2$

i) partial derivatives
ii) total derivative - or Jacobian derivative,
iii) directional derivative (in the direction of some two dimensional vector in $\mathbb{R}^2$).

For the functional Φ we introduce the concept of a gradient of Φ. All these concepts have approprate generalizations in Hilbert space. We can also introduce the appropriate concepts of integration in $\mathbb{R}^2$. Given any simple arc Γ, whose endpoints are p_1 , p_2, we can introduce the Riemann integral

$\int_{\Gamma(p_1,p_2)} \langle \underset{\sim}{f}(x,y), \underset{\sim}{ds} \rangle$. If $\underset{\sim}{f}(x,y)$: $\mathbb{R}^2 \to \mathbb{R}^2$ has

the property that for any p_1, $p_2 \in \Omega \subset \mathbb{R}^2$ an integral is independent of the path, we introduce the idea of a potential functional:

$$\Phi(x_0,y_0) = 0, \quad \Phi(x,y) = \int_{(x_0,y_0)\,\Gamma}^{(x,y)} \langle \underset{\sim}{f}(x,y) \cdot \underset{\sim}{ds} \rangle .$$

We note that the independence of path condition requires no properties apart from Riemann integrability.

The more commonly used condition: curl $f(x,y) = 0$ in $\mathbb{R}^3$ or

$$\frac{\partial f_1(x,y)}{\partial y} = \frac{\partial f_2(x,y)}{\partial x} \text{ in } \mathbb{R}^2, \ \underset{\sim}{f} = (f_1, f_2) ,$$

assume the differentiability of f, and assume the simple connectivity of the region Ω.

Each of these criteria can be shown to have important generalizations in Hilbert spaces.

A.1.2 Definitions of derivatives of maps in Hilbert spaces.

By a map in Hilbert space we mean a map (a transformation, an operator...)

$$f: \Omega \subset H_1 \to H_2$$

where Ω is some subset of a Hilbert space H_1, and the range of f is a subset of a Hilbert space H_2.

The simplest case when $H_1 = \mathbb{R}$ is easily disposed, since division by a scalar is defined in H_2. We say that f: $\mathbb{R} \to H_2$ is differentiable at the point $t_0 \in \mathbb{R}$ if

$$\lim_{t \to t_0} \frac{f(t) - f(t_0)}{t - t_0} \text{ exists.}$$

Unless otherwise qualified the symbol lim. will stand for limit in the norm that is, strong limit. We postulate the existence of a vector $f'(t_0) \in H_2$, such that

$$\lim_{t \to t_0} \frac{f(t) - f(t_0)}{t - t_0} - f'(t_0) = 0 \qquad \text{(A(1.1)}$$

We can rewrite this in the form (1.1) as an equation in H_2:

$$f(t) - f(t_0) = (t - t_0)\, f'(t_0) + r(t;t_0), \qquad \text{A(1.2)}$$

$$\text{where} \quad \lim_{t \to t_0} \frac{r(t;t_0)}{t - t_0} = 0, \qquad \text{A(1.3)}$$

and where we deliberately avoid complications at this point by insisting on this limit being defined, instead of possible alternate conditions (one sided limit, lim sup,...etc.)

If we replace convergence in the norm by weak convengence, we have the equivalent definition of a weak derivative: f: $\mathbb{R} \to H_2$ has a weak derivative at the point $t_0 \in \mathbb{R}$, if there exist

a vector $f'(t_0) \in H_2$ such that

$$f'(t_0) = \text{weak limit}_{t \to t_0} \left(\frac{f(t) - f(t_0)}{t - t_0} \right), \qquad A(1.4)$$

with the formula A(1.2) still valid, and condition A(1.3) replaced by

$$\left(\frac{r(t;t_0)}{t-t_0} \right) \xrightarrow{\text{weakly}} \emptyset, \qquad A(1.5)$$

where $\emptyset$ is the zero vector in H_2. These definitions don't make much sense if f is an operator $f: H_1 \to H_2$ where H_1 is an infinite dimensional space. (See [4].) Next, let us offer the definition of the Gateaux (directional) derivative in $\mathbb{R}^2$. We consider a continous map $f: \Omega \subset H_1 \to H_2$. Let x_0 be an interior point of Ω. Pick a fixed vector $h \in H_1$. For fixed x_0, $h \in H_1$, we consider the vector $(f(x_0 + th) - f(x_0)) \in H_2$, where the constant t is picked to be sufficiently small in absolute value to make sure that $x_0 + th \in \Omega$. This is always possible since x_0 was an interior point of Ω. For a fixed x_0 and h the difference $\delta f(t) = (f(x_0 + th) - f(x_0))$ is a function of the real variable t only. Hence, we are back in the previously discussed case $\delta f(t): \mathbb{R} \to H_2$, and we can define the <u>directional derivative</u> of f in the direction of h, <u>computed at x_0</u>, if there exists a vector $f(x_0;h) \in H_2$ such that for all sufficiently small values of t, the following formula holds:

$$\delta f(t;x_0,h) = f(x_0+th) - f(x_0) = t \cdot f'(x_0;h) + r(t;x_0,h), \qquad A(1.6)$$

where $\lim_{t \to 0} \dfrac{r(t;x_0 h)}{t} = 0$; (See [1]), A(1.7)

or $\frac{1}{t} r(t;x_0 h) \xrightarrow{w} \emptyset$ as $t \to 0$ in case of a weak derivative. We should point out that the directional derivative $f'(x_0;h)$ does not have to be linear in h. In general, it turns out that $f'(x_0;h_1+h_2) \neq f'(x_0;h_1) + f'(x_0;h_2)$. It is however homogeneous of degree one in h, as can be easily checked from the definition, and $f'(x_0;ch) = cf'(x_0;h)$, for any $c \in R$.

A. 1.3 The Gateaux derivative.

We assume that $f'(x_0;h)$ exists and is linear in h. Then $f'(x_0;h)$ is called the Gateaux derivative of f in the direction of h, computed at the point x_0. The Gateaux derivative is not necessarily continuous with respect to h either in the strong, or even in the weak topology of H_2. It only implies the existence of a linear operator $L_{(x_0)}$ such that for a fixed $h \in H_2$ and for sufficiently small (in absolute value) $t \in R$,

$$f(x_0 + th) - f(x_0) = tL_{(x_0)}h + r(x_0;th)$$

where

$$\lim_{t\to 0} \frac{r(x_0;th)}{t} = 0.$$

(or weak $\lim_{t\to 0} \frac{1}{t} r(x_0;th) = \emptyset$.) We would like to point out that

$$\lim_{t\to 0} \frac{r(x_0;th)}{t} = 0$$

does not imply $\lim_{\|h\|\to 0} \left(\frac{\| r(x_0;h) \|}{\|h\|}\right) = 0$,

for arbitrary $h \in H_2$ of sufficiently small norm. We note that the Gateaux derivative may exist in the direction of a vector h_1, but may fail to exist in the direction of a vector h_2. This is easily checked to be true even in some two dimensional cases, that is for the mappings: $f: \mathbb{R}^2 \to \mathbb{R}^2$; or $f : \mathbb{C} \to \mathbb{C}$.

A. 1.4. Fréchet differentiation.

Assume that the operator L_{x_0} in (A1.8) is linear and bounded, i.e. for fixed $x_0 \in H_2$ we have

i) $L_{x_0}(\alpha h_1 + \beta h_2) = \alpha L_{x_0} + \beta L_{x_0} h_2$, and there exists $M > 0$ such that

ii) $\| L_{x_0} h\| \leq M(x_0) \|h\|$ for all h of sufficiently small norm.

We also assume that for all such $h \in H_2$ it is true that

(iii) $f(x_0+h) - f(x_0) = L_{x_0} h + r(x_0;h)$ where

$$\lim_{\|h\|\to 0} \frac{\|r(x_0;h)\|}{\|h\|} = 0, \quad \text{(see [7].}$$

Then L_{x_0} is called the Fréchet derivative of the operator f evaluated at x_0. The following condition can replace condition (iii)

(iv) $\lim\limits_{t\to 0} \dfrac{r(x_0,th)}{t} = 0$ uniformly with respect to h on all bounded subsets of H_2.

The fact that (iii) and (iv) turn out to be equivalent is not trivial. The proof of it can be found in [2] or [3] .

Theorem. Fréchet derivative exists if and only if the Gateaux derivative exists and satisfies condition (iii). The proof is easy. See [1] , [2] or [6] .

References

[1] M.M. Vainberg, Variational methods for study of nonlinear operators, Gostekhizdat, Moscow, 1956.

[2] M.M. Vainberg, Variational methods and techniques of monotone operators, Nauka, Moscow, 1972.

[3] M.Z. Nashed, Differentiability and related properties of nonlinear operators: some aspects of the role differentials in nonlinear functional analysis, in "Nonlinear Functional Analysis and Applications" (L.B. Rall, Ed.), pp. 103-309, Academic Press, New York, 1971.

[4] M. Fréchet, Sur la notion differentielle, Jour. de Math., 16 (1937), p. 233-250.

[5] R. Gateaux, Sur les functionelles continues et les functionelles analitiques, Comptés Rendues, 157, (1913), p. 325-327.

[6] A. Alexiewicz and W. Orlicz, on differentials in Banach Spaces, Ann. Soc. Polon. Math. 25, (1952), p. 95-99.

[7] M. Fréchet, La notion de differentielle dans l'analyse generale. Ann. Soc. de l'Ecole Norm. Super., 42, (1925), p. 293-323.

VARIATIONAL PRINCIPLES OF CONTINUUM MECHANICS

Appendix B

An introduction to analysis in Banach and Hilbert Spaces - Selected topics

B. 2.1. The inner product.

Let $\mathbf{V}$ be a linear vector space over the complex numbers, such that for any pair of vectors f, g, ϵ $\mathbf{V}$ there exists a complex number $\langle f,g\rangle$ and the maps $\mathbf{V}\times\mathbf{V}\to\mathbf{C}$ called the inner (or scalar) product $\cdot\{f,g\} \to \langle f,g\rangle$ satisfies the following axioms

i) $\langle f,g\rangle = \overline{\langle g,f\rangle}$ (bar denotes complex conjugate)

ii) for any α_1, $\alpha_2 \in \mathbf{C}, f_1, f_2, g \in \mathbf{V}$

$$\langle \alpha_1 f_1 + \alpha_2 f_2, g\rangle = \alpha_1\langle f_1, g\rangle + \bar{\alpha}_2\langle f_2, g\rangle .$$

iii) $\langle f,f\rangle$ is a nonnegative real number (for any $f \in V$) and $\langle f,f\rangle = 0 \Leftrightarrow f = \emptyset$.

iv) It follows from i) and iii) that

$$\langle \alpha f, g\rangle = \alpha\langle f, g\rangle = \langle f, \bar{\alpha} g\rangle .$$

Exercise: Show that the usual definition of a vector dot product in $\mathbf{R}^3$ satisfies the axioms i)-iv).

$$(\vec{A}\cdot\vec{B} = A_x B_x + A_y B_y + A_z B_z = |A|\cdot|B|\cdot \cos(\vec{A},\vec{B})\cdot)$$

Example a: The space of polynomials over $\mathbf{C}$ of order less than or equal to K is an inner product space with the product defined below:

$$\langle P_1, P_2\rangle = \sum_{i=0}^{K} a_{i(1)}\overline{a_{i(2)}}$$

where $P_1 = \sum_{i=0}^{K} a_{i(1)} x^i$

$$P_2 = \sum_{i=0}^{K} a_{i(2)} x^i$$

($a_i \in \mathbb{C}$). (Check the validity of axioms (i) -(iv))

Example b: Consider the linear vector space C[0,1] of real valued continuous functions on the interval [0,1],with usual pointwise addition and pointwise scalar multiplication by a real number. Check that

$$\int_0^1 w(x)f(x)g(x)dx = \langle f,g \rangle$$

defines the inner product on C[0,1],if w(x) is a given function which is positive on [0,1] .
Problem: Suggest additional examples of inner product vector spaces.

B. 2.2 The Cauchy-Schwartz inequality:

Let **V** be a vector space with an inner product < , >. We shall prove that for any $f \in \mathbf{V}$, $g \in V$

$$|\langle f,g \rangle|^2 \leq \langle f,f \rangle \cdot \langle g\cdot g \rangle$$

Proof: If <f,g> = 0 the statement of this theorem is trivially correct. We shall therefore consider the case <f,g> ≠ 0. Denote by α the complex number

$$\alpha = \frac{\langle f,g \rangle}{|\langle f,g \rangle|}$$ (recall that $|c| = (c\bar{c})^{\frac{1}{2}}$ for any $c \in \mathbb{C}$).

By axiom (iii) we have

$$0 \leq \langle \bar{\alpha}f + \beta g, \bar{\alpha}f + \beta g \rangle = |\beta|^2\langle g,g \rangle + 2\beta|\langle f,g \rangle| +$$

$$+ \langle f,f \rangle \quad \text{for any complex number } \beta.$$

Regarding this as a quadratic inequality in β, we conclude that the discriminant is non-positive, or

$$|\langle f,g\rangle|^2 \leq \langle f,f\rangle \cdot \langle g,g\rangle ,$$

which was to be proved. Moreover the strict equality is correct only if $\bar{\alpha}f + \beta g = \emptyset$, that is if f and g are linearly dependent vectors.

B. <u>2.3 The norm.</u>

A vector space $\mathbf{V}$ with a map $\|\ \ \| : \mathbf{V} \to \mathbb{R}_+$ is called a <u>normed space</u>. The <u>norm</u> $\|\ \ \|$ is a generalization of the concept of length of a line segment. It satisfies the following axioms

a) $\|f\| \geq 0$ and $\|f\| = 0 \iff f = \emptyset$.

b) $\forall\ \alpha \in \mathbb{C}$ the map $\|\ \ \| : \mathbf{V} \to \mathbb{R}_+$ satisfies:

$$\|\alpha f\| = |\alpha| \cdot \|f\| .$$

c) $\|f + g\| \leq \|f\| + \|g\|$. (This is called the triangular inequality)

Assertion: If $\mathbf{V}$ is an inner product space, then $\langle f,f\rangle^{\frac{1}{2}}$ defines a norm on $\mathbf{V}$.

Proof: We shall only prove that it satisfies axiom (c). Axioms (a) and (b) follow trivially from the properties of inner product. We use the Cauchy-Schwartz inequality.

$$\begin{aligned} &\langle f + g,\ f + g\rangle \\ &= \langle f,f\rangle + \langle f,g\rangle + \langle g,f\rangle + \langle g,g\rangle \\ &\leq \langle f,f\rangle + 2\sqrt{\langle f,f\rangle}\ \sqrt{\langle g,g\rangle} + \langle g,g\rangle \\ &= (\sqrt{\langle f,f\rangle} + \sqrt{\langle g,g\rangle})^2, \text{ or} \end{aligned}$$

denoting $\langle f,f\rangle^{\frac{1}{2}}$ by $\|f\|$, we have

$$\|f + g\|^2 \leq (\|f\| + \|g\|)^2$$

Hence $\|f + g\| \leq \|f\| + \|g\|$.

<u>Problem</u>: Check that $\|f-g\|$ is a metric $\rho(f,g)$. The norm $\langle f,f\rangle^{\frac{1}{2}}$ is called the <u>norm generated by</u>

the inner product, and the topology induced by the metric $\rho(f,g) = \| f-g \|$ is called the norm topology or the strong topology in inner product spaces. The concepts of convergence and continuity are then defined with respect to the norm topology. We shall use the notation:

$$\{\underset{\sim}{x}_n\} \xrightarrow{\text{(strongly)}} \underset{\sim}{x} \ (\epsilon V) \text{ if}$$

$$\lim_{n\to\infty} \| \underset{\sim}{x}_n - \underset{\sim}{x} \| = 0.$$

The "strongly", or "strong" will be frequently omitted, and $\{\underset{\sim}{x}_n\} \to \underset{\sim}{x}$ will mean that the sequence of vectors $\{\underset{\sim}{x}_n\}$ converges to the vector $\underset{\sim}{x}$ in the norm topology.

A function f: $V_1 \to V_2$ (where V_1, V_2 are normed spaces) is called continuous at the point $x \in V_1$ if: a) some open neighborhood of x is in the domain of f. (i.e., $\exists\ \varepsilon>0$ s.t. $\| x-y \| < \varepsilon \Rightarrow y$ is in the domain D_f of f). b) $\{x_i\} \to x$ implies that $f(x_i) \to f(x)$, for any such sequence of vectors x_i in the domain of f.

Problem: Prove the continuity of the inner product.

$$x_n \to x \text{ and } y_n \to y \quad \text{(in the norm topology)}$$

$$\Rightarrow (x_n,y_n) \to (x,y) \in \mathbb{C}.$$

Exercise: Show that in any inner product space $\| x + y\|^2 + \| x - y\|^2 = 2(\| x\|^2 + \|y\|^2)$. (The parallelogram law.)

The converse is also true. In any normed space the parallelogram law implies the existence of an inner product.

B. 2.4 Hilbert Space, Subspaces, Projection theorem.

Definition: An inner product space is complete if any Cauchy sequence converges to an element of the space.

Definition: A complete inner product space is called a Hilbert Space.

Example: Space ℓ_2 of all sequences of real numbers $\{a_i\}$ with the inner product

$$<\{a_i\}, \{b_i\}> \sum_{i=1}^{\infty} a_i b_i \text{ such that}$$

$$\sum_{i=1}^{\infty} a_i^2 < \infty, \text{ i.e. } \lim_{N\to\infty} \sum_{i=1} a_i^2 \text{ exists.}$$

We need to show that this inner product is defined, and then show that ℓ_2 is complete.

$$|<\{a_i\}, \{b_i\}>|^2 \leq <\{a_i\}, \{a_i\}> \cdot <\{b_i\}, \{b_i\}>$$

by the Cauchy-Schwartz inequality. Since $\{a_i\} \varepsilon\, \ell_2$, $\{b_i\} \varepsilon\, \ell_2$ it follows that $<\{a_i\}$, $\{b_i\}>^2$ exists.

To show that the space ℓ_2 is complete we choose a Cauchy sequence of sequences in ℓ_2. That is we choose a sequence $\{a_i\}_K$ of sequences in ℓ_2 such that

$$\lim_{K_1, K_2\to\infty} \| \{a_i\}_{K_1} - \{a_i\}_{K_2} \| = 0 \quad .$$

Using our definition of inner product and of the norm in ℓ_2, we have

$$\lim_{K_1, K_2\to\infty} \sum_{m=1}^{\infty} [(a_m)_{K_1} - (a_m)_{K_2}]^2 = 0$$

This is possible only if for each m=1,2,...

$\lim_{K_1, K_2\to\infty} [(a_m)_{K_1} - (a_m)_{K_2}] = 0$. But $\{a_m\}_{K_2}$ is a sequence of real numbers, and on the real line every Cauchy sequence converges. Hence,

$\lim_{K\to\infty} \{a_i\}_K = \{A_i\}$ is defined. It remains to show that $\{A_i\} \in \ell_2$, i.e., $\sum_{i=1} A_i^2 < \infty$. We leave this as an exercise.

We observe that the space ℓ_2 is <u>separable</u>, that is it contains a countable dense subset. An example of such a subset of ℓ_2 is the set of all sequences containing only finitely many non-zero entries, and with the additional requirement that the non-zero entries are rational numbers.

The following unit vectors are clearly the basis of ℓ_2:

$$\mathbf{e}_1 = \{1,0,0,\ldots,0,\ldots\}$$

$$\mathbf{e}_2 = \{0,1,0,0,\ldots,0,\ldots\}$$

$$\mathbf{e}_3 = \{0,0,1,0,\ldots,0,\ldots\}$$

Moreover the basis $\mathbf{e}_i$ are mutually orthogonal with respect to the inner product $<\ ,\ >$ on ℓ_2. That is $<\mathbf{e}_i, \mathbf{e}_j> = 0$ if $i \neq j$.

2.5. <u>Orthogonal subsets of Hilbert spaces</u>.

The subsets of S_1 and S_2 of a Hilbert H space are called mutually orthogonal if for any $x \in S_1$, $y \in S_2$, $<x,y> = 0$, i.e., if any elements of S_1 is orthogonal to any element of S_2. We note that the concept of orthogonality of vectors is a generalization of the definition of two vectors being perpendicular to each other in $\mathbb{R}^n$.

In $\mathbb{R}^3$ vectors are perpendicular if their dot product is equal to zero. Here we replace this notion by the vanishing of the inner product in H.

<u>Exercise</u>: Define the following inner product:

$$<f,g> = \int_0^1 f(x)g(x)dx$$ for functions which are bounded an measurable on $[0,1]$. Give some non-trivial examples of orthogonal functions. (non-trivial means that $f(x) \equiv 0$ a.e., is not allowed in the example.)

Definition: A subset M of a Hilbert space H (over $\mathbb{C}$) is called a linear manifold if for any f_1, $f_2 \in M$ and $\alpha_1, \alpha_2 \in \mathbb{C}$ it is true that $\alpha_1 f_1 + \alpha_2 f_2 \in M$. A closed linear manifold will be called a subspace of H.

Claim: Any subspace G of a Hilbert space H is itself a Hilbert space.

Proof: We need to show that G is complete. Since H is complete, every fundamental sequence in G must have a limit H. But G is closed. Hence that limit is in G.

Theorem: Let G be a subspace of a Hilbert space H. Choose any $h \in H$. Then there exists a unique $\tilde{g} \in G$, such that

$$\|\tilde{g} - h\| < \|g - h\| \text{ for any } g \in G; \quad g \neq \tilde{g}.$$

Idea of the proof: If $h \in G$ it is trival. If $h \notin G$ $\inf_{g\in G} \|g-h\|$ exists, and therefore there exists a sequence $\{g_i\} \in G$ such that $\lim_{i\to\infty} \|g_i - h\| = \inf_{g\in G} \|g-h\| = D$. It only remains to be shown that $\{g_i\}$ is a Cauchy sequence, which is left as an exercise. To show that $\tilde{g} = \lim_{i\to\infty} \{g_i\}$ is the unique vector such that $\|g-h\| = D$, assume to the contrary that two such vectors exist, say $\tilde{g}_1$ and $\tilde{g}_2$, and use the triangular inequality: $\|h - \alpha_1 g_1 - \alpha_2 g_2\| \leq \alpha_1 \|h - \tilde{g}_1\| + \alpha_2 \|h - \tilde{g}_2\|$ for any positive numbers α_1, α_2 such that $\alpha_1 + \alpha_2 = 1$, $\alpha_1 \neq 0$, $\alpha_1 \neq 1$.

However $\alpha_1 \|h - \tilde{g}_1\| + \alpha_2 \|h - \tilde{g}_2\| = \alpha_1 D + \alpha_2 D = D$. Hence, we must have strict equality (D was the infimum of all possible distances!), and

$\|\alpha_1(h-\tilde{g}_1) + \alpha_2(h-\tilde{g}_2)\| = \|\alpha_1(h-\tilde{g}_1)\| + \|\alpha_2(h-\tilde{g}_2)\|$; so that $\alpha_1(h-\tilde{g}_1)$ and $\alpha_2(h-\tilde{g}_2)$ are linearly dependent. Therefore

$$(h-\tilde{g}_1) = K\frac{\alpha_2}{\alpha_1}(h-\tilde{g}_2) = K_1(h-\tilde{g}_2) \text{ or } \tilde{g}_1 +$$

$K_1\tilde{g}_2 = (1-K_1)h$, which is impossible because this implies that h lies in (the linear manifold!) G. This completes the proof. It is easy to check that $\tilde{g}$ is orthogonal to h.

3.25a <u>The projection theorem</u>: Let G be a subset of a Hilbert space H. Then every vector $X \in H$ can be represented uniquely in the form: $X = x_1 + x_2$ where $x_1 \in G$, $x_2 \in G^{\perp}$. ($G^{\perp}$ denotes the set of all vectors which are orthogonal to G). To prove this theorem consider only the non-trivial case $x \notin G$, $x \notin G^{\perp}$. Hint: Choose as x_1 the unique vector satisfying conditions of the preceding theorem $\| x-x_1 \| = \min_{g\in G} \|x-g\|$, and use the fairly obvious arguments to show that x_1 (and therefore, $x_2 = x-x_1$) is unique.

<u>Problem</u>: Show that for non-zero x,y $\| x-y\|$ is minimized if and only if

$$\langle x-\frac{(y,x)}{\|y\|} y, y\rangle = 0 .$$

<u>An Approximation problem</u>: Consider the problem of finding the best approximation to an element x in a subspace M generated by the orthonormal vectors $\{e_1, e_2, \ldots, e_n\}$, all contained in the Hilbert space H. This problem may be restated as follows: choose constants $c_1, c_2, \ldots, c_n$ such that

$\| x - \sum_{i=1}^{n} c_i e_i \|^2$ is minimized. By the projection theorem the c_i should be chosen to satisfy

$(x-\sum_{i=1}^{n} c_i e_i, e_k) = 0$, $k = 1,2\ldots,n$, i.e., $c_k = (x,e_k)$, $k = 1,\ldots,n$. This can be shown also by direct expansion of $\| x-\sum_{i=1}^{n} c_i e_i\|^2$ to get

$$\|x\|^2 + \sum_{i=1}^{n} c_i{}^2 - \sum_{i=1}^{n} c_i{}^*(x,\mathbf{e}_i) - \sum_{i=1}^{n} c_i(x,\mathbf{e}_i)^*,$$

where * denotes conjugation.

Problem: Interpret the best approximation problem above in the following cases: (1) H = $\mathbf{e}_2$,{$\mathbf{e}_1$, $\mathbf{e}_2$, ...,$\mathbf{e}_n$...}are the basis {1,0,0,...},{0,1,0,...}... as given in section 2.4.(2) H is the Hilbert space obtained by completion of C[-1,1] with the product $\langle f,g\rangle = \int_{-1}^{1} f(x)g(x)dx$. $\mathbf{e}_1 = 1$ (i.e., is identically equal to one on [-1,+1], $\mathbf{e}_2 = x$, $\mathbf{e}_3 = x^2$. (3) H is the same space of functions as in (2) with additional restriction f(x) = 0 if $x \leq 0$. (Check that it is a Hilbert space!).

Following the standard physics notation, we could adopt the bra-ket symbolism.

If $|\psi\rangle$ denotes a vector in a space H, $\langle\eta|$ denotes a vector in the dual space, and $\langle\eta|\psi\rangle$ denotes the inner product of η and ψ. The notation $\langle\eta,\psi\rangle$ can also be used.

2.5c Find the projection of the vector $f(x) = |x| \in L_2[-\pi,+\pi]$ on the three dimensional subset of $L_2[-\pi,+\pi]$ spanned by

$$\mathbf{e}_1 = \frac{1}{(2\pi)^{\frac{1}{2}}}, \quad \mathbf{e}_2 = \frac{\cos x}{\pi^{\frac{1}{2}}}, \quad \mathbf{e}_3 = \frac{\sin x}{\pi^{\frac{1}{2}}}.$$

$(x \in [-\pi,+\pi])$.

The inner product is the usual $L_2[-\pi,+\pi]$ product:

$$\langle f,g\rangle \triangleq \int_{-\pi}^{+\pi} f(x)\, g(x)\, dx.$$ Check that

$\mathbf{e}_1$, $\mathbf{e}_2$, $\mathbf{e}_3$ are unit vectors in $L_2[-\pi,+\pi]$, which are mutually orthogonal.

25.d Change the inner product to :

$$<f,g> = \int_{-\pi}^{+\pi} \{(\pi^2-x^2)^{-\frac{1}{2}} \; f(x)\cdot g(x)\} dx \; ,$$

and show that $\mathbf{e}_1$, $\mathbf{e}_2$, $\mathbf{e}_3$ are no longer orthonormal. Find some orthonormal vectors.

3.26 <u>The Gramm-Schmidt technique</u>:

H will denote a Hilbert space in all future discussion. A set of vectors $M \subset H$, such that any two (different) vectors are orthogonal, and each vector is of norm one, is called <u>orthonormal</u>. (That is, $f \in M$, $g \in M$ implies $<f,g> = 0$ if $f \neq g$, and $<f,g> = 1$ if $f=g$.)

Suppose that some collection of vectors ϕ_i, $i=1,2,\ldots,n$ is given. The Gramm-Schmidt procedure offers a technique for constructing linear combinations of $\{\phi_i\}$ which form an orthonormal set. Without any loss of generality let us assume that $\{\phi_i\}$ form a linearly independent set. Our construction proceeds as follows.

Take $\mathbf{e}_1 = \dfrac{\phi_1}{\|\phi_1\|}$, Take $\psi_2 = \phi_2 - <\phi_2,\mathbf{e}_1>\mathbf{e}_1$

and $\mathbf{e}_2 = \dfrac{\psi_2}{\|\psi_2\|}$, $\psi_3 = \phi_3 - <\phi_3,\mathbf{e}_2>\mathbf{e}_2 - <\phi_3,\mathbf{e}_1>\mathbf{e}_1$,

$$\mathbf{e}_3 = \frac{\psi_3}{\|\psi_3\|}, \quad \psi_n = \phi_{n-1} - \sum_{j=1}^{n-1} <\phi_{n-1},\mathbf{e}_j>\mathbf{e}_j \quad \mathbf{e}_n = \frac{\psi_n}{\|\psi_n\|}.$$

It is easy to check that $<\mathbf{e}_k,\mathbf{e}_\ell> = 0$ if $k \neq i$ and

$<\mathbf{e}_k,\mathbf{e}_k> = 1$.

<u>Exercises</u>: Use the Gramm-Schmidt procedure to generate an orthonormal set which is a linear combination of: (a) $1, x, x^2, x^3, x^4$, $-1 \leq x \leq +1$ on the inner product space with the inner product

$<f,g> \overset{\text{def}}{=} \int_{-1}^{+1} f(x)g(x)dx$. (b) $\{1,0,0,\ldots,0\ldots\}$, $\{1,1,0,\ldots,0\ldots\}$, $\{1,1,1,0,\ldots,0\ldots\}$ in ℓ_2.

(c) 1, $\cos x$, $\cos 2x$, $\cos 3x$, $<f,g> = \int_0^1 f(x)\cdot g(x)\,dx$.

(Observe that the interval is [0,1] not [0,2π]!).

B.2.7 <u>Complete orthonormal systems:</u>

Following our previous discussion we shall offer a proof of the following statement. Let h be any vector in H, and $\{\mathbf{e}_i\}$ some (possibly infinite) orthonormal sequence of vectors.

Then $\sum_{i=1}^{\infty} \langle h,\mathbf{e}_i\rangle^2 \leq \|h\|^2$.

(This is known as Bessel's inequality.)

<u>Proof</u>: Take a linear combination of $\mathbf{e}_i$-s of the form $\sum_{i=1}^{n} h_i \mathbf{e}_i$. Then there exists a vector s_n nearest to h in the subspace spanned by $\mathbf{e}_i$, i = 1,2,...,n. The distance from h to that subspace is given by

$$d_{(n)} = \min_{h_i} \| h-\sum h_i\mathbf{e}_i\| = \|h\|^2 - \sum_{i=1}^{n} \langle h,\mathbf{e}_i\rangle^2$$

(see [1]) Since $\{d_n\}$ is a sequence of montone non-increasing positive numbers, $\lim_{n\to\infty} d_n$ exists and $\lim_{n\to\infty} d_n = d \geq 0$, which was to be proved. If for every $h \in H$, $\|h\|^2 = \sum_{i=1}^{\infty} \langle h,\mathbf{e}_i\rangle^2$, the orthonormal system $\{\mathbf{e}_i\}$ is called closed in H.

<u>Definition</u>: If a set $\{\mathbf{e}_\alpha\}$ (α in some index set) of orthonormal vectors in H has the property that no (non-zero) vector can be found in H orthogonal to every element of $\{\mathbf{e}_\alpha\}$ then the set $\{\mathbf{e}_\alpha\}$ is called <u>complete</u>.

<u>Theorem</u>: Every closed subset of a Hilbert space is a complete, subspace and in a separable Hilbert space every complete subset is closed. The proof is left as a problem. The relation $\sum_{i=1}^{\infty} \langle h,\mathbf{e}_i\rangle^2 = \|h\|^2$ is called Parseval's equality.

Example: The space L_2. The space $L_2[a,b]$ consists of eignevalence classes of complex valued functions defined on $[a,b] \subset \mathbf{R}$ such that

$\int_a^b |f(x)|^2 dx$ exists. Two functions belong to the same equivalence class if they differ only on a set measure zero. The inner product on $L_2[a,b]$ is given by $\int_a^b f(x)\overline{g(x)}dx = \langle f,g \rangle$. The proof that L_2 is complete will be delayed. For the time being we shall accept the completeness of L_2 without proof.

Two basic theorems of Hilbert space theory.

B.2.8 The Riesz representation theorem.

Let Φ be a continuous linear functional Φ: $H \to R$ (or $H \to C$). Then there exist a unique element $f \in H$ such that $\Phi(g) = \langle f,g \rangle$ for all $g \in H$, and $||\Phi|| = \sup_{||h||=1} \Phi(h) = ||f||$.

Proof: Let $\eta(\Phi)$ denote the null space of Φ, i.e., $x \in \eta(\Phi)$ if $\Phi(x) = 0$. It is easy to show that $\eta(\Phi)$ is a subspace of H. Take an arbitarary vector $y \in \eta^{\perp}$(recall that $\eta^{\perp} = H - \eta$). Then, for any $z \in H$, $\Phi(z)y-\Phi(y)z$ is an element of $\eta(\Phi)$. (Check that $\Phi(\Phi(z)y - \Phi(y)z) = 0$!) But y was chosen to be an element of $\eta^{\perp}$. Hence $\langle y, \Phi(z)y-\Phi(y)z \rangle = 0$ or $\Phi(z)\langle y,y \rangle-\Phi(y)\langle y,z \rangle = 0$ and therefore for any $z \in H$ we have

$$\Phi(z) = \frac{\Phi(y)}{\langle y,y \rangle} \langle y,z \rangle = \langle Ky,z \rangle = \langle f,z \rangle, \quad f = Ky,$$

where the constant K is given $K = \dfrac{\Phi(y)}{\langle y,y \rangle}$.

This proves the "representation" part of our theorem. The uniqueness of f is proved by contradiction. Assume, that for all $z \in H$, $\Phi(z) = \langle f^1,z \rangle = \langle f^2,z \rangle$ $f^1 \neq f^2$. By choosing $z = f^1 - f^2$ we arrive at an easy contradiction: $||f^1-f^2||^2 = 0$ (but $f^1 \neq f^2$, therefore $f^1 - f^2 \neq \emptyset$) We omit the proof of the last statement.

B.2.8[a] The Lax-Milgram theorem:

Every bilinear functional $\Phi(f,g)$, Φ: $H \times H \to \mathbb{C}$ (or $\mathbb{R}$), linear in each argument has a unique representation of the form $\Phi(f,g) = \langle Af,g\rangle$ where A is a bounded linear map A: $H \to H$ which is uniquely determined by Φ. Moreover,

$\|\Phi\| = \|A\|$, where

$$\|\Phi\| = \sup_{\|x\|=1} |\Phi(x)| \qquad \|A\| = \sup_{\|x\|=1} \|Ax\| .$$

B.2.9 The Generalized Fourier Series:

Let $\{e_i\}$ be a set of complete orthonormal vectors in H. A representation of a vector $h \in H$ of the form $h = \sum_{i=1}^{\infty} c_i e_i$ is called the Fourier series representation of h. In classical analysis $\{e_i\}$ are taken to be

$$\frac{\sin nx}{\|\sin nx\|} \quad n = 1,2,\ldots, \frac{\cos nx}{\|\cos nx\|}, \quad n = 1,2,\ldots,$$

which can be shown to form a complete orthonormal subset of L_2.

The Fourier Series representation problem is clearly related to our previous discussion on a best approximation.

We shall consider the problem of finding the best approximation to an element $\underset{\sim}{x}$ in a subspace M generated by the orthonormal vectors $e_1, e_2, \ldots e_n$ all contained in the Hilbert space H. This problem may be restated as follows: choose constants $c_1, c_2, \ldots c_n$, such that

$$\|\underset{\sim}{x} - \sum_{i=1}^{n} c_i e_i\|^2 \text{ is minimized.}$$

By the projection theorem the c_i should be chosen to satisfy

$$<(\underset{\sim}{x} - \sum_{i=1}^{n} c_i \mathbf{e}_i, \mathbf{e}_k)> = 0, \; k = 1,2,\ldots,n, \text{ i.e.}$$

$c_k = (\underset{\sim}{x}, \mathbf{e}_k)$ $k=1,2,\ldots,n$. This can be shown also by direct expansion of $\| \underset{\sim}{x} - \sum_{i=1}^{n} c_i \mathbf{e}_i \|^2$ to get

$$\| \underset{\sim}{x} \|^2 + \sum_{i=1}^{n} |^2 - \sum_{i-1}^{n} c_i^k (\underset{\sim}{x}, \mathbf{e}_i) - \sum_{i=1}^{n} c_i (\underset{\sim}{x}, \mathbf{e}_i)^*$$

Adding constant terms $|(\underset{\sim}{x}, \mathbf{e}_i)|^2$ $i = 1,\ldots,n$ it can be seen that minimizing $\| \underset{\sim}{x} - \sum_{i=1}^{n} c_i \mathbf{e}_i \|^2$ is equivalent to minimizing $\| \underset{\sim}{x} \|^2 + \sum_{i=1}^{n} |c_i - (\underset{\sim}{x}, \mathbf{e}_i)|^2$

which occurs when $c_i = (\underset{\sim}{x}, \mathbf{e}_i)$ $i = 1,\ldots,n$.

Bessel's Inequality: Let x be an element of a Hilbert space H and suppose $\{\mathbf{e}_i\}$ is an orthonormal sequence in H. Then $\sum_{n=1}^{\infty} |(\underset{\sim}{x}, \mathbf{e}_i)|^2 \leqq \| \underset{\sim}{x} \|^2$.

We have claimed that the sequence $\{\mathbf{e}_i\}$ is complete if equality holds in the Bessel inequality for all $\underset{\sim}{x} \in H$. It is clear that in this case the Fourier Series of any element $\underset{\sim}{x} \in H$ converges in H (not pointwise!) to that element.

Example: The space $L_2(0,2\pi)$ of square integrable functions (real) is a Hilbert space with inner product

$$(f,g) = \int_0^{2\pi} f(x)g(x)dx \; , \quad \|f-g\|^2 = \int_0^{2\pi} |f(x)-g(x)|^2 dx$$

It can be shown that the functions $\frac{1}{\sqrt{2\pi}}$, $\frac{1}{\sqrt{\pi}}\sin nx$, $\frac{1}{\sqrt{\pi}}\cdot\cos nx$ is a complete orthonormal set in $L_2(0,2\pi)$.

It follows therefore that if $f \in L_2(0,2\pi)$ then

$$\frac{a_0}{2} + \sum_{k=1}^{n} a_k\cos(Kx) + \sum_{K=1}^{n} b_K\sin(Kx) \underset{(\text{as } n\to\infty)}{\to} f(x)$$

(in the L_2 sense!) In general, the above series fail to converge pointwise to $f(x)$.

A classical engineering example of application of Fourier Series is in the analysis of vibrating systems subjected to a periodic disturbance. Consider for example a crankshaft of an internal combustion engine. For simplicity we can consider a single cylinder driving a crankshaft. If the engine runs at some frequency f revolutions per minute the angular velocity is $\omega=2\pi f$, and the torque transmitted to the crankshaft is periodic with a period of $2\pi\omega^{-1}$. The graph of torque transmitted may look almost like a randomly drawn continuous cruve, but is periodic.

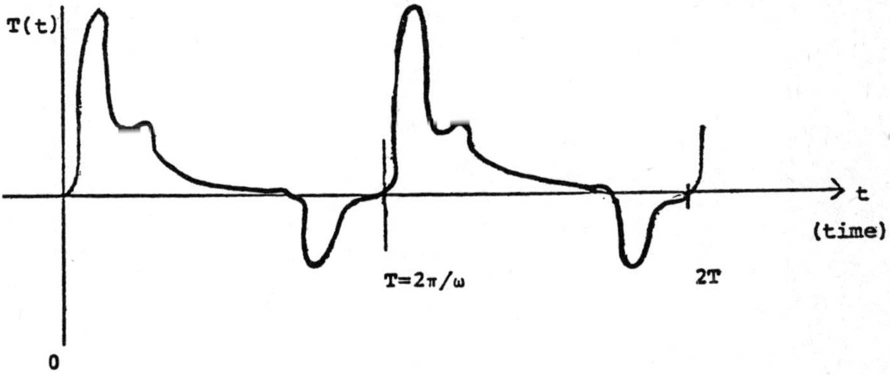

Figure B1

The engineers suggest at this point a simple formula for approximating the torque curve by sine and cosine functions

$$T(t) = a_0 + a_1\cos \omega t + a_2\cos 2\omega t + \dots$$
$$+ b_1\sin \omega t + b_2\sin 2\omega t + \dots$$

and interpret physically the meaning of the coefficients $a_0, a_1, a_2, \dots, b_1, b_2, \dots$ as amplitudes of the "harmonic" components of the torque T(t).

Moreover should one of the coefficients (say a_5) be a relatively large number (in absolute value) they will check if any of the natural frequencies of the mechanical system matches $5\omega/2\pi$. Again there is a mathematical justification for validity of their approach. The energy considerations show that displacement and velocity functions are elements of the space $L_2[0,2\pi\omega]$ and it remains only to be shown that the sines and cosine functions of the form $\sin(n\omega t)$, $\cos(n\omega t)$, $n = 0,1,2,\dots$ are dense in $L_2[0,2\pi\omega]$. For a more complete account see the monograph of Akhiezer and Glazman [1]. A relation between Fourier series and Fourier integrals is analyzed in the 1932 monograph of S. Bochner [2]. For a modern insight including the relation between functional analysis and partial differential equations read Hörmander [3], Hille and Phillips [4], Yosida [5], or Kantorovich and Akilov [6].

References and Sources for Appendix B

[1] N.I. Akhiezer and I.M. Glazman, The theory of linear operators in Hilbert spaces. Ungar, 1963.

[2] S. Bochner, Vorlesungen über Fourierische Integrale, Akademie Verlag, Berlin, 1932.

[3] L. Hörmander Linear Partial differential operators, Vol. I, II and III, Springer Verlag, Berlin, First edition of Vol. I in 1963.

[4] E. Hille and R. Phillips, Functional analysis and semigroups, Colloquia of American Mathematical Scoeity, first edition, A.M.S., Providence, Rhode Island, 1948.

[5] K. Yosida, Functional analysis, revised second edition, Springer Verlag, New York, 1968.

[6] L.V. Kantorovich and G.P. Akilov, Functional analysis in normed spaces, Pergammon Press, Oxford and New York, 1964.

General References for Volume 1

[1] V.I. Arnol'd, Mathematical methods of classical mechanics, Springer Verlag, Berlin, and New York, 1978.

[2] R. Becker, Introduction to Theoretical Mechanics, McGraw-Hill, New York, 1954.

[3] Y. Choquet Bruhat, Problems and Solutions in Mathematical Physics, Holden Day, San Francisco, 1967.

[4] J.P. den Hartog, Mechanical Vibrations, McGraw-Hill, 1950.

[5] H. Goldstein, Classical Mechanics, Addison Wesley, Reading, Mass., 1950.

[6] Mikhlin, Linear Equations of Mathematical Physics, Holt, Rinehart, and Winston, New York, 1967.

[7] P.M. Morse and H. Feshbach, Methods of Theoretical Physics, McGraw-Hill, New York, 1953, Vol. I and II.

[8] N.I. Muskhelishvili, Some Basic Problems of the Mathematical Theory of Elasticity, Noordhoff, Gronigen, Holland, 1953.

[9] W. Prager, Introduction to Mechanics of Continua, Ginn, & Co., Boston, 1961.

[10] A.N. Tikhonov and A.A. Samarskii, Equations of Mathematical Physics (in Russian), "Nauka", Moscow, 1966.

[11] S. Timoshenko and J.N. Goodier, Theory of Elasticity, McGraw-Hill, New York, 1951.

[12] K. Yosida, Lectures on differential and integral equations, Interscience, New York, 1960.

[13] R. Courant and D. Hilbert, Methods of mathematical physics, Interscience, New York, vol. 1,1953, vol. 2 1962.

[14] L. D. Landau and E.M. Lifshitz, Mechanics, Vol. 1 of Course in theoretical physics, Pergammon Press, Oxford, London and New York, 1960.

[15] S. G. Mikhlin, Mathematical physics - an advanced course, North Holland Publ. Co. Amsterdam, 1970.

[16] F. Riesz and Sz. Nagy, Lectures on functional analysis, Akad. Krado, Budapest 1952, Ungar Publ. 1962, English translation.

[17] S.L. Sobolëv, Some applications of functional analysis to mathematical physics, Leningrad, Univ. 1945, English translation, A.M.S. monograph translation #7, Providence, Rh. I., 1963.

[18] N. Dunford and J. Schwartz, Linear Operators, Vol. 1, Interscience, 1958.

[19] Yosida, Functional Analysis, 2nd edition, Springer, New York, 1968.

Index